HISTORY OF MATHEMATICAL SCIENCES

HISTORY OF MATHEMATICAL SCIENCES

Editor

Dominga Abascal

scitus
academics

History of Mathematical Sciences
Edited by **Dominga Abascal**

This Book is Printed on Acid-Free Paper. ♾

ISBN: 9781681179612
Library of Congress Control Number: 2017949092

SCITUS Academics LLC,
616, Corporate Way, Suite 2, 4766,
Valley Cottage, NY 10989

www.scitusacademics.com

Notice

Printed in United States of America.

Preface

In the 19th Century, though, mathematics expanded to include mathematical or symbolic logic, and therefore came to be viewed increasingly as the science of relations or of portraying necessary conclusions. The history of mathematics is approximately as old as humanity itself. Since antiquity, mathematics has been ultimate to developments in science, engineering, and philosophy. It has grown from simple measurement and calculation, and the methodical study of the shapes and motions of physical matters, through the application of abstraction, imagination and logic, to the broad, complex and often abstract discipline we know today. Mathematics displays much more robustness in its devotion to concepts and theories than do other sciences. These days history of mathematics is a powerful tool for a disseminating an understanding of mathematics. We look at history as a way of motivating the learner to see the significance of the area being studied. We consider to history as a route to help the learner understand the path of development to a mathematical concept or process. With the history of mathematics, students will come to know that mathematics and science is a work of all civilizations and teachers will find more confidence in teaching.

In this book, topics are introduced in their historical contexts will assist students to better gain and appreciate the mathematical ideas involved. It covers number theory, calculus/analysis, and several historically oriented courses and includes chapters on simple groups and combinatorics, and some topics, including the Poincare conjecture. It presents new insights on topics of current interest from broad perspectives and/or reviewing all major developments in an important area over many years. This is not intended as a comprehensive and definitive guide to all of mathematics, but as an easy-to-use summary of the major mathematicians and the developments of mathematical thought over the centuries. This is intended to introduce some of the major thinkers and some of the most important advances in mathematics.

Table of Contents

CHAPTER 1

The Mathematics of Harmony. Proclus' Hypothesis and New View on Euclid's Elements and History of Mathematics Starting since Euclid

Alexey Stakhov

International Club of the Golden Section, Bolton, Ontario, Canada

ABSTRACT

We are discussing one of the most unlikely hypotheses in the history of mathematics—Proclus' hypothesis, which overturns a traditional view on Euclid's Elements and the history of mathematics, starting since Euclid. According to Proclus, the main goal of Euclid, when writing the Elements, was to build a complete geometric theory of Platonic solids (Book XIII), associated in the ancient philosophy (Pythagoras, Plato) with the Universe harmony. To construct this theory, Euclid introduced in Book II the problem of dividing a segment into extreme and mean ratio (the "golden sec- tion"). It follows from Proclus' hypothesis that Euclid's Elements are the first attempt to create the "Mathematical Theory of the Universe Harmony", based on Platonic solids and the "golden section".

Keywords: Proclus' Hypothesis, Mathematics of Harmony, Euclid's Elements, Pythagoras, Plato, Golden Section, Platonic Solids

1. PREFACE

Algebra and Geometry have one and the same fate. The very slow successes followed after the fast successes at the beginning. Mathematicians left science in a state very far from perfect. It happened, probably, because Mathematicians drew all their attention on the Advanced Parts of Analysis, by neglecting the origin of Mathematics and not wishing to dig in the field, which they already went through and had left behind.

2. INTRODUCTION

Euclid's Elements is the greatest mathematical work of the ancient Greek era. This scientific work was created by Euclid (Figure 1) in the 3d century BC and contained the foundations of ancient mathematics: elementary geometry, number theory, algebra, theory of proportions and relations, methods of determining areas and volumes, and others. In this work Euclid presented the result of the Greek mathematics development during 3 centuries and created a powerful foundation for further development of mathematics.

Euclid's Elements is an inexhaustible source of new mathematical findings. Lobachevsky's geometry, which arose as a result of the deep mathematical analysis of Euclid's V-th postulate (the "postulate about parallel lines") is a brilliant example of this.

This article reveals another secret of Euclid: for what purpose Euclid introduced the "problem of the division of a line segment in extreme and mean ratio"? In modern science, this division is called the "golden section". Euclid introduced the "golden section" in the Book II (Proposition II.11). Euclid then considered the "golden section" in the Book VI. The most often the Propositions, concerning to the "golden section," are found in the Book XIII, which is devoted to the geometric theory of regular polyhedra ("Platonic solids").

Modern historians of mathematics are not very often discuss the role of the "golden section" in Euclid's Elements and prefer to keep silent about this and, unfortunately, they do not answer the question: what was the purpose of Euclid to introduce the "golden section" in his famous mathematical work. The purpose of this article is to uncover one of the greatest mysteries of Euclid's Elements, associated with the "golden section".

Greek philosopher and mathematician Proclus Diadochus (410-485) was one of the first brilliant commentators of Euclid's Elements. He was first thinker, who tried to uncover this secret. He put forward the most unusual hypothesis in the history of mathematics. The essence of Proclus' hypothesis consists in the fact that the main purpose of Euclid when writing the Elements was to build a complete geometric theory of Platonic solids. This theory has been placed in the Book XIII, that is, in the concluding Book of the Elements. Because the most important material of scientific book is usually placed in the final part of the book, the fact that the theory of Platonic solids was placed by Euclid in the Book XIII, that is, the concluding book of the Elements, is an indirect confirmation of Proclus' hypothesis.

Figure 1. Euclid.

Because the Platonic solids were associated in ancient Greek science with the "Universe Harmony", it follows from this fact that Euclid's Elements were written under the direct influence of the "harmonic ideas" of Pythagoras and Plato. This is the main secret of Euclid's Elements. From these reasonings, it follows another interesting conclusion: Euclid's Elements are historically the first attempt to create the "Mathematical Principles of the Universe Harmony". And since mathematics as a science was formed under the powerful influence of Euclid's Elements, we must admit that the main goal of mathematics is to discover the mathematical laws of the Universe Harmony. From these considerations it follows another important conclusion: the role of the "golden section" and the Platonic solids should be increased in mathematics and especially in mathematics education. Modern mathematicians have not yet grasped these important conclusions.

According to the author, the "golden section" was introduced by Euclid in his Elements not accidental. Euclid was in need of the "golden section" to construct the geometric theory of "dodecahedron", one of the main Platonic Solids. That's why the Book XIII, the concluding Book of the Elements, contains the largest number of Proposals relating to the "golden section".

In modern science, interest in the "golden section" and the Platonic solids increased significantly. Two scientific discoveries of modern science (fullerenes and quasi-crystals), based on the "golden section" and Platonic solids, has been awarded the Nobel Prize in chemistry. That is, the modern theoretical natural sciences make an extensive use of the well-known geometric discoveries of the ancient Greeks—the "golden section" and Platonic solids.

The author's book The Mathematics of Harmony. From Euclid to Contemporary Mathematics (World Scientific, 2009) http://www.worldscientific.com/worldscibooks/10.1142/6635, which in its origins dates back to Euclid's Elements, is a reflection of one of the most important trends in the development of modern science— the revival of "harmonic ideas" Pythagoras, Plato and Euclid. Proclus hypothesis plays a central role in this revival.

3. THE IDEA OF THE UNIVERSE HARMONY IN THE ANCIENT GREEK SCIENCE

3.1. What Is the Harmony?

As pointed out by the Russian philosopher VP Shestakov in the book "Harmony as an Aesthetic Category" [1] "the history of aesthetic doctrines puts forward a variety of types of understanding harmony. The very concept of "harmony" is used very widely and has many meanings. It signified the natural construction of nature and cosmos, the beauty of the physical and moral world of a man, the principles of the construction of art works, and the laws of aesthetic perception."

Shestakov identifies three basic understandings of harmony, prevailing in the development of science and aesthetics:

1) Mathematical understanding of harmony or mathematical harmony. In this sense, harmony is understood as equality or proportionality of parts with each other and part with the whole. In the Great Soviet Encyclopaedia, we find the following definition of harmony, which expresses the mathematical understanding of harmony:

   "Harmony is a proportionality of parts with the whole, the confluence of the various components of the object into a single organic whole. In the harmony, the internal ordering and the measure of being get the external manifestation."

2) Aesthetic harmony. In contrast to the mathematical understanding of the harmony, the aesthetic understanding is not only quantitative but also qualitative, and expresses the inner nature of things. The aesthetic harmony associated with aesthetic experiences, with an aesthetic evaluation. Most clearly this type of harmony is manifested in the perception of the nature beauty.

3) Artistic harmony. This type of harmony is associated with art. The artistic harmony is the actualization of the principle of harmony in the material of art itself.

Most importantly, it follows from the arguments, given in [1] , that the "harmony" is a universal concept that is relevant not only to mathematics and science but also to arts.

3.2. Numerical Harmony of the Pythagoreans

Pythagoras (Figure 2) and Heraclitus (Figure 3) are Greek thinkers whose names are usually attributed to the beginning of the philosophical doctrine of the harmony. According to many authors, the key idea of the harmony as a commensurate unity of opposites belongs to Pythagoras. Pythagoreans first proposed the idea of the harmonic organization of the world, including not only man and nature, but also the entire cosmos. According to the Pythagoreans, "the harmony represents the internal relationship of things, without which the cosmos could not exist." Finally, according to Pythagoras, the harmony has a numeric expression, that is, it is integrally linked to the concept of number.

Pythagoreans created the doctrine about constructive essence of a number. Aristotle in his "Metaphysics" notes this particular feature of the Pythagorean doctrine as follows:

"The so-called Pythagoreans, by exploring the mathematical sciences, for the first time advanced them forward and, by taking them as the basis, began to consider them the principle of all things... Since all the rest is explicitly likened to the numbers by the essence, therefore the numbers occupied the first place in the whole nature, therefore the elements of numbers, named the elements of all things and the whole universe, were recognized by harmony and number."

Figure 2. Pythagoras (approx. 570-c. 500 BC).

Figure 3. Heraclitus (535-475 BC).

3.3. Heraclitus' Contribution to the Development of the Harmony Doctrine

From antiquity up to the present day, the name of Heraclitus remains one of the most popular in the history of philosophy.

Heraclitus believed that everything is constantly changing. Heraclitus connects an assertion about the general variability with the idea of the inner division of things and processes on opposite parts, with their constant interaction. Heraclitus believed that everything in life arises from opposites and is cognized through them. The idea of perpetual motion was embodied by Heraclitus in the image of ever-flowing river. The postulate about the general changeability of the world, as one of the cornerstones of the entire dialectics, was formulated by Heraclitus in the famous formula: "One can not enter twice in the same river."

The moment of evaluation is characteristic for Heraclitus' aesthetics. Especially clearly such evaluation expressed in the doctrine about the two kinds of harmony: the "hidden" and "explicit". Heraclitus gives preference for the "hidden" harmony. It is widely known the following saying of Heraclitus: "The hidden harmony is stronger explicit".

Cosmos, as the supreme and perfect beauty, is an example of the "hidden" harmony. Only at first glance the world seems chaos. In fact, "the marvelous harmony" is hiding for the game of elements and randomness.

3.4. Pythagorean Musical Harmony and Music of the Spheres

Pythagoreans made remarkable discoveries in music. Pythagoras discovered that the most pleasant hearing harmonies—consonances—are obtained only in the case where the length of the strings, which produce these sounds, relate as the first natural numbers 1, 2, 3, 4, 5, 6, that is, 1:1, 1:2 (an unison and an octave), 2:3, 3:cv 4 (a fifth and a fourth), 4:5, 5:6 (a third), etc. This discovery shocked Pythagoras. This discovery first pointed out on the existence of numerical regularities in nature and was the starting point in the development of Pythagorean philosophy, based on the thesis: "Everything is a number". Therefore, the day when Pythagoras discovered the law of consonances, the German physicist Arnold Sommerfeld called the birthday of theoretical physics.

The discovery of mathematical regularities in musical consonances became the first "experimental" proof of Pythagoras' doctrine of numbers. From that moment, the music and the associated with it doctrine of the harmony begin to occupy a central place in the Pythagorean system of knowledge. The Pythagorean idea of musical relations soon acquired the "cosmic scale" and grew into the idea of the universal or global, harmony.

Pythagoreans began to claim that the Universe is arranged on the basis of simple numerical relations and that the moving planets produce "music of the heavenly spheres", but the traditional music is merely a reflection of the prevailing everywhere "universal harmony". Thus, music and astronomy were reduced by the Pythagoreans to the analysis of numerical laws, that is, to the arithmetic and geometry. Thus, all four disciplines are mathematical and called by one word—"Mathematics".

3.5. Plato's Cosmology

The Greeks made the first attempt to "mathematize harmony", that is, to express the harmony in numerical and geometric form. According to

Edward Soroko [2] , "the ancient Greeks associated the idea about the "universal harmony" of the Universe with its implementation in the Platonic solids." In other words, the regular polyhedra, called Platonic solids after the prominent Greek philosopher Plato (Figure 4), were considered as the indicators of the geometric harmony of the Universe.

There are only five Platonic solids (this was proved in Euclid's Elements): tetrahedron, hexahedron or cube, octahedron, dodecahedron, and icosahedron (Figure 5).

According to Plato, the first four of the five regular solids (tetrahedron, icosahedron, cube and octahedron) embodied in themselves the main four "elements" (fire, water, land, air). Tetrahedron symbolized the fire, because its apex directed upwards; icosahedron symbolized the water, because it is the most "streamlined" polyhedron; cube symbolized the land, because it is the most "sustainable" polyhedron; octahedron symbolized the air as the most "air" polyhedron. The fifth polyhedron, the dodecahedron, embodied in itself "all things", "Universal Mind", was a symbol of the ether and was considered the main geometrical figure of the Universe.

Pythagorean doctrine about the numerical harmony of the Universe had a huge creative force and had a great influence on the development of all subsequent doctrines on the nature and essence of harmony, in particular, it was the basis of Plato's cosmology. In his works, Plato develops the Pythagorean doctrine, especially emphasizing the cosmic significance of harmony. He is firmly convinced that the harmony of the world can be expressed in numerical proportions. The influence of the Pythagoreans is especially evident in Plato's "Timaeus", where after the Pythagoras, Plato developed the theory of proportions and analyzed the role of regular polyhedra ("Platonic solids"), from which, in his opinion, God created the world.

The main conclusion, which follows from Pythagoras and Plato's doctrines, consists in the fact that the harmony is objective phenomenon, it exists independently of our consciousness and is expressed in the harmonious organization of everything, starting with the Cosmos and ending by microcosm. But if the harmony is objective, it should be the subject of mathematical research.

This idea underlies Alexey Stakhov's (2009) book "The Mathematics of Harmony. From Euclid to Contemporary Mathematics and Computer Science" [3] . The main purpose of this book is to develop and generalize the mathematical concepts that are relevant to the harmony and emerged in science for several millennia. The next goal of the book is to show perspectives of the applications of the "mathematics of harmony" for the solution of the vital problems of modern science.

Figure 4. Plato (428/427 or 424/423 BC-348/347 BC).

3.6. About the Term of "the Mathematics of Harmony"

For the first time, the term of "the Mathematics of Harmony" was used in the article "Harmony of spheres", published in "The Oxford dictionary of philosophy" [4] :

Harmony of spheres. A doctrine often traced to Pythagoras and fusing together mathematics, music, and astronomy. In essence the heavenly bodies being large objects in motion, must produce music. The perfection of the celestial world requires that this music be harmonious, it is hidden from our ears only because it is always present. The mathematics of harmony was a central discovery of immense significance to the Pythagoreans.

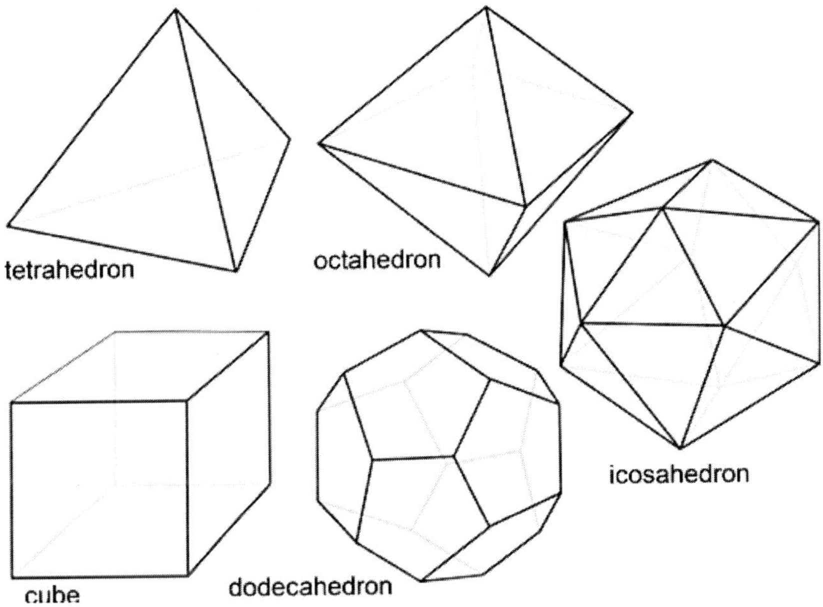

Figure 5. Platonic solids: tetrahedron, octahedron, cube, dodecahedron, icosahedrons.

Thus, the concept of the "Mathematics of Harmony" is associated here with the "harmony of the spheres", which is also called the "harmony of the world" (Latin "harmonica mundi") or world music (Latin "musical mundane"). The harmony of the spheres [4] is the ancient and medieval doctrine about the musical-mathemati- cal organization of Cosmos that goes back to the Pythagorean and Platonic philosophical tradition.

Another mention about the "Mathematics of Harmony", as the ancient Greek great discovery, we find in the book by Vladimir Dimitrov, A new kind of social science. Study of self-organization of human dynamics [5] . Let us consider the quote from the book:

Harmony was a key concept of the Greeks, a conjunction of three strands of meaning. Its root meaning was aro, join, so harmonia was what joined. Another meaning was proportion, the balance of things that allowed an easy fit. The quality of joining and proportion then came to be seen in music and other arts.

The precondition for harmony for the Greeks was expressed in the phrase "nothing to much". It also had a mysterious positive quality, which became the object of enquiry of their finest minds. Thinkers such as Pythagoras sought to capture the mystery of harmony as something both inexpressible yet also illuminated by mathematics. The mathematics of harmony explored by the ancient Greeks is still an inspiring model for contemporary scientists. Crucial to it is their discovery of its quantitative expression in astonishing diversity and complexity of nature through the golden mean (golden ratio), Φ (phi): $\Phi = \frac{1+\sqrt{5}}{2}$, which is approximately equal to 1.618. It is described by Euclid in book five of his Elements: "A straight line is said to have been cut in extreme and mean ratio when, as the whole line is to greater, so is greater to the less".

Thus, in the book [5] the concept of the "Mathematics of Harmony" is directly associated with the golden ratio—the most important ancient mathematical discovery in the field of harmony, which at that time was called the division of the line segment in extreme and mean ratio.

Finally, it is pertinent to mention that this term was used in Stakhov's speech "The Golden Section and Modern Harmony Mathematics", made at the 7th International Conference "Fibonacci Numbers and Their Applications" (Austria, Graz, 1996) [6] .

4. THE GOLDEN SECTION IN EUCLID'S "ELEMENTS"

4.1. Proposition II.11

In Euclid's Elements we meet the task, which later had played a great role in the development of science. This task is called the "division of line segment in extreme and mean ratio". In the Elements this task occurs in two forms. The first form is formulated in the Proposition 11 of the Book II.

Proposition II.11. Divide a given line segment AD into two unequal parts AF and FD so that the area of the square, which is built on the larger segment AF would be equal to the area of the rectangle, which is built on the segment AD and the smaller segment FD.

Depict this problem geometrically (Figure 6).

Thus, according to the Proposition II.11, the area of the square AGHF should be equal to the area of the rectangle ABCD. If we denote the length of the larger segment AF through b (it is equal to the side of the square AGHE), and the side of the smaller segment FD through a (it is equal to the vertical side of the rectangle ABCD), then the condition for the Proposition II.11 can be written as follows:

$$b^2 = a \times (a+b) \quad (1)$$

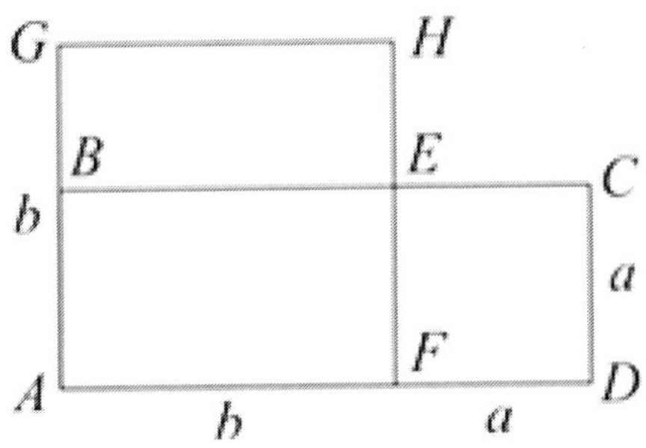

Figure 6. A division of a line segment in extreme and mean ratio (the "golden ratio").

In Euclid's Elements, we meet another form of the "golden ratio". This form follows from the first one, given by (1), if we make the following transformations. Dividing both parts of (1) at first by a, and then by b, we get the following proportion:

$$\frac{b}{a} = \frac{a+b}{b} \quad (2)$$

The proportion (2) has the following geometric interpretation (Figure 7). Divide the segment AB in the point C in such a relation when the larger part CB relates to the smaller part AC, as the segment AB to its larger part CB (Figure 7), that is,

$$\frac{AB}{CB} = \frac{CB}{AC} \quad (3)$$

This is the definition of the "golden ratio", used in modern science.

We denote the proportion (3) through x. Then, taking into consideration, that $AB = AC + CB$, the proportion (3) can be written as follows:

$$x = \frac{AC + CB}{CB} = 1 + \frac{AC}{CB} = 1 + \frac{1}{\frac{CB}{AC}} = 1 + \frac{1}{x} \quad (4)$$

The following algebraic equation follows from (4):

$$x^2 - x - 1 = 0. \quad (5)$$

From the "physical meaning" of the proportion (5), it implies that we should use the positive root of the Equation (5). We denote this root through F:

$$\Phi = \frac{1 + \sqrt{5}}{2}. \quad (6)$$

4.2. Mordukhai-Boltovsky's Comments about the "Golden Section"

Euclid's Elements are translated into many languages. The most authoritative Russian edition [7] -[9] is Euclid's Elements in the translation and commentaries of the famous Russian historian of mathematics Prof. D.D. Mordukhai-Boltovsky (Figure 8).

Consider Mordukhai-Boltovsky's comments about the "golden section":

Now look at the place of the golden section in Euclid's Elements. First of all, it should be noted that it arises in two forms, the difference between them is almost imperceptible to us, but was very significant in the eyes of the Greek mathematician of V-VI-th century BC. The first form, the prototype of which we have seen in Egypt, appears in the Book II of the Elements, namely in the Proposition 11 together with its preceding Proposals 5 and 6; here the golden ratio is defined as such ratio, for which the square, built on the larger segment, is equal to the rectangle, built on the whole line and the smaller segment. We can find the second form in the definition 3 of the Book VI, wherein the golden ratio is determined as the proportion of the whole line to the larger segment, and a larger segment to the smaller one and is called the division in extreme and mean ratio; in this form the golden section could be known only since Eudoxus' time...

In the book XIII, the golden ratio is presented in the above two forms, namely in the first form in the Propositions 1 - 5, and in the second dorm in the Propositions 8 - 10...

Figure 7. The golden ratio.

Figure 8. D.D. Mordukhai-Boltovsky (1876-1952).

All this allows to think that Propositions 4, 7, 8 of the Book II and Propositions 1 - 5 of the book XIII represent the remains of one of the oldest in the history of Greek geometry document, which dated probably to the first half of the V century and emerged in the Pythagorean school, on the grounds of the material, which was brought from Egypt. Relative antiquity of this document can be ascertained from the fact that Propositions 4 and 7 of the Book II are used in it to prove a generalization of the Pythagorean theorem [the square of the side against acute and obtuse angle

(Propositions 12 and 13 of the book II)], which, of course, was known to Hippocrates of Chios."

Valentin Bunin in the book [10] also draws attention to "the Egyptian trace" in the origin of the "golden section": "It is worth recalling that the original geometric meaning of the 'golden section' is very simple and has been associated with the annual redistribution of land area, flooded by Nile. In this case, the following problem was considered: to find the side of the rectangle, the area of which was equal to the area of the square, and the ratio of sides would provide ease of layout."

4.3. The Origin of the Term of the "Golden Section"

Sometimes the introduction of the name of the "golden section" ("sectio aurea") attributed to Leonardo da Vinci. However, it is believed that the great Leonardo was not the first. According to the statement of Edward Soroko [2] , the term goes back to Ptolemy's book "On the harmony". However, in the book [11] , devoted to the history of the "golden number", there is stated that the German mathematician Martin Ohm first coined the term "golden section" ("goldene shnitt") in 1835 in the book "Die reine Elementar-Mathematik".

The designation of the "golden section" by the Greek letter F (the number of PHI) is not accidental. This letter is the first letter in the name of the famous Greek sculptor Phidias, who is widely used the "golden section" in his sculptural works.

4.4. Citations of Alexey Losev and Johannes Kepler

What is the main idea, which underlies the ancient Greek science? The vast number of researchers is inclined to the following answer: the idea of Harmony, associated with the "golden section". As is known, in ancient Greek philosophy the Harmony is opposed to the Chaos and means organization of the Universe.

The outstanding Russian philosopher Alexey Losev (Figure 9), a researcher at the aesthetics of antiquity and the Renaissance, assesses the main achievements of the ancient Greeks in this field as follows:

"From Plato's point of view, and in general from the point of view of the whole ancient cosmology, the world is presented as a certain proportional whole, which obeys to the law of harmonic division, the golden section... Their system of cosmic proportions in the literature often is presented sometimes as a curious result of unrestrained and wild imagination. In such explanations we can see the antiscientific helplessness of those, who this assert. However, the understanding of this historical and aesthetic phenomenon can only be in connection with a holistic understanding of history, that is, by using a dialectical materialist view on culture and by looking for answers in the peculiarities of ancient social life."

Figure 9. Alexey Losev (1893-1988).

In this statement Alexey Losev convincingly formulated the "golden" paradigm of ancient cosmology. It is based on the most important ideas of ancient science, which in modern science is sometimes interpreted as a "curious result of unrestrained and wild imagination". First of all, this is the Pythagorean idea of the numerical harmony of the Universe and Plato's cosmology, based on the Platonic solids. By turning to the geometric structure of the Universe and the arithmetic relations, expressing the harmony, the Pythagoreans anticipated the emergence of mathematical natural sciences, which began to develop rapidly in the 20th century. The idea of Pythagoras and Plato about the numerical harmony of the Universe proved to be immortal.

Thus, in the center of the created by the ancient Greeks mathematical doctrine of Nature was "the concept of harmony", and mathematics itself of the ancient Greeks was the "Mathematics of Harmony", which was directly related to the "golden section", the most important mathematical discovery of ancient science in the field of harmony.

Here is another well-known statement, concerning the "golden section". It belongs to the brilliant astronomer Johannes Kepler (Figure 10), the author of the three famous "Kepler's laws".

Kepler expressed his admiration for the "golden section" in the following words:

"Geometry has two great treasures: one is the Theorem of Pythagoras; the other, the division of a line into extreme and mean ratio. The first, we may compare to a measure of gold; the second we may name a precious stone."

Recall that the antique problem of the division of the line segment into extreme and mean ratio, which is referred to in this statement, is the "golden section"!

4.5. The Golden Section in the Dodecahedron and Icosahedron

The dodecahedron and the dual to it icosahedron occupy a special place among the Platonic solids (Figure 5). First of all, it is necessary to emphasize that the geometry of the dodecahedron and icosahedron is directly related to the "golden section".

Indeed, the regular pentagon, based on the "golden section", are the faces of the dodecahedron (Figure 5). If we look closely at the icosahedron (Figure 4), we can see that in each of its vertices the five triangles come together and their outer sides form the regular pentagon. Already these facts are enough to make sure that the golden ratio plays a decisive role in the construction of these two Platonic solids.

But there are the deeper mathematical confirmations of the fundamental role, played by the golden section in the icosahedron and dodecahedron. It is known that these Platonic solids have three characteristic spheres. The first (inner) sphere is inscribed in the Platonic solid and touches its faces. We denote the radius of the inner sphere through R_i. The second or middle sphere touches its ribs. We denote the radius of this sphere through R_m.

Finally, the third (outer) sphere is described around the Platonic solid and passes through its vertices. We denote its radius through. In geometry, it is proved that the values of the radii of these spheres for the dodecahedron and icosahedron, which have the ribs of unit length, expressed in terms of the golden ratio (Table 1).

Note, that the ratio of the radii $\frac{R_c}{R_i}=\frac{\sqrt{3(3-\Phi)}}{\Phi}$ is the same for the icosahedron and dodecahedron. Thus, if the dodecahedron and icosahedron have the same inscribed spheres, then their described spheres are equal. The proof of this mathematical result is given in Euclid's Elements.

4.6. Ancient Greeks' Mathematical Doctrine about Nature

According to the eminent American historian of mathematics Morris Kline [12] , the main ancient Greeks' contribution, "which had a decisive influence on all subsequent culture, consists in the fact that they begun to study the laws of Nature." The main conclusion to be drawn from Morris Kline's book [12] is the fact that the ancient Greeks offered the innovative concept of the Cosmos, where everything was subordinated to mathematical laws. The question arises: when this concept was developed? The answer to this question is also found in the book [12] .

Figure 10. Johannes Kepler (1571-1630).

In the book [12] it is assumed that the innovative concept of the Cosmos has been developed by the ancient Greeks during VI-III c. BC.

But according to the prominent Russian mathematician A.N. Kolmogorov [13] , in the same period (VI-III c. BC.) in the ancient Greece "there arises mathematics as an independent science with a clear understanding of the uniqueness of its method and the need for systematic development of its basic concepts and proposals in a rather general way."

But then the question arises: whether there is any relationship between the process of creating a mathematical doctrine about the Nature what is considered as the most important achievement of the ancient Greek science [12] , and the process of creation of mathematics, which was developed in the ancient Greece in the same period? Or those are different processes? It turns out that such a relationship existed definitely. Moreover, it can be argued that these processes actually coincided. This means that mathematics, created by the ancient Greeks, and their doctrine of Nature, based on mathematical and "harmonic" principles, are one and the same scientific objects. Euclid's Elements, written in the III c. BC, is the most striking embodiment of the process of the "Mathematization of Harmony" in the ancient Greece

Table 1. The Golden Ratio in the spheres of the dodecahedron and icosahedrons.

Polyhedron	R_c	R_m	R_i
Icosahedron	$\frac{1}{2}\Phi\sqrt{3-\Phi}$	$\frac{1}{2}\Phi$	$\frac{\frac{1}{2}\Phi^2}{\sqrt{3}}$
Dodecahedron	$\frac{\Phi\sqrt{3}}{2}$	$\frac{\Phi^2}{2}$	$\frac{\Phi^2}{2\sqrt{3-\Phi}}$

5. PROCLUS' HYPOTHESIS

5.1. What Was the Purpose of Euclid to Create His Elements?

At first glance, it seems that the answer to this question is very simple: The main objective of Euclid was to present the main achievements of the Greek mathematicians 300 years before Euclid, using the "axiomatic method". This view of the Elements is the most common in modern mathematics.

However, apart from "axiomatic" point of view, there is another point of view on Euclid's motives in writing the Elements. This point of view is expressed by the Greek philosopher and mathematician Proclus Diadoxus (412-485), one of the first commentators of Euclid's Elements.

Among Proclus' mathematical works the "Commentary on the first book of Euclid's Elements" is the most known. In his Commentary he suggested the following unusual hypothesis.

As is known, the Book XIII, that is, the final book of the Elements, is devoted to the geometric theory of the five regular polyhedra, which played a major role in Plato's Cosmology and are known in modern science as Platonic solids. This fact drew Proclus' attention. As pointed out by well-known Belarusian philosopher Eduard Soroko [2] , according to Proclus, Euclid "created the Elements not for the purpose to present geometry itself, but to give a full systematic theory for constructing the five 'Platonic solids', highlighting in passing some of the latest achievements of ancient Greeks' mathematics."

5.2. How Did Euclid Use the "Golden Section"?

The question arises: why Euclid introduced in the Elements different forms of the "golden ratio", which we can find in the Books II, VI and XIII? To answer this question, we again return back to the Platonic solids (Figure 5). As is known, only three types of regular polygons can be the faces of the Platonic solids: equilateral triangle (tetrahedron, octahedron, icosahedron), square (cube) and a regular pentagon (dodecahedron). In order to construct the Platonic solids, we must, first of all, be able to build the faces of Platonic solids geometrically (that is, by using a ruler and compass).

Euclid knew, how to construct an equilateral triangle and a square, but he met some difficulties at constructing of the regular pentagon, which underlies the dodecahedron (Figure 11(a)). The icosahedron in Figure 11(b) is geometric figure dual to the dodecahedron.

It is for this purpose, Euclid had introduced in the Book II the "golden ratio" (Proposition II.11), which is presented in the Elements in two forms. By using the "golden ratio", Euclid constructs the "golden" isosceles triangle, whose angles at the base are equal to the doubled angle at the vertex (Figure 12(a)).

At first, we construct the "golden" isosceles triangle by using a ruler and compass (see Figure 12(a)). The triangle ABD has the equal sides AB and AD and the equal angles B and D at the base BD. These angles are equal to the doubled angle at the vertex A.

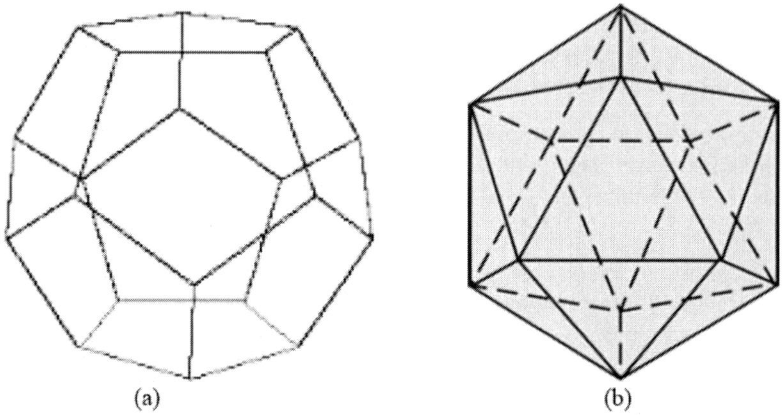

Figure 11. Dodecahedron (a) and icosahedron (b).

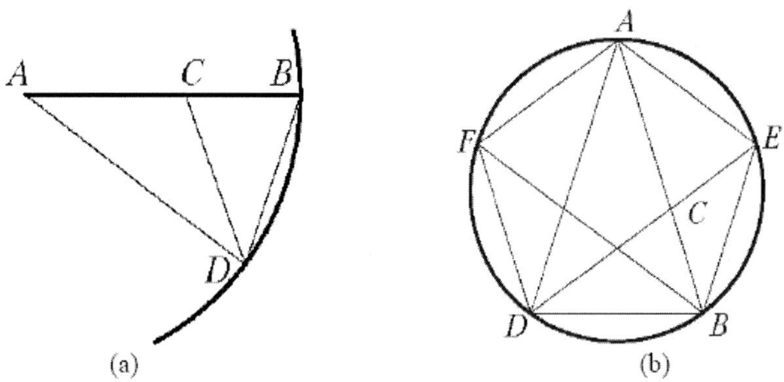

Figure 12. The "golden" isosceles triangle (a) and the pentagon (b).

By using the "golden" isosceles triangle ABD (Figure 12(a)), we construct the regular pentagon in Figure 12(b).

And then we should do only one step to the geometric construction of the dodecahedron (Figure 11(a)), one of the most important regular polyhedra, which symbolized the Ether or Universal Mind in Plato's cosmology.

Pentagon and pentagram. Note that the regular pentagon (Figure 13), which is the base of the dodecahedron, was very important geometric figure of Greek science because the golden ratio is a basis of the regular pentagon. The word of "pentagon" is derived from the Greek word of "pentagonon".

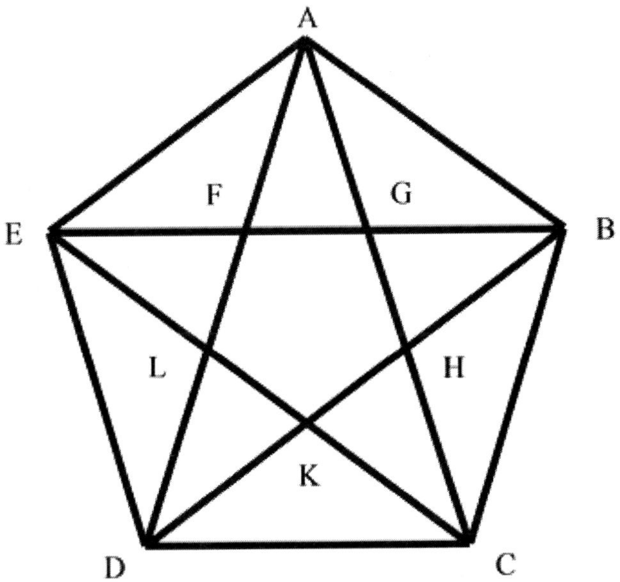

Figure 13. The regular pentagon and pentagram.

If we draw in a pentagon all diagonals, then we get the pentagonal star, also known as a pentagram or pentacle. The name of "pentagram" is derived from the Greek word "pentagrammon" (pente—five and grammon— line).

It is proved that the points F, G, H, K, L of the intersection of pentagon's diagonals are always the points of the golden ratio. At the same time, they form a new regular pentagon FGHKL. In the new pentagon we can draw new diagonals. The points of their intersection adduce to the new regular pentagon and this process can be continued indefinitely. Thus, the pentagon ABCDE consists of an infinite number of regular pentagons, which are formed each time by the points of the diagonals intersection. This endless repetition of the same geometric figure (the regular pentagon) creates a sense of rhythm and harmony, which is fixed by our unconscious mind.

The Pythagoreans were delighted with the pentagram, which was considered as their main distinctive symbol.

5.3. Three "Key" Problems of the Ancient Mathematics

The prominent Russian mathematician academician Kolmogorov in the book [13] had identified the two main, that is, "key" problems, which stimulated the development of mathematics at the stage of its origin: the counting problem and the measurement problem. However, it follows from "Proclus' hypothesis" that one more "key" problem, the problem of harmony, which underlies Euclid's Elements, stimulated the development of mathematics at the stage of its origin.

Thus, not two but three "key" problems, the counting problem, the measurement problem, and the harmony problem, underlie mathematics at the stage of its origin (see Figure 14).

The first two "key" problems resulted in the creation of two fundamental notions of mathematics—natural number and irrational number, which underlie the Classical Mathematics. The problem of harmony, connected with the division of line segment in extreme and mean ratio (the "golden ratio") (Proposition II.11 of Euclid's Elements) resulted in the Mathematics of Harmony, which is interpreted now as a new interdisciplinary direction of contemporary science and mathematics [3] .

This approach is unexpected for many mathematicians. It turns out, that in parallel with the Classical Mathematics, another mathematical direction, the Mathematics of Harmony, was developing in mathematics starting since ancient science.

Similarly to the Classical Mathematics, the Harmony Mathematics takes its origin in Euclid's Elements. However, the Classical Mathematics focuses on the axiomatic approach and other ancient achievements (number theory, theory of irrationalities and so on), while the Mathematics of Harmony is based on the golden ratio (Proposition II.11) and Platonic solids, described in Book XIII of Euclid's Elements. Thus, Euclid's Elements is a source of the two independent mathematical directions: the Classical Mathematics and the Mathematics of Harmony.

5.4. The Historical Significance of Proclus' Hypothesis for Mathematics

The main conclusion from Proclus' hypothesis consists in the fact that Euclid's Elements, the greatest Greek mathematical work, was written by Euclid under the direct influence of the ancient idea of Harmony, which was associated with the Platonic solids.

Thus, the Proclus' hypothesis allows suggesting that the well-known in the ancient science Pythagorean doctrine of numerical harmony of the Universe and Plato's cosmology, based on regular polyhedra, were embodied in the greatest mathematical work of Greek mathematics, Euclid's Elements.

From this reasoning's, we can consider Euclid's Elements as the first attempt to create the "Mathematical Theory of the Universe Harmony", associated in ancient science with Platonic solids. This is the main secret of

Euclid's Elements that leads us to the revision of the history of the mathematics, starting since Euclid.

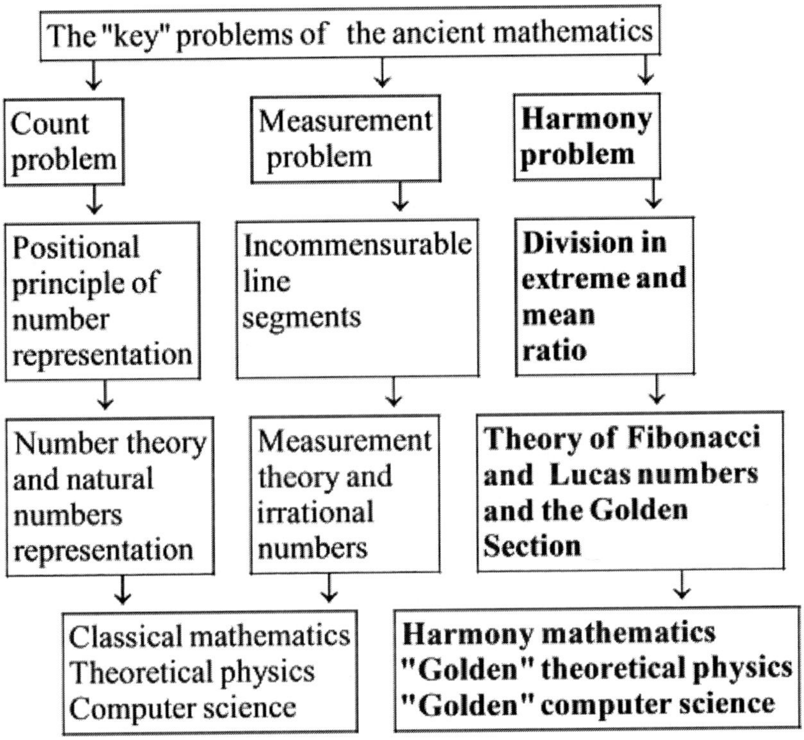

Figure 14. Three "key" problems of the ancient mathematics.

5.5. A Role of the Mathematics of Harmony for School Mathematical Education

For many centuries, the creation of the Classical Mathematics, the Tsarina of science, was the main focus of mathematicians. However, starting from Pythagoras, Plato, Euclid, Fibonacci, Pacioli, Kepler, the intellectual forces of many prominent mathematicians and thinkers were directed towards the development of the basic concepts and applications of the Mathematics of Harmony. We don't have a right to ignore this important historical fact.

Unfortunately, these two important mathematical directions (Classical Mathematics and Mathematics of Harmony) evolved separately from one another. A time came to unite these important mathematical directions. This unusual union can lead to new scientific discoveries in mathematics and theoretical natural sciences.

A new approach to the origin of mathematics (see Figure 11) is very important for school mathematical education. The prominent Ukrainian mathematician academician Yuri Mitropolsky (Figure 15) wrote in his review [14] :

"One may wonder what place in the general theory of mathematics is occupied by Mathematics of Harmony created by Prof. Stakhov? It seems to me, that in the last centuries, as Nikolay Lobachevsky said, 'mathematicians turned all their attention to the Advanced Parts of Analytics, neglecting the origins of Mathematics and not willing to dig the field that already been harvested by them and left behind.' As a result, this created a gap between 'Elementary Mathematics'—basis of modern mathematical education, and 'Advanced Mathematics'. In my opinion, the Mathematics of Harmony developed by Prof. Stakhov fills up that gap. i.e, 'Mathematics of Harmony' is a big theoretical contribution, first of all to the development of 'Elementary Mathematics' and as such should be considered of great importance for mathematical education."

Mitroplsky's proposal introduces, in very natural manner, the idea of harmony and the golden ratio into school mathematical education. This provides for pupils the natural access to ancient science and to its main achievement—the idea of Harmony—and to tell them about the most important architectural and sculptural works of the ancient arts based upon the golden ratio (including pyramid of Khufu (Cheops), Nefertiti, Parthenon, Doryphoros, Venus and so on).

5.6. A Discussion of Proclus' Hypothesis in the Historical-Mathematical Literature

The analysis of Proclus' hypothesis is found in many mathematical sources. Consider some of them [15] -[17] . In the book [15] we read: "According to Proclus, the main objective of the 'Elements' was to present the geometric construction of the so-called Platonic solids."

In the book [16] , this idea got a further concrete development: "Proclus, by mentioning all previous mathematicians of Plato's circle, said: Euclid lived later than the mathematicians of Plato's circle, but earlier than Eratosthenes and Archimedes... He belonged to Plato's school and was well acquainted with Plato's philosophy and his cosmology; that's why he put a creation of the geometric theory of the so-called Platonic solids as the main purpose of the Elements."

Figure 15. Academician Yuri Mitropolsky (1917-2008).

This comment is very important and draws our attention to the connection of Euclid with Plato. Euclid fully shared Plato's philosophy and cosmology, based on Platonic solids, that is why, Euclid put forward the creation of the geometric theory of Platonic solids as the main purpose of the Elements.

In the book [17] , there is discussed the influence of Plato and Euclid' ideas on Johannes Kepler at designing of Kepler's Cosmic Cup (Figure 16) in his first book "Mysterium Cosmographicum":

"Kepler's project in 'Mysterium Cosmographicum' was to give 'true and perfect reasons for the numbers, quantities, and periodic motions of celestial orbits.' The perfect reasons must be based on the simple mathematical principles, which had been found by Kepler in the Solar system, by using multiple geometric demonstrations. The general scheme of his model was borrowed by Kepler from Plato's Timaeus, but the mathematical relations for the Platonic solids (pyramid, cube, octahedron, dodecahedron, icosahedron) were taken by Kepler from the works by Euclid and Ptolemy.

At that, Kepler followed to Proclus and believed that 'the main goal of Euclid was to build a geometric theory of the so-called Platonic solids.' Kepler was completely fascinated by Proclus, he often quotes him and calls him 'Pythagorean'."

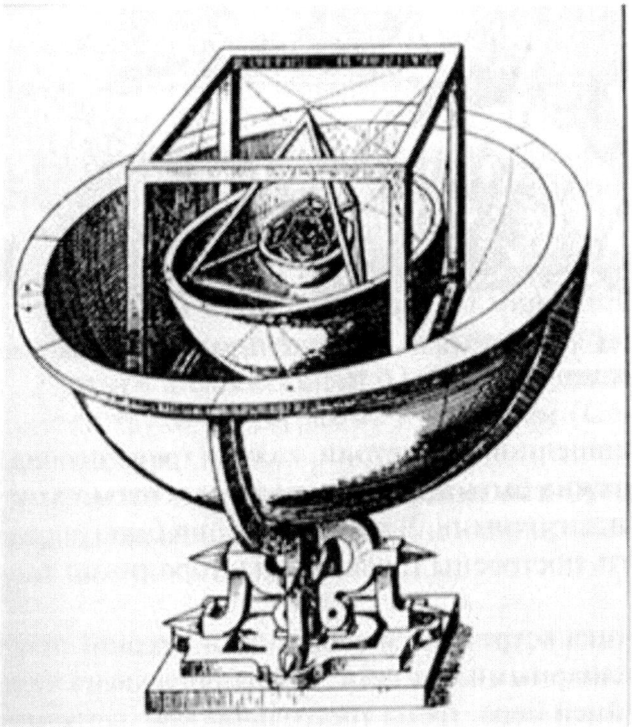

Figure 16. "Cosmic Cup"—Kepler's model of the Solar system based on Platonic solids.

From this quote, we can conclude that Kepler used the Platonic solids to create the Cosmic Cup, but all the mathematical relations for the Platonic solids were borrowed by him from Book XIII of the Elements, that is, he united in his studies Plato's Cosmology with Euclid's Elements.

At that, he fully believed in Proclus' hypothesis that the main goal of Euclid was to give the complete geometric theory of Platonic solids, which were used by Kepler in his geometric model of the Solar system (Figure 16).

5.7. Icosahedron as the Main Geometrical Object of Mathematics

In the late of the 19th century, the great German mathematician Felix Klein (Figure 17) drew attention to the Platonic solids. He predicted an outstanding role of the Platonic Solids, in particular, the icosahedron for the future development of science and mathematics. In 1884 Felix Klein published the book "Lectures on the Icosahedron" [18] , dedicated to the geometric theory of the icosahedron.

According to Klein, the tissue of mathematics widely and freely extends like the sheets of the different mathematical theories. But there are geometric objects, which unites many mathematical theories. Their geometry binds these mathematical theories and allows embracing a general mathematical sense of the miscellaneous theories. In Klein's opinion, the icosahedron is precisely such mathematical object. Klein treats the regular icosahedron as the mathematical object, from which the branches of the five mathematical theories follow, namely, geometry, Galois' theory, group theory, invariants theory and differential equations.

Figure 17. Felix Klein (1849-1925).

Thus, the prominent mathematician Felix Klein after Pythagoras, Plato, Euclid, and Johannes Kepler could estimate a fundamental role of the Platonic Solids, in particular, of the icosahedron, for the development of science and mathematics. Klein's main idea is extremely simple [18] : "Each unique geometrical object is somehow or other connected to the properties of the regular icosahedron".

Unfortunately, Klein's contemporaries could not understand and appreciate a revolutionary importance of Klein's idea, suggested by him in 19th century. However, its significance was appreciated one century after, when the Israeli scientist Dan Shechtman discovered in 1982 a special alloy called quasi-crystals and the outstanding researchers Robert F. Curl, Harold W. Kroto and Richard E. Smalley discovered in 1985 a special kind of carbon called fullerenes. It is important to emphasize that the quasi-crystals are based on the Platonic icosahedron and the fullerenes on the Archimedean truncated icosahedron. This means that Felix Klein in the 19th century made a brilliant prediction of the two outstanding scientific discoveries of modern science: fullerenes (Nobel Prize in chemistry-1996) and quasi-crystals (Nobel Prize in chemistry-2011).

6. CONCLUSIONS

1) For several millennia the prominent thinkers and scientists: Pythagoras, Plato, Euclid, Fibonacci, Pacioli, Kepler, Binet, Lucas, Klein, and in the 20th century *Coxeter*, Vorobyev, Hoggatt, Vajda and other famous mathematicians participated in the development of the "Mathematics of Harmony".

2) It is clear that, if the truth will prevail and Proclus' hypothesis will be recognized by modern historians of mathematics as it was done in [15] -[17] , it will be a prerequisite for the introduction of the Ancient Concept of Harmony into modern mathematics and mathematical education.

3) In this case, the Idea of Harmony introduces into mathematics the main applied goal, for which the ancient Greeks have created the Mathematical Theory of Nature [12] , namely, the creation of the useful mathematical models of the Universe Harmony.

4) It is clear that Platonic solids, "golden section", Fibonacci numbers and their generalizations should underlie these models [3] . These outstanding mathematical discoveries of ancient and medieval mathematics should no longer play the role of a sort of "rogue" in mathematics and mathematical education and they will occupy here the worthy place, which they certainly deserve.

Alexey Stakhov (19,35)">19] the "Harmonization of Mathematics" and has been embodied in author's book "The Mathematics of Harmony" (World Scientific, 2009) [3] , can become the beginning of the "golden" revolution in mathematics and can lead to the merger of mathematics with theoretical natural sciences. Morris Klain in the final chapter, "The Authority of Nature," of his wonderful book [12] is proclaiming this goal.

REFERENCES

1. Shestakov, V.P. (1973) Harmony as an Aesthetic Category. Science, Moscow (Russian).
2. Soroko, E.M. (1984) Structural Harmony of Systems. Science and Technology, Minsk (Russian).
3. Stakhov, A.P. (2009) The Mathematics of Harmony. From Euclid to Contemporary Mathemartics and Computer Science. World Scientific, New Jersey, London, Singapore, Beijing, Shanghai, Hong Kong, Taipei, Chennai, 748 p.
4. Harmony of Spheres. The Oxford Dictionary of Philosophy, Oxford University Press, 1994, 1996, 2005.
5. Dimitrov, V. (2005) A New Kind of Social Science. Study of Self-Organization of Human Dynamics. Morrisville Lulu Press, Richmond, Australia.
6. Stakhov, A.P. (1998) The Golden Section and Modern Harmony Mathematics. Applications of Fibonacci Numbers, Kluwer Academic Publishers, 7, 393-399.
7. Euclid's Elements. Books I-VI. Translated from the Greek and Comments D.D. Mordukhai-Boltovsky. Moscow-Leningrad: GITTL 1948 (Russian).
8. Euclid's Elements. Books VII-X. Translated from the Greek and Comments D.D. Mordukhai-Boltovsky. Moscow-Leningrad: GITTL 1949 (Russian).
9. Euclid's Elements. Books XI-XV. Translated from the Greek and Comments D.D. Mordukhai-Boltovsky. Moscow-Leningrad: GITTL 1950 (Russian).
10. Bunin, V.A. (2009) The Code of Bio-Similarity. Ternary Code of Meta-Harmony as as Bio-Similarity of Technological Systems by the Criterion of the Objective Function. "Academy of Trinitarism", Moscow, № 77-6567, publ.15669, 24.11.2009 (Russian).
11. Herz-Fischler, R. (1998) A Mathematical History of the Golden Number. Dover Publications, Inc., New York.
12. Kline, M. (1984) Mathematics: The Loss of Certainty. (transl. from Engl.). Mir, Moscow. (Russian)
13. Kolmogorov, A.N. (1991) Mathematics in Its Historical Development. Science, Moscow. (Russian)
14. Academician Mitropolsky's Commentary on the Scientific Research of the Ukrainian Scientist Doctor of Engineering Sciences Professor Alexey Stakhov. Preface to the book Alexey Stakhov "The Mathematics of Harmony. From Euclid to Contemporary Mathematics and Computer Science", World Scientific, New Jersey, London, Singapore, Beijing, Shanghai, Hong Kong, Taipei, Chennai, 2009, 748 p.
15. Kahn, C.H. (2001) Pythagoras and Pythagoreans: A Brief History. Hackett Publishing Co, Inc., Indianapolis.
16. Zhmud, L. (2006) The Origin of the History of Science in Classical Antiquity. Walter de Gruyter, Berlin.
17. Smorinsky, C. (2008) History of Mathematics: A Supplement. Springer, Berlin.
18. Klein, F. (1956) Lectures on the Icosahedron. Courier Dover Publications, Mineola.
19. Stakhov, A.P. (2011) Mathematization of Harmony and Harmonization of Mathematics. Academy of Trinitarism, Moscow. № 77-6567, publ.16897, 16.10.2011 (Russian).

CHAPTER 2

Amelija Mažylytė (1900-1972)—The First Lithuanian Female Graduate in Mathematics

Juozas Banionis

Lithuanian University of Educational Sciences, Vilnius, Lithuania

ABSTRACT

Amelija Mažylytė (1900-1972) was the first Lithuanian woman, graduate of The Lithuanian University in Department of Mathematics and also philosophy, and a participant of the catholic wo- men movement. She started dissemination of knowledge of the history of mathematics in the Lithuanian language. The article presents enhanced biography of Amelija Mažylytė and the work of her research interests. Special attention is given to publications about the history of arithmetic, the development of geometry and the universally known specialist of mathematics C. Gauss.

Keywords: C.F. Gauss, Arithmetic, Calculus, Geometry, History of Didactics, Women's Movement in Lithuania

1. INTRODUCTION

In the second decade of the 20th century cultural foundations were actively laid in Lithuania which had been rebuilt as a modern republic. Considerable attention was devoted to educating the nation by introducing the achievements of various civilizations and doing it in the Lithuanian language. Amelija Mažylytė (1900-1972), the first Lithuanian woman graduate in mathematics, was very actively participating in this field. Lithuanian researchers Algirdas Ažubalis and Bronius Riauba had introduced her as a Lithuanian University graduate and a mathematician (Ažubalis, 2000, 2012 ;Riauba, 2011). This article adds new details to Amelija Mažylytė's biography by focusing on the merits of her social activities and her contribution to the development of education and the history of mathematics in Lithuania.

On the basis of archive material and the textual legacy the author of this article aims to record the most important facts of A. Mažylytė's biography and to introduce her most significant scientific 1928-1939 works by employing the methods of analysis, synthesis and description. A particular focus is provided to A. Mažylytė's works concerning the history of mathematics. In addition, this article emphasizes her role in promoting women's education as well as their involvement in academic activities.

2. A BIOGRAPHICAL TIMELINE

Amelija Mažylytė was born on the 17th of March (the 4th of March according to the old calendar) in 1900 to a Mažyliai family of a large farmer Kazimieras and a young noblewoman Jankevičiūtė living in the village of Vebriai near the town Panemunėlis in Rokiškis region (Ažubalis, 2000 ; Riauba, 2011). They already had an older daughter Ona and a few years later their youngest daughter Veronika was born (Figure 1).

The Mažyliai were bright people, for they not only socialized with the Panemunėlis priest Jonas Katelė (1831- 1908) who was well known as an organizer of Lithuanian education and a contributor to publishing as well as introducing "Uždauotinas" (Book of Problems, Tilsit, 1885), but they also decided to educate their three daughters. Amelija started her primary education in Panemunėlis school and then moved to Kalkūnai school (VVPI: p. 2).

In 1912 she continued her education as a 4th form student in the Vilnius Gymnasium for Girls which, after the break of the First World War, was transferred to Voronezh in Russia. There she was a member of the Ateitininkai circle, a catholic youth organization, and together with other girls she read and analyzed Lithuanian literature and learnt by heart Lithuanian folklore and poems. She was also broadening her knowledge about Christianity by reading the Bible and by writing essays about the way she saw the world (Mažylytė, 1933b: p. 315). She was also developing her literary skills by translating fairytales into the Lithuanian language together with her classmates. In Voronezh she spent time among famous Lithuanian people such as the mathematicians Zigmas Že- maitis (1884-1969), Pranas Mašiotas (1863-1940) and Marcelinas Šikšnys (1874-1970), the linguists Jonas Jablonskis (1860-1930) and Juozas Balčikonis (1885-1969) (Ažubalis, 2012: p. 123).

In 1918 she graduated the Gymnasium with honors. When the Lithuanian State was restored Amelija Ma- žylytė together with her sisters returned back to the home country and joined a group of Lithuanian teachers. In the school year 1919-1920 she started working as a teacher in the State Gymnasium in Panevėžys. However, in January 1920 when the Higher Courses, the start of the Lithuanian University, were established in Kaunas Amelija Mažylytė came to the capital city and entered the department of mathematics and physics. In June 1929 she started working at the Ministry of Education in the administer office, later she worked there as a secretary of the Primary Education Department and finally as a member of the Publishing Committee. During her studies she had an opportunity to attend the lectures of such Lithuanian mathematicians as Zigmas Žemaitis, Viktoras Bir- žiška (1886-1964), Otto T. Volk (1892-1989) and Julijonas Graurogkas (1885-1968). She graduated in 1925 after the successful defense of the theses "Astronomical Time Setting" on 28th of October under the tutorship of Bernardas Kodatis (1879-1957) (LU MGF: p. 205).

After graduating the Lithuanian University with a diploma in mathematics Amelija Mažylytė continued work- ing as a lecturer in the private teacher-training college for girls that belonged to the Congregation of the Sisters of St. Kazimieras where she had been teaching mathematics and its methodology since the 9th of February 1925 (LR: p. 3). Unwilling to limit herself to the current subject and teaching she decided to continue her education at the Lithuanian University and entered the faculty of Theology and Philosophy choosing studies in the Philoso- phy department for her second degree. At that time at the university there were teaching such well known Li- thuanian personalities as Pranas Dovydaitis (1886-1942), Pranas Kuraitis (1883-1964), Leonas Bistras (1890- 1971), Vincas Mykolaitis (1893-1967), Mečislovas Reinys (1884-1953) and Stasys Šalkauskis (1886-1941). Studying for her second degree she chose the philosophy of natural sciences as a major and the system of phi- losophy and sociology as a minor. On the 15th of June 1929 Amelija Mažylytė successfully passed her final ex-ams and was awarded a Diploma in Philosophy with honors (LU TFF: p. 17).

Figure 1. Amelija Mažylytė (1900-1972).

Without a doubt the study of philosophy had contributed to forming her firm catholic attitudes. She repre- sented these attitudes properly as an organizer of the educated women's movement in Lithuania and as a developer of women's education. Since the time at gymnasium she had belonged to the Ateitininkai organization, and since university studies she had been cooperating with the catholic press. In 1928-1929 she had been editing Ateitininkai girls' magazine "Naujoji Vaidilutė" which in 1931 developed into the educated women's monthly magazine. In her published works Amelija Mažylytė was promoting women's education and participation in politics embodying the image of a modern woman.

On the 15th of February 1929, as a recognition of her pedagogical and subject qualifications, the ministry of Education awarded her with a title of a Teacher of a Higher Education which gave her the right to teach mathematics, physics and cosmography; it also allowed her to teach the Russian language, history, geography and natural sciences at the secondary school level (LR: pp. 5-6).

In 1930-1936 Amilija Mažylytė continued her pedagogical career by teaching mathematics in Kaunas "Saulės" Seminary of teachers for girls and from 1931 cosmography at the S. Daukantas Seminary of Teachers. In 1936 when teacher seminaries were closed she was sent to work as a teacher in "Saulės" gymnasium (VVPI: p. 4). In 1940 Sovietization started and she was moved to the gymnasium in Zarasai. Later, already during the Second World War, she returned back to Kaunas and was teaching mathematics and its methodology in a restored teachers seminary. In 1943 she fell seriously ill and till 1944 stayed in the Red Cross hospital in Kaunas. In 1944-1945, when the Germans were retreating, she was employed at Braunsberg hospital. In 1945-1946 she re- turned back to Vilnius and worked at a hospital (Figure 2).

In 1946 Amelija Mažylytė was employed as a head of the books sector in the library of The Science Academy. She worked in this position till her retirement in 1956. The choice of a job outside teaching activities was most probably determined by the realities of the soviet education system which was based so different beliefs and values to those of her own. Though she made some attempts to return back to her well-liked mathematics and didactics. For two summers she taught mathematics in Kaunas at courses for teachers and in September 1947 she was also employed as a lecturer by Vilnius Pedagogical Institute in the Physics and Mathematics faculty at the Department of Mathematical Analysis and Geometry. However, she stayed in this job for only one academic year (VVPI: p. 6). In the remaining records she is described as a good mathematician and a serious teacher (LR: p. 6).

Amelija Mažylytė died in Vilnius on the 30th of April 1972.

3. PUBLISHED WORKS WITH A SPECIAL FOCUS ON THE HISTORY OF MATHEMATICS

After graduating Amelija Mažylytė dedicated quite a lot of her time to writing and publishing various works. One of the reasons for being so active in this field was her high command of the Lithuanian language. When still at the gymnasium she socialized with Jonas Jablonskis and Juozas Balčikonis and when Lithuania restored its independence she got acquainted with Kazimieras Būga (1879-1924). These prominent linguists helped her to enrich the native language and broaden humanitarian education. The important factor was also the knowledge of foreign languages; she knew the German, French, Russian and Polish languages (LR: p. 4). In 1933 Amelija

Mažylytė spent some time at Grenoble University where she broadened her experience in Western European science. Her broad outlook and erudition directed her scientific interests towards the history of mathematics.

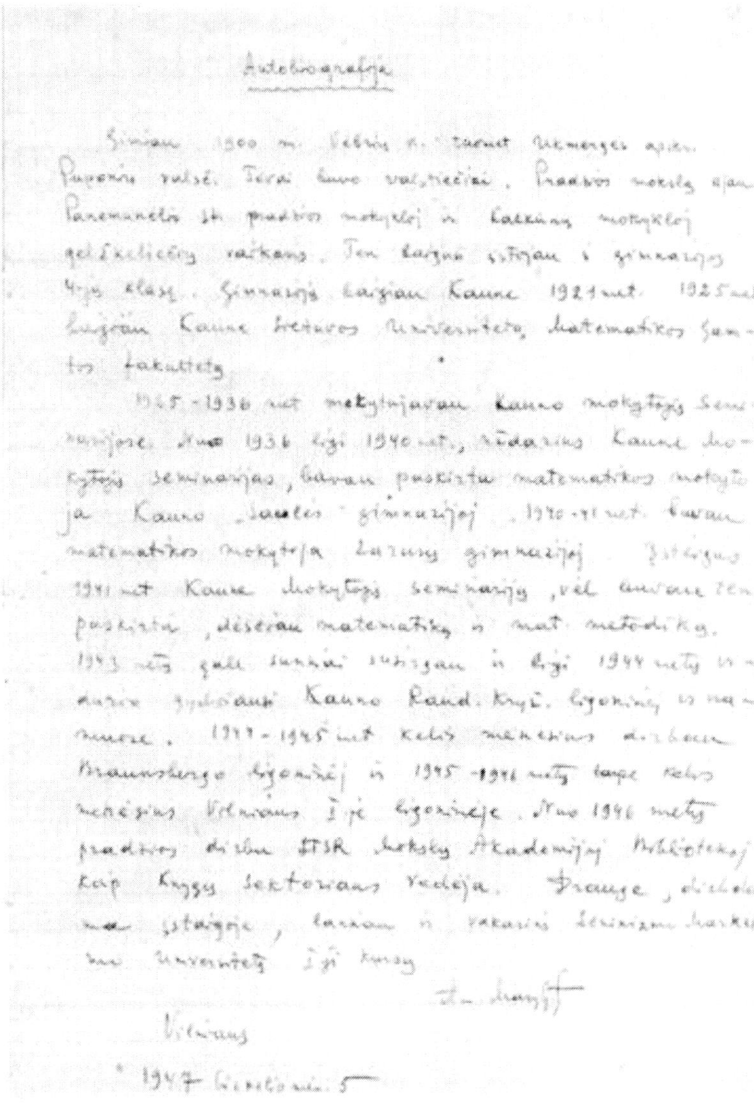

Figure 2. CV of A. Mažylytė (Vilnius, 1947).

She supported women's active participation in society by promoting education for girls and encouraging young women to study; she herself was actively participating in public life by taking part in such organizations as the association for the women with higher education, the catholic teachers' union and the circle of teachers of mathematics and physics (LR: p. 5 ; VVPI: p. 5). Mažylytė published her works on a wide variety of subjects but the main area of her interest was the subject of mathematics. Beside the articles dedicated to the development of the fields of mathematics, she also published articles about the didactics of mathematics which was again very closely interrelated to the history of mathematics. The best five articles on the latter subject were published in the period between 1928 and 1939.

Amelija Mažylytė, most probably encouraged by professor Pranas Dovydaitis with whom she had been socializing since her activities in the Federation of Ateitininkai through her second-degree studies at the university, took a challenge to write an article about Carl F. Gauss (1777-1855) "the Prince of Mathematicians" (Mažylytė, 1928). With references to the biographical researches carried out by A. Kitner, F. Mathe and W. Ahrens she revealed the personality of Carl F. Gauss not only as of a great mathematician but also as of a man who experienced family joy and tragedies. Presenting his biography she focused on the most important stages of his life. After mentioning a special talent for mathematics that had become obvious in early childhood and his readings of L. Euler, J. L. Lagrange, and I. Newton, when he was still a schoolboy, she focused on his first achievements in the theory of mathematics, i.e. a simple method of errors and the prime numbers theorem. Such success encouraged him to further study mathematics and to abandon the idea of getting deeper into philosophy. Carl F. Gauss became even more known as mathematician during his studies in Goettingen when in 1796 he presented a regular heptagon inscribed in a unit circle by using a ruler and calipers. Another stage of Carl F. Gauss life was related to the acquaintance with Professor John Friedrich Pfaff and the preparation of theses on the existence of polynomial equation roots. Evaluating his most important work "Arithmetical Investigations" Amelija Mažylytė wrote that he had described the theory of numbers since the times Diophantus and "must be considered to be the most important in the science of numbers" (Mažylytė, 1928: p. 422). Her article also covered other works published by Carl F. Gauss in the period between 1808 and 1828. She wrote about his achievements in astronomy, his contribution to works on triangulation and his interests in electromagnetic phenomena. She described Carl F. Gauss's working style as "make little but complete" and according to her "his works did not require correction". Describing the importance of his legacy to future generations Amelija Mažylytė used Ernst Kummer words "those are the masterful works with a stamp of exemplar" (Mažylytė, 1928: p. 424). Though the article was occasional, written for the Carl F. Gauss 150th anniversary, it was insightful and can be justly compared to the works written on the same topic by professors Pranas Dovydaitis and Zigmas Žemaitis.

Amelija Mažylytė, with the references to the major works by the Russian and German authors, also wrote two articles analyzing the development of arithmetic and geometry, the two oldest fields of mathematics. Both articles are similar in the way that they discuss the teaching of those subjects at different stages of history. The article "Evolution of Teaching Calculus" (Mažylytė, 1933a) covers the arts of calculating through the whole history of civilizations, starting with the Achmes Papyrus dated back to 1800 BC Egypt, and according to the author "the oldest known textbook of arithmetic", and closing with the German philosophers and mathematicians of the 19th century Johann Friedrich Herbart (1776-1841) and August W. Grube. The article elaborated on the most important facts of the science of arithmetic: Nicomachus textbook "Introduction to Arithmetic" that was used in Europe for a thousand years, the books propagating practical counting "The Rules of Abacus Counting" and "Definition of the Thing" by Gerbert d'Aurillac who became Pope Sylvester II in 999 and the work by Luca Pacioli "The Totality of Arithmetic". She devoted special attention to the work "The Calculation" by Adam Riese admiring "the absolute truth of number ratio" and evaluating it as "the thing important for trading" (Mažylytė, 1933a: p. 335). Because the development of arithmetic is closely related to its teaching in different stages of history in this article, the author intentionally names the 18th century as the century of Padagogy. The article showed how separate events of teaching mathematics in the past lead to the history of mathematics in general. By describing the development of arithmetic, Amelija Mažylytė gave meaning to the fact that in the 19th century arithmetic had become a compulsory subject in education (Figure 3).

The other article "The Most Important Thoughts about the Maturation of Geometry" (Mažylytė, 1939a ; Mažylytė, 1939b) covers a long and interesting history of this field in mathematics. The work was published in two different parts corresponding to two important stages of the development of geometry. The first part of the article covers the rise of geometry as a science, starting with the Rhind Papyrus that dates to 1650 BC through Assyro-Babylonian Clay tablets and finishing with the heyday of this science in Ancient Greece. After introduction to the famous Greek scientists and philosophers like Thales, Pythagoras, Hippocrates, Plato, Eudoxus, Menaechmus, Archimedes and Apollonius, she focuses on the work "Elements" by Euclid, especially on the problem of the 5th postulate. According to Amelija Mažylytė that problem in geometry was influenced by three "methodological trends". The first one was seeking to prove the rightness of the postulate, the second tried to refute the cases that had contradicted the postulate and the third one was demonstrating the independence of the postulate (Mažylytė, 1939a: p. 233). The second part of the article introduces the representatives of the second "methodological trend": Sacher, G. (1667-1733), Lambert, J. (1728-1777) and Lagendre, A. (1752-1833). While evaluating their contribution it was pointed out that they not only had a critical view on the foundations of geometry but also "prepared the basis for the non-Euclidean geometries" (Mažylytė, 1939b: p. 294). Later it was recognized that the geometry created by the Russian mathematician Nikolai Ivanovich Lobachevsky proved the independence of the 5th postulate from other axioms and postulates. By the end of the article the author also elaborated on the ideas of David Hilbert's work "Foundations of Geometry" by introducing five groups of axioms. In both parts of the article the author comes across as a smart mathematician; she used proofs of various triangle theorems to illustrate her points. She made an important conclusion that the geometry created by David Hilbert "does not have a tangible basis but is a matter of agreement" that may be any existing patterns (Mažylytė, 1939b: p. 298). The article was written with references to the works of famous Russian and German authors, such as V. Kagan, V. Deputatov, H. Wieleitner and D. Hilbert.

Such an insightful article on the development of geometry was a result of long and hard work. The epilogue is considered to be her other paper "The First Non-Euclideans" (Mažylytė, 1935) that was written in a popular but a very informative style providing the whole history of the analysis of the 5th Euclidean Postulate and revealing the achievements of such famous mathematicians as G. Sacher, N. Lobachevsky, J. Bolyai and B. Riemann. Comparing different geometries the author notices that Euclidean Geometry is based on imagination, Non-Euclidean Geometry (N. Lobachevsky and J. Bolyai) introduces analysis next to the imagination and Riemann's concept of space is considered to be a "variety of numbers" (Mažylytė, 1935: p. 680).

A. Mažylytė initiated the women's movement in Lithuania and was encouraging them to pursue university studies, including mathematics. She introduced Lithuanian women to role models of female mathematicians in the history of world science. By that she suggested the idea that education improves the social status of women and makes their lives more balanced. One more article by A. Mažylytė that often remains unnoticed by researchers is "Girls and Mathematics" (Mažylytė, 1936). The article focuses on the issues of teaching mathematics in gymnasiums for girls, talks about women with notable achievements in the science of mathematics and also stresses the importance of mathematics in the process of education. She disagrees with a stereotypical notion that "girls are not good at mathematics" and introduces well-known women mathematicians Hypatia from Alexandria, Maria Gaetana Agnesi from Italy (the 18th century), Sophie Germaine from France and Grace Chisholm from England (the 19th century). She also discussed Sofia Kovalevskaya's life and achievements in mathematics (Mažylytė, 1936: p. 382). The author is drawing a conclusion that "mathematics is not a kind of princess in the world of sorcery that is beyond the comprehension of girls" (Mažylytė, 1936: 383). By promoting an educated women's movement A. Mažylytė also had published the article about Nobel Prize winner Maria Curie-Sklodowska.

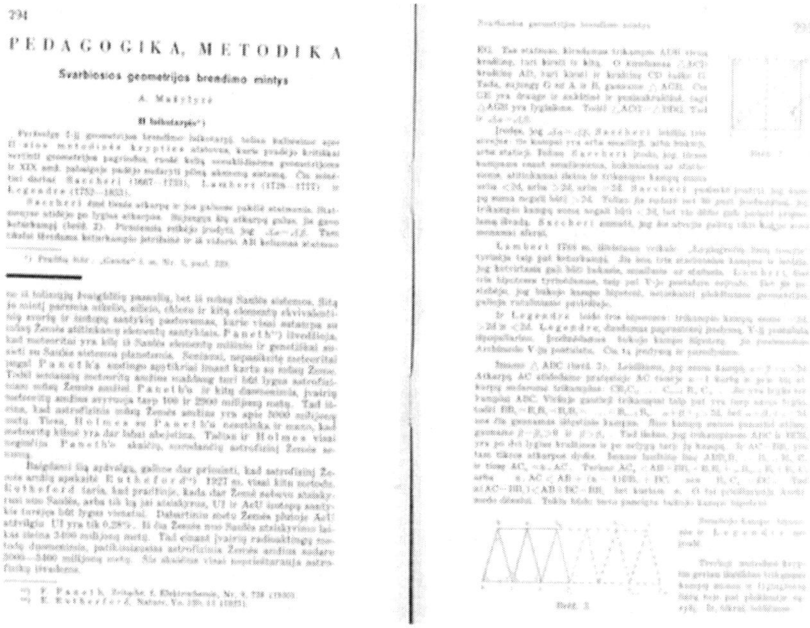
PEDAGOGIKA, METODIKA

Svarbiosios geometrijos brendimo mintys

Figure 3. Fragmen of A. Mažylė article "The Most Important Thoughts about the Maturation of Geometry".

She also published articles about the origins and development of the metric system in order to advertise its benefits in Lithuania (Mažylytė, 1939c). It contributed to the popularization of the system in the country as well as to reinforcing the foundation of a modern state.

4. CONCLUSION

Amelija Mažylytė (1900-1972) was brought up in an intelligent family of Lithuanian farmers who were open to learning. Not only her activities, but also her entire life had been about promoting education, especially that of women, in the young state of Lithuania. A. Mažylytė was one of the first

Lithuanian women seeking university education in the restored state of Lithuania . She is the first Lithuanian woman graduate in mathematics. A. Mažylytė also got a second university degree in philosophy and was actively involved in educational work especially by promoting the rich mathematical heritage of different civilizations in the Lithuanian language for the first time in the history of the country. She was one of the organizers of the educated women's movement inLithuania and a role model of a modern woman. In Soviet times she abandoned mathematics and social activities because of differences in views and values. The published works discussed in this article demonstrate her great erudition and the depth of her knowledge of the history of mathematics. The mentioned works can be separated into two groups. The first one discusses purely the history of mathematics and the second one proves the correctness of thesis by employing the historical facts.

ACKNOWLEDGEMENTS

Juozas Banionis Translated by Jūratė Marchertaitė, Lithuanian University of Educational Sciences.

REFERENCES

1. Ažubalis , A. (2012). Pirmoji Lietuvos universiteto diplomantė matematikė. Mokslo ir technikos raida Lietuvoje, Vilnius, 122-128.
2. Ažubalis, A. (2000). Pirmoji Lietuvos universiteto diplomantė matematikė. Mokslo Lietuva, 2000.05.18.
3. LR Švietimo ministerija (Ministry of Education of Rep. Lithuania). A. Mažylytės byla (Case of A. Mažylytė). LCVA[1], F.391, Ap.7, B.3653.
4. LU MGF tarybos protokolai (Lithuanian University Council on the Faculty of Sciences. VUB RS[2], F.96.VDU3.
5. LU TFF tarybos protokolai (Lithuanian University Council on the Faculty of Theology and Philosophy). LVCA, F. 631, Ap.13, B.155.
6. Mažylytė, A. (1928). Carl Friedrich Gauss (1777-1855). Kosmos, Nr. 9, 422-424.
7. Mažylytė, A. (1933a). Skaičiuotės mokymo raida. Lietuvos mokykla, Nr. 18, 333-340.
8. Mažylytė, A. (1933b). Organizuotas moterų veikimas At-kų Federacijoje. Naujoji Vaidilutė, Nr. 7.
9. Mažylytė, A. (1935). Pirmieji neeuklidininkai. Šviesos keliai, Nr. 11, 675-680.
10. Mažylytė, A. (1936). Mergaitės ir matematika. Naujoji Vaidilutė, Nr. 10, 379-385.
11. Mažylytė, A (1939a). Svarbiausios geometrijos brendimo mintys (I). Gamta, 1939, Nr. 3, 229-234.
12. Mažylytė, A. (1939b). Svarbiausios geometrijos brendimo mintys (II). Gamta, 1939, Nr. 4, 294-301.
13. Mažylytė, A. (1939c). Metras. Lietuvos mokykla, Nr. 12, 766-768.
14. Riauba, B. (2011). Amelija Mažylytė—Matematikė nuo Panemunėlio. Prie Nemunėlio, Nr. 2, 12-14.
15. VVPI (Vilnius State Pedagogical Institute) A. Mažylytės byla. LEUA[3], F.K2744.

APPENDIX

List of A. Mažylytė (1900-1972) research (1928-1939) papers

1. Carl Friedrich Gauss (1777-1855), Kosmos, 1928, Nr. 9, pp. 422-424.
2. Kilimas moterų švietimo/Rise of Womens Education, Naujoji Vaidilutė, 1928, Nr. 10, pp. 255-260.
3. Matematikos mokymo tikslai/Proposes to Teaching Mathematic, Lietuvos mokykla, 1932, Nr. 7, pp. 151-155.
4. Skaičiuotės mokymo raida/Evolution of Teaching Calculus, Lietuvos mokykla, 1933, Nr. 18, pp. 333-340.
5. Mergaičių švietimas/Girls Education, Naujoji Vaidilutė, 1934, Nr. 1, pp. 14-16; Nr. 2, pp. 77-81.
6. Pirmieji neeuklidininkai/The First Non-Euclideans, Šviesos keliai, 1935, Nr. 11, pp. 675-680.
7. Mergaitės ir matematika/Girls and Mathematics, Naujoji Vaidilutė, 1936, Nr. 10, pp. 379-385.
8. Matematikos žurnalai mokykliniais klausimais/Mathematics magazine for Scholl Questions, Gamta, 1939, Nr. 1, pp. 86-87.
9. Marija Curie (1867-1934), Naujoji Vaidilutė, 1939, Nr. 2, pp. 89-98.
10. Svarbiausios geometrijos brendimo mintys/The Most Important Thoughts about the Maturation of Geometry, Gamta, 1939, Nr. 3; 4, pp. 229-234; 294-301.
11. Studentės ir studijos/Students (Females) and Studies, Naujoji Vaidilutė, 1939, Nr. 8-9, pp. 401-405.
12. Metras/Metre, Lietuvos mokykla, 1939, Nr. 12, pp. 766-768.

NOTES

1. LCVA—Central State Archive of Lithuania.
2. VUB RS—Vilnius University Library, Departament of Manuscrips.
3. LEUA—Archive of Lithuanian University of Educational Sciences.

CHAPTER 3

Epistemology of Greek Harmonics

Danilo Capecchi

Dipartimento di ingegneria strutturale e geotecnica, University of Rome, Rome, Italy

ABSTRACT

Greek harmonics is the theory of the organization of notes with respect to the highness or lowness of their pitches. This was a mathematical science, a thorough knowledge of which was not necessary for the professional musician and was thus cultivated essentially by mathematicians and philosophers. For them, harmonics was mixed mathematics, the mix being between experimental assertions derived from hearing and propositions derived from reason, that was, mathematics. This classification was maintained through the Middle Ages and the Renaissance. Despite the important role played by harmonics in ancient science, little consideration has been given by historians of science. This study shows that harmonics merits a careful study and that its methods are the same as other sciences. Particular attention was devoted to Ptolemy who applied the same approach to harmonics as he did to astronomy, astrology and geography.

Keywords: Harmonics, Epistemology, History of Science, Greek Music, Mixed Mathematics

1. INTRODUCTION

The theory of organization of notes with respect to the highness or lowness of their pitches was named harmonics by the Greeks. Modern harmonics deals with concords, that is notes played simultaneously. Greek harmonics dealt with notes played in sequence or, in other words, musical scales. Even though the concept of concord was fundamental in it, to formulate a theory of a musical interval, Greek music was essentially melodic, with no use for concords (Comotti, 1991) . Ptolemy and Aristoxenus, two of the most important theoreticians of the time, each provided a definition of harmonics. Ptolemy defined harmonics as a "perceptive function of the differences in sounds between high and low" (Ptolemy, 2000: p. 2) . By extension, harmonics is the discipline that deals with the way in which sounds should be organized so that they sound "melodic", which is pleasurable to the ear.

Aristoxenus definition was more articulated:

Harmonics is only a part of the musician's accomplishment, as are the sciences of rhythm, metre and instruments. We must therefore give an account of harmonics and its parts. It is to be understood, speaking generally, as the science which deals with all melody, and enquires how the voice naturally places intervals as it is tensed and relaxed. For we assert that the voice has a natural way of moving, and does not place intervals haphazardly (Barker, 1989: p. 149) .

Harmonics was a mathematical science, a thorough knowledge of which was not necessary for a professional musician and was thus cultivated essentially by mathematicians and philosophers. There were two main traditions, usually referred to as Pythagoreans and Aristoxenians. The former assumed that pitches of notes could be represented by integers and their relations by ratios (usually simple ratios). This was the dominant position of music theorists, namely Philolao, Archytas of Tarentun, Plato, Euclid, Theon, Nicomachus, and Ptolemy. The Aristoxenian school, on the other hand, denied the fundamentality of the mathematical representation of the pitches of sound and developed an essentially empirical approach, albeit quantitative and thus associated with mathematics.

Mathematicians and philosophers considered harmonics as a mixed mathematics (Renaissance terminology), the mix being between experimental assertions derived from hearing, and propositions derived from reason, or, rather, mathematics. This classification was maintained throughout the Middle Ages, where it was part of the quadrivium (arithmetic, geometry, music and astronomy), and the Renaissance (Palisca, 1985) . In any scientific discussion of the 17th century, harmonics was involved. These discussions included, among others, René Descartes, Marin

Mersenne, Johannes Kepler, Christiann Huygens, Athanasius Kirker, and John Wallis. Galileo Galilei considered several fundamental musical questions in his scientific writings and he was not only a lutanist himself, but also a son, brother, and father of musicians, in short, a member of a musical dynasty (Palisca, 1961: p. 92) .

Despite the important role of harmonics in ancient science, little consideration has been given to this by historians of science. The great synthesis of the 20th century was carried out by Duhem, Koyré, Dijksterhuis, etc., practically ignores harmonics. This is not so for the historians of mathematics; most textbooks mention harmonics, because of the reference to the theory of proportions as well as the discussion on irrational numbers. Harmonics has been and is of course the object of historians of music that has written widely on the theme. Bibliographical references can be found in (Barker, 1989) . However, they are not generally interested in science, apart from the fundamental works by Andrew Barker (Barker, 1978, 1989, 1994, 2006) .

It is the purpose of the present paper to cover part of this gap, which to my opinion is inexcusable. In particular, the paper intends:

1) To clarify scientific aspects and the epistemological implication of various approaches;
2) To relate the epistemology of harmonics to that of other sciences.

A particular attention was devoted to conceptions of Ptolemy who used the same approach for harmonics as he did for astronomy, astrology, and geography.

2. FUNDAMENTS OF GREEK MUSIC

According to a tradition going back to Aristotle, sounds were classified as discrete or continuous. Continuous sounds are such that no single part is equal-toned over a perceptible interval of time, that is there is a continuous variation in pitch. Discrete sounds are such that their parts remain equal-toned over a perceptible interval of time (Ptolemy, 2000: p. 15) . Music and harmonics are concerned with discrete sounds only; a sound of a given pitch is called a note.

And now we call such sounds notes, since a note (φθόγγον) is a sound which has one and the same tone (τόνος) (Ptolemy, 2000: p. 15) .

But Greeks did not possess the concept of an absolute value for the pitch of a note; they only could say that one pitch was higher than another (today we know that pitches are associated with the frequencies of vibration of air, a physical magnitude measured in Hz?cycles per second. A note corresponds to a given frequency; for instance we associate 440 Hz with the note La). In a Greek context a note alone had no sense, it was "irrational" (Ptolemy, 2000: p. 15) [1]. A note thus should be associated with other different notes. Such associations are made up of emmelic (melodic) notes, which when joined together are acceptable to the ear; otherwise they are made up of ecmelic (unmelodic) notes. They are termed concordant if they create a homogeneous impression on the hearing, discordant if they do not (Ptolemy, 2000: p. 17) .

Two different notes are said to be divided by a (musical) interval. How to measure an interval differed for Aristoxenians and for Pythagoreans. The former introduced something similar to a unit of measure that was based on hearing only. The pitches of notes were admitted to vary continuously as points in a straight line. Here clearly a geometric analogy was involved; as the interval is what keeps two points on a straight line apart. When the pitches of two notes are compared, one of which is maintained as fixed while the other is varied, hearing recognizes concordance for certain values of the varying notes, in the sense that the two notes blend and appear as a unique pleasant sound. The first interval for which this occurs was called a fourth, the second a fifth, the third an octave. The interval between the fifth and the fourth was called a tone and was assumed as a unit of measure of intervals.

Pythagoreans operated in a very different way that could be classified as arithmetic. They associated a note with the length of a vibrating string, which could be measured and expressed by a number. When comparing numbers corresponding to different lengths of vibrating strings they realized that the interval, as it is felt by the hear, could be represented by the ratio of the numbers associated with the lengths. In particular a fourth corresponds to a ratio of 4: 3, a fifth to a ratio of 3: 2, an octave to a ratio of 2: 1 (today we know that the length of a string, ceteris paribus, is inversely proportional to its frequency of vibration, and a fourth corresponds to two notes whose frequencies have a ration o 4/3 and so on). The composition (sum) of intervals is the product of ratios, their division (difference) the division of ratios. Of course an interval in such an instance has lost all reference to a length, but the name was maintained. Pythagoreans also defined the tone as the difference between a fifth and a fourth, which has the ratio obtained by dividing the ratios of the two intervals, that is (3:2):(4:3) = 9:8[2].

The two concept of the measure of intervals to some extent do clash. For instance Aristoxenians, in their geometric approach, maintained that the fourth measured two and half tones, the fifth three and half and the octave six tones. For the Pythagoreans, however, six tones spanned an interval of $(9:8)^6 \simeq 2.002:1$, greater than an octave. Moreover, the Aristoxenians could divide the tone in equal parts, hearing and geometry allows this; the Pythagoreans could not; arithmetics forbids this because dividing the ratio of 9:8 in two n equal parts requires evaluating the n-square root (modern meaning) of 9:8, which is not rational, and thus not an interval.

The tetrachord, an association of notes, is the basis of Greek music. It is made up of a set of four notes of various heights ordered from the lowest to the highest. The lowest and highest notes are separated by an interval of fourth. The intervals among the intermediate notes and the lowest note may vary and defines the genus of the tetrachord. There are three main genera: enharmonic, chromatic and diatonic. In the enharmonic genus the interval between the top of the tetrachord and the highest of the movable notes is high; in the diatonic the interval is relatively small and in the chromatic genus its size is intermediate between those of the other two genera. This is illustrated in Figure 1 where the notes are indicated with the name of the chords in a Lyre: ὑπάτη, the farthest chord from the singer; παρυπάτη, the closest to ὑπάτη; λιχανός, the chord touched by the forefinger and μέση, the middle chord.

Tetrachords of the same genus can be joined to constitute a system of tetrachords. They are in conjunction if the lowest note of the upper tetrachord is the same as the highest note of the lower tetrachord; this system is comprised of fourths, with seven notes and six intervals. They are in disjunction if the two tetrachords are separated by an interval; here the system is made up of a fourth and a fifth alternate, with eight notes and seven intervals; two disjoint tetrachords form an octave.

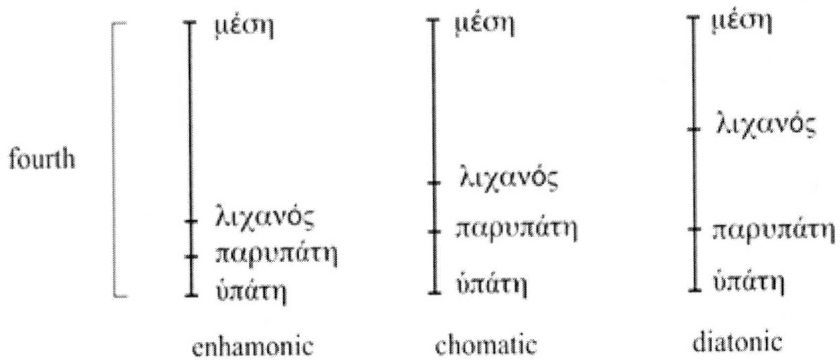

Figure 1. Genera of tetrachords.

The position of the notes in a disjoint tetrachord justifies the names, fourth, fifth and octave, assigned by the Greeks to the intervals: the interval between the first and fourth note is indeed a fourth (by definition of the tetrachord), the interval between the first and the fifth note is a fifth and the interval between the first and the eighth note an octave.

3. EMPIRICIST AND RATIONALIST THEORETICIANS

As already mentioned, in classical and hellenistic Greece there were two basic approaches to the study of harmonics, the Pythagorean and the Aristoxenian. It must be said that neither school was monolithic and that the work of each did not develop independently of the other. In the following I will present two typical expressions of the two schools from two of the oldest treatises of musical theory. One of these is the Euclidean Sectio canonis, and the other the Aristoxenian Elementa harmonica.

3.1. Euclid

The attribution of the Sectio canonis to Euclid (3rd century BC) is controversial, as often occurs for ancient texts; but this is not my concern. The text to which I refer to is that used by Barker who translated it into English (Barker, 1989: pp. 191-207) . It consists of a short premise and nineteen propositions (theorems). The premise refers to general physical and epistemological considerations. Sound was associated with motion, "If there were stillness and no movement there would be silence" (Barker, 1989: p. 191) . From the nature of motion derives the nature of notes, that are not explicitly defined. Euclid, somewhat vaguely classified motion accordingly to its frequency:

And since of movements some are closer packed, others more widely spaced, those which are closer packed producing higher notes and those which are more widely spaced lower ones, it follows that some notes must be higher, since they are composed of closer packed and more numerous movements, and others lower, since they are composed of movements more widely spaced and less numerous (Barker, 1989: p. 192) .

Here packing could be interpreted as occurring in time, so that a closer packed motion has a greater frequency of oscillation. If this interpretation is correct, Euclid was asserting a vision of sound similar to a modern one, a vision not common in ancient Greece, where the association of pitches with speed of air was more common.

Euclid assumed that packing varies in a discrete way, so that notes do, too, and could be represented by numerical discrete values, that is integer numbers. Notes are classified as concordant or discordant. Concordant notes are those that blend and seem almost to form a unique sound, discordant those that do not.

Among notes we also recognize some as concordant, others as discordant, the concordant making a single blend out of the two, while the discordant do not. In view of this, it is to be expected that the concordant notes, since they make a single blend of sound out of the two, are among those numbers which are spoken of under a single name in relation to one another, either being multiple or epimoric [superparticular][3] (Barker, 1989: p. 192) .

This quotation expresses, with no particular emphasis, the fundamental principle of the treatise, and belongs to the Pythagorean school. This is the principle by which the notes in a concord must be in whole or super-particular ratios of the numbers that represent them. For many Pythagoreans this principle was of a metaphysical nature and derives from the idea that the phenomena of the physical world are governed by numbers and numerical relations that obey particular rules of "symmetry". For other Pythagoreans the principle, albeit maintaining a metaphysical basis, is accepted on the basis of musical practice. In both cases the principle was taken for granted and assumed as a premise of more geometrico argument, typical of mixed mathematics.

Euclid belonged to the second category of Pythagoreans; his mixed position, though not explicitly stated, is clear because he devoted large parts of the Sectio canonis (in particular most of his theorems) to prove that notes considered as concordant by hearing were (by chance?) represented by simple ratios; whereas pure Pythagoreans would have said that notes corresponding to simple ratios should be concordant a priori.

After the brief preface, Euclid presented the nineteenth propositions. The first nine were purely mathematical covering the theory of proportions. However, in these propositions musical terminology is evident; the term interval is used to indicate a ratio of integers; the union of intervals is an interval given by the product of the ratios of the constituent intervals; the difference of two intervals is the division of the corresponding ratios.

The following list reports sixteen of the nineteen propositions: the last three are left out for their lack of interest or pertinence.

1) If a multiple interval put together twice makes some interval, this interval too will be multiple;

2) If an interval put together twice makes a whole that is multiple, then that interval will also be multiple;

3) In the case of an epimoric [super-particular] interval, no mean number, neither one nor more than one, will fall within it proportionally;

4) If an interval which is not multiple is put together twice, the whole will be neither multiple nor epimoric;
5) If an interval put together twice does not make a whole that is multiple, that interval itself will not be multiple either;
6) The duple interval is composed of the two greatest epimoric intervals, the hemiolic (3:2) and the epitritic (4:3);
7) From the duple interval and the hemiolic, a triple interval is generated;
8) If from a hemiolic interval an epitritic interval is subtracted, the remainder left is epogdoic (9:8);
9) Six epogdoic intervals are greater than one duple interval;
10) The octave interval is multiple;
11) The interval of the fourth and that of the fifth are each epimoric;
12) The octave interval is duple;
13) It remains to consider the interval of a tone, to show that it is epogdoic;
14) The octave is less than six tones;
15) The fourth is less than two and a half tones, and the fifth is less than three and a half tones;
16) The tone will not be divided into two or more equal intervals.

Proposition 10 refers to the octave, which is defined from a musical point of view, as the interval between two notes that shows the maximum concordance according to hearing sole. Nothing is said about its numerical ratio, whose evaluation is the object of the proposition. The octave is also presented as the interval that separates the higher note in a two-octave system from the μέση, the intermediate note, located an octave below, or the μέση from the lower note of the system. The statement of the proposition and its proof follows (see Figure 2):

Proposition 10 The octave interval is multiple Let A be nēte hyperbolaiōn [the highest note], let B be mese and let C be proslambanomenos [the lowest note]. Then the interval AC, being a double octave, is concordant. It is therefore either epimoric [super-particular] or multiple. It is not epimoric, since no mean falls proportionally within an epimoric interval. Therefore it is multiple. Thus since the two equal intervals AB and BC put together make a whole that is multiple, AB is therefore multiple too (Barker, 1989: p. 199) .

Euclid began with an empirical observation, hearing, that the two-octave interval is concordant. From his architectonic principle he could deduce that the interval AC must be either multiple or super-particular. But it can not be super-particular because the μέση B divides the interval AC into two equal intervals (one octave from the one hand and another octave from the other hand). By proposition 3, however, it is known that a super-particular ratio cannot be divided into two equal ratios, thus the ratio AC is multiple. By proposition 2 it follows that the ratio AB (an octave) is multiple. The demonstration appears somewhat contorted and in my opinion contains a petitio principii. That is, it confuses the fact that the μέση is halfway musically with the fact that it is halfway numerically. Nonetheless, it remains quite interesting, as it illustrates how the experimental data is analyzed with the theory of the ratio between numbers.

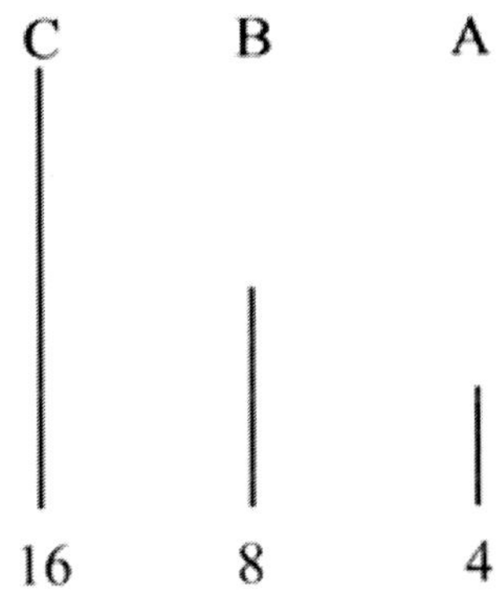

Figure 2. Two joined octave intervals[4].

Demonstrations of other propositions proceed in much the same way. Proposition 13 is worth noting that shows that the range of a tone is expressed by the ratio 9: 8, that is it is epogodic.

Proposition 13. It remains to consider the interval of a tone, to show that it is epogodic. It is known that if an epitritic [4:3] interval is subtracted from a hemiolic [3:2] interval, the remainder left is epogdoic [9:8]. And if the fourth is taken from the fifth, the remainder is the interval of a tone. Therefore, the interval of a tone is epogodic (Barker, 1989: p. 201) .

The starting point is the implicit definition of the interval as the distance between a fifth and a fourth (their difference). As proposition 11 (and 12) show that the intervals of fourth and fifth have ratios of 4:3 and 3:2 respectively, the proof of proposition 13 is immediate because the subtraction of the intervals is equal to their division, so: (3:2): (4:3) = 9:8.

Worthy of note are propositions 14 and 15, later emphasised by Ptolemy, to counteract the thesis that he attributed to Aristoxenus, according to which the octave is six tones, the sum of a fourth of two and a half tones, and a fifth of three and a half tones. Proposition 16 shows the limit of the arithmetic approach to harmonics. A tone cannot be divided into two equal intervals because a super-particular ratio cannot be divided into two equal ratios; this would correspond to the evaluation of the square root (modern term), which is irrational and thus not a ratio.

3.2. Aristoxenus

Aristoxenus (c. 360-c. 320 BC) was an important student of Aristotle; he even aspired to succeed him in Liceum, however with no success. He is considered (one of) the most important figures in the history of Greek musical theory. Ingemar Düring, the editor of Ptolemy's Harmonics, observes that "Aristoxenus's theory of music is revolutionary. It rests on the fundamental principle that the human ear is the sole arbiter of the correctness of pitches and harmonic function" (Düring, 1956) . He was a prolific writer in philosophy, although little interested in metaphysics. He wrote several treatises on music and, among these, the Elementa harmonica, that has come down to us relatively intact. At least two editions exist in English; one, older, which also contains the Greek text(Aristoxenus, 1902) , and a more recent version (Barker, 1989) . There is also a large commentary(Laloy, 1904) and quite good Italian edition (Aristoxenus, 1954) .

The Elementa harmonica is the most ancient work known on music and is a repository of ancient thought and accomplishments. The extant text is divided into three books: the first two are partially overlapping and report the main elements of theory and epistemology. The third book contains the demonstration of some propositions of a musical nature. Aristoxenus' epistemology is strongly influenced by that of Aristotle; harmonics must be based on principles considered true and evident. Like Aristotle he assumed that during a demonstration it was not possible to move from one genus to another (prohibition of metabasis):

Since every science that consists of more than one proposition should adopt first principles from which the things dependent on these principles can be demonstrated, we must adopt such principles, bearing the following two points in mind. We must ensure, first, that each of the fundamental propositions is true and evident [emphasis added], and secondly that each is such as to be accepted by perception as belonging among the primary parts of harmonic science: for whatever demands demonstration is not fundamental. And in general we must be very careful, as we set out, not to slip into extraneous territory by beginning from a conception of sound as a movement of the air, and equally not to turn back too soon and leave out many things which belong to the subject (Barker, 1989: p. 159) .

Principles of harmonics should be derived directly from harmonics itself and not from other sciences such as acoustics, for instance.

Both the first two books try to clarify what is harmonic; meanwhile, it does not coincide with music but only affects the organization of musical intervals:

Harmonics is only a part of the musician's accomplishment, as are the sciences of rhythm, metre and instruments. We must therefore give an account of harmonics and its parts. It is to be understood, speaking generally, as the science which deals with all melody, and enquires how the voice naturally places intervals as it is tensed and relaxed. For we assert that the voice has a natural way of moving, and does not place intervals haphazardly (Barker, 1989: p. 149) .

Harmonics depends on two things, hearing and reason. And the sense perception, hearing, plays a major role, which is not the case with mathematics:

The geometer makes no use of the faculty of perception: he does not train his eyesight to assess the straight or the circular or anything else of that kind either well or badly: it is rather the carpenter, the wood-turner, and some of the other crafts that concern themselves with this. But for the student of music accuracy of perception stands just about first in order of importance, since if he perceives badly it is impossible for him to give a good account of the things which he does not perceive at all (Barker, 1989: pp. 150-151) .

Aristoxenus also stated the importance of memory to compare what was past with what is present:

Comprehension of music comes from two things, perception and memory: for we have to perceive what is coming to be and remember what has come to be. There is no other way of following the contents of music (Barker, 1989: p. 155) .

He divided harmonics into seven parts. The first concerns the distinction between genera (inside the tetrachord), the second the distinction between intervals, the third notes; the other parts relate to more complex organizations, such as the union of more tetrachords.

In Book I, sound is presented as a process in which pitches may vary continuously from top to bottom and vice versa, with an image that evokes the movement of a point on a line. There is obviously an upward and downward physical limit to pitches both for the generation and for the ear. However, Aristoxenus rather than the absolute value of the pitch of a note, refers to the interval of two notes. For example, according to him, the ear cannot clearly distinguish an interval under a fourth of a tone. After this introduction he went on to define the note. A note in essence is the stabilization of the sound on a single pitch.

Now that this is understood we must say what a note is. To put it briefly, a note is the incidence of the voice on one pitch: for it is when the voice appears to rest at one pitch that there seems to be a note capable of being put into a position in a harmonically attuned melody. That, then, is the sort of thing a note is (Barker, 1989: p. 13) .

The interval is what is limited by two notes that do not have the same pitch, in analogy to the interval of a straight line which is defined by two points. The word Aristoxenus used for interval is διάστεμα, the same used by Euclid in the Elements (Barbera, 1977) . Intervals are classified with respect to their size, if they are concord or discord, if they are composite or in-composite, with respect to the genus to which they belong.

Aristoxenus said that there were eight magnitudes of concords, though he referred only to three of them. The smallest is the fourth, the second the fifth, the third is the sum of a fourth and a fifth, that is an octave. Other concordant intervals are obtained by adding an octave to a concordant interval.

We must explain first that the addition of any concordant interval to the octave makes the magnitude resulting from them concordant. This is a quality intrinsic and peculiar to the concord of the octave, for whether the concord added is smaller than it or equal or greater, the result of the combination is concordant (Barker, 1989: p. 160) .

That by which the fifth is greater than the fourth is named a tone. Moreover the fourth is two and half tones and the fifth three and a half tones. Note that Aristoxenus never explicitly said that an octave consists of six tones, though this can be deduced from his writings. A tone may be divided in a manner that intervals result melodic into two parts (halftone), three parts, four parts. However he accepted that from a purely abstract point of view there is no limit in the division, as occur in a geometric line. A tone defines a metric in Aristoxenus' harmonics, so that it becomes mathematical.

At the end of Book II Aristoxenus laid down the principle on which his whole theory of harmonics was based; how the intervals should be organized so that they result melodic:

We must next give an account of the first and most indispensable of the conditions that bear upon the melodic combination of intervals. Let it be accepted that in every genus, as the melodic sequence progresses through successive notes both up and down from any given note, it must make with the fourth successive note the concord of a fourth or with the fifth successive note the concord of a fifth. Any note which fulfils neither of these conditions must be considered unmelodic relative to all the notes with which it fails to form concords in the numerical relations mentioned(Barker, 1989: p. 167) .

Alongside this architectonical principle there is another, which concerns how two tetrachords are joined:

Similar to it, in a way, is the principle which governs the positioning of tetrachords in relation to one another. Tetrachords that belong to the same system must have one or other of two properties. Either they must be concordant with one another, so that each note of the one forms some concord with the corresponding note of the other, or they must both be concordant with the same tetrachord, each of them being continuous with the one with which they are concordant, but not in the same direction (Barker, 1989: p. 167) .

These two principles are largely experimental in character, even if they are mediated by reason. They do not reflect, however, the Aristotelian epistemology he declared. Indeed its principles can be considered as true but they were certainly not obvious.

From a reading of the first two books of Aristoxenus, what emerges is a scholar who adopted a rigorous and fairly concise language, closer to that of mathematicians than to philosophers; his arguments were based on natural logic and not on syllogism, and in most cases they are sufficiently stringent.

Aristoxenus did not repudiate the use of mathematics; he only maintained that it could not be of any help in establishing the principles of harmonics, as the Pythagoreans did. Having introduced a metric, the tone, he could measure musical intervals and can prove theorems with the aid of geometry, and thus, as already noted, could move in the root of mixed mathematics. Barker in(Barker, 1989: p. 16) criticizes Aristoxenus for his claim that tone could be divided into many parts—Aristoxenus also mentioned intervals as small as one twelfth of a tone, much smaller than the smallest melodic interval of one fourth of tone—because he did not dispose of any empiric method to verify this division; and because he did not possess a criterion to hear an interval of one fourth of tone he was not justified in claiming that it was the smallest melodic interval. I do not share Barker's criticism and believe that, in any mixed mathematics, aspects are suggested that cannot be verified experimentally, but this notwithstanding can still be classified as empiric.

How Aristoxenus presented his measure of intervals, always made reference to the three concords, the fourth, the fifth and the octave, the only ones that the ear can distinguish very clearly. For instance to find a note E, two tones below a given note A (see Figure 3), one can move by a fifth and a fourth alternatively, as explained in the quotation:

Thus, if we have the task of constructing from a given note a discord such as a ditone downwards (or any of those which can be constructed by means of concords) we should construct from the given note [A] a fourth upwards [B], from there a fifth downwards [C], then another fourth upwards [D], and then another fifth downwards [E]. In this way the ditone [AE] downwards from the given note will have been constructed (Barker, 1989: p. 167) .

Using a variant of this approach Aristoxenus could measure the interval of a semitone (see below), but in no way could he measure smaller intervals.

One important reason for which Aristoxenus thought that mathematics could not exhaust the study of harmonics, was that the musical value of a set of notes is not only given by the measure of their intervals, but it is also given by its function in a melody, by its dynamics. For instance, the intermediate notes of a tetrachord of the same genus could be described by different (infinite) mathematical ratios and thus they could be considered as different notes. However, the ear, if the variation of the pitches is not very high, hears the notes as belonging to the same genus, and this could not be explained by mathematics (Barker, 1978; Barker, 1989: pp. 12, 161-163) .

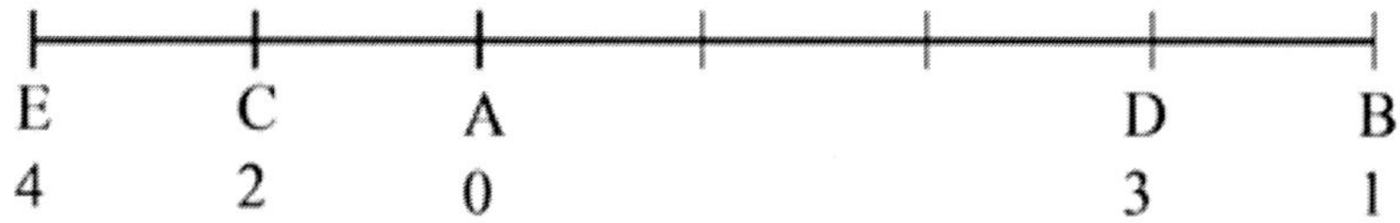

Figure 3. Construction of a ditone interval.

Book III concerns propositions (theorems) to be proved starting from assumed principles. It parallels Euclid's Elements. There are 27 propositions. Some of the most relevant are listed here.

1) Successive tetrachords are either conjunct or disjunct.

3) In variations of genus it is only the parts of the fourth that change, while the interval peculiar to disjunction is unaltered.

8) Two ditones will not be placed in succession.

9) In the enharmonic and chromatic, two tones will not be placed in succession.

11) In this same genus, two semitones will not be placed in succession.

To show how Aristoxenus' argues this, below is a proof of proposition 1 and the proposition considered in Book II, according to which a fourth is composed of two and half tones.

Successive tetrachords are either conjunct or disjunct.

Let it be called 'conjunction' when two tetrachords, successive in the melodic series and with the same arrangement [schema], have in the middle a common note; and let it be called "disjunction" when between two tetrachords, successive in the melodic series and with the same arrangement, there lies a tone. That one or the other of these must apply to successive tetrachords is clear from the assumptions we laid down. When the notes four places apart in a continuous series form the concord of a fourth they will produce conjunction, and when those five places apart from a fifth they will produce disjunction. But the notes must fulfil one or other of these conditions, so that successive tetrachords must have one or the other of the properties mentioned (Barker, 1989: p. 170) .

Aristoxenus used his principle by which, in the conjunction of two tetrachords, as in any musical scale, the distance of four successive notes should have an interval of a fourth, that of five successive notes an interval of a fifth. However the first case is that of a conjunct tetrachord, the second that of a disjunct tetrachord. Considered otherwise one could turn the proposition into the following problem: how to merge two tetrachords so that their union respects the principle of melody set by Aristoxenus? The answer is: two tetrachords must be either conjunct or disjunct, tertium non datur.

The fourth consists of two and a half tones.

The question whether we were correct in the assumption, which we made in our introduction, that the fourth consists of two and a half tones, can be investigated most accurately in the following way. Take a fourth, and starting from each of its extremes in turn, mark off a ditone by means of concords. It is clear that the remainders are equal, since equals have been taken from equals. Next, take a fourth upwards from the lower bounding note of the upper ditone, and take another fourth downwards from the upper bounding note of the lower ditone. It is clear that next to each of the notes which bound the resulting systema there will be two consecutive remainders which must be equal, for the reasons stated before. When this construction has been set up, we must bring to the judgement of perception the outermost of the notes that have been located. If they appear to perception as discordant, it will be evident that the fourth is not two and a half tones: but if they sound the concord of a fifth, it will be evident that the fourth is two and a half tones. For the lowest of the notes constructed was tuned to make the concord of a fourth with the upper bounding note of the lower ditone, and the highest of the notes constructed has turned out to form the concord of a fifth with the lowest; so that since the difference is a tone, and since it is divided into equal parts, each of which is a semitone and is also the excess of a fourth over a ditone, it is clear that the fourth consists of five semitones (Barker, 1989: pp. 168-169) .

Aristoxenus's proof assumes that intervals could be added together to obtain larger intervals, just as with the line segments. This is not only developed on the basis of the premises and established principles but, at a certain point, it also requires experience to prove a crucial point. The demonstration can be better understood if reference is made to geometry. In Figure 4, the segment AB represents the interval of a fourth, which is concordant by definition. Point C is a pitch distant a ditone from B (that is the segment CB is a ditone)[5]; similarly the point D is a pitch distant two ditones from A. Obviously AC = BD. Aristoxenus then considered pitches E and F a fourth away from D and C respectively. At this point he declared the need for a musical experiment, to see if the interval EF is concordant. And without being particularly clear, he stated that EF is in fact concordant. Thus it is easy to show that EF is a fifth because it is a concord closer to a fourth than it is to an octave (this also should be an experimental result) and there are no other nearby concords. But if EF is a fifth, by definition of tone, DF is the length of the tone (difference between EF and AB), DB is a semitone (half of a tone) and AB, the sum of the ditone AD and of the half tone DB, is two tones and a half.

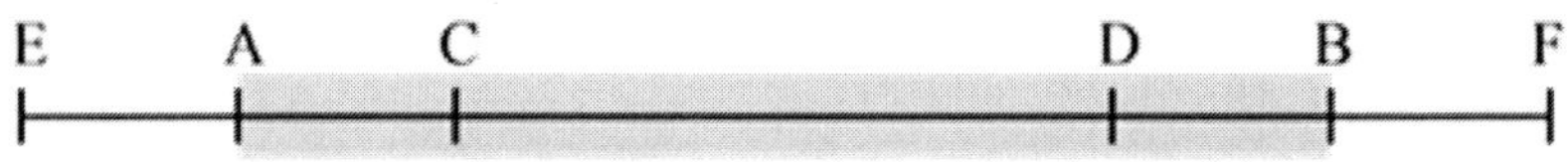

Figure 4. A fourth [AB] is composed of two [AD] and half [DB] tones.

Aristoxenus's proof, though very interesting is certainly questionable. In order to be considered a true demonstration more geometrico, the result of experimental observations should appear in the premise. However, this is only a defect of a formal kind and could be overlooked. What perplexes the modern reader is that the proposition to be proved is false if interpreted with modern categories and instruments, where pitches are considered as frequencies and measured in Hz. This issue is discussed in (Litchfield, 1988) . However, it is this author's opinion that to judge Aristoxenus's demonstration as wrong is questionable and the error, if any, can can be found in the experimental results; indeed a very accurate experimentation, carried out by using electronic device, shows that EF is not a fifth, so the error in Aristoxenus's proof is of an experimental and not logical nature. However, an experiment only using the ear could not give a definitive verdict. Therefore, Aristoxenus's results, instead of being wrong should simply be considered approximate[6].

4. PTOLEMY'S HARMONICA

Among the great treatises of Ptolemy (c. 100 AD-c. 175 AD), such as the Almagestum and the Geographia, the Harmonica is the least fortunate. The existing text is not reliable. In particular the last three chapters are missing and authenticity of the third as a whole is doubtful (Ptolemy, 2000: p. XXX) . When it became available in Europe, it was no more than a historical document, and musical theory had evolved quite differently[7].

The treatise is composed of three books each divided into sixteen chapters. In Book I Ptolemy tackled the fundamentals of the subject, defined epistemology, and criticized previous approaches, in particular those of the schools of Pythagoras and Aristoxenus. It was here that he put forward his theory on the tetrachord. Book II fully develops his theory of music. Book III describes the concepts of musical harmony and their relation to the human soul and to the heavens.

References concerning epistemology are scattered throughout the treatise. From the very first pages Ptolemy's efforts to indicate that his epistemology is not only about music but about all of sciences are quite clear. However, in the Harmonica the epistemological positions—surprising for us today—are fully explicit, more than they were for example in the Almagestum, which could appear to be a much more scientific text. Ptolemy seeks an agreement between a mathematical theory, made explicit by a number of assumptions on numerical relationships, and experience, which concerns the consent of the human ear to sounds, in the sense that it finds them somehow melodious. Ptolemy was convinced that this agreement should exist because nature itself has an intrinsic harmony. It is man's purpose to recognize this through the use of reason and senses.

For in every subject it is inherent in observation and knowledge to demonstrate that the works of nature have been crafted with some reason and prearranged cause and completed not at all in random or as it happened, particularly in its most beautiful constructions which are simply those of the sense perceptions more closely related to reason-sight and hearing (Ptolemy, 2000: p. 8) .

The roles of reasons and hearing are distinct "hearing is criterion for matter and conditio, while reason is the criterion for form and causes" (Ptolemy, 2000: p. 3) . Ptolemy specified that the role of hearing is "discovering what is approximate and accepting what is exact", while the role of reason is "accepting what is approximate and discovering what is exact". That is, given two or more sounds, the role of hearing is to discover the approximate ratio of their pitches for which the sounds appear concordant. Then reason examines the ratios and corrects them if they appear irrational, that is contrasting with the hypothesis of harmonics. Subsequently these rationally modified ratios and mathematical inference from them are subject to hearing that should give its assent. Ptolemy specified that sense perception is characterized by incertitude:

Perception, on the other hand, is of constantly mixed and fluctuating matter, so that on account of its instability neither the perception of all men nor that of the same men always observes the same thing in what has remained the same. Perception needs as its crutch, as it were, the educational assistance of reason (Ptolemy, 2000: p. 4) .

and that it can distinguish between similar concords even though the difference is minimal at a first glance.

Just as a circle described by one's eye alone has often seemed to be accurate until one drawn by reason redirects the eye to recognize what is truly accurate, if some specific difference of sounds is perceived by the hearing alone, it will then for a time seem neither to fall short nor to be excessive; yet it will frequently be proved not to be so when one selected according to the appropriate reason is heard (Ptolemy, 2000: p. 4) .

In any case there is a limit to human perception. But a continuous hearing can help perceive even very small differences.

In the comparison of smaller parts, more of the deficiency will be concentrated and now made detectable in them, and even more so in still smaller ones. The reason is that something which is just once quite short of the truth and cannot make the amount of its shortness perceivable by few comparisons, will when a number of comparisons are made be most noticeable and comprehensible to all (Ptolemy, 2000: p. 5) .

This quotation is interesting as it refers to actual experimentation, similar to that carried out in a modern laboratory, where an experience is repeated many times to evidence the regularity of a phenomenon.

4.1. Ptolemy's Musical Hypotheses

Ptolemy defined his musical hypotheses first in the negative sense, stating how they differed for those of the Pythagoreans and Aristoxenians. Those regarding Pythagoreans are of greater interest and only these will be remarked on[8]. Ptolemy said that their rational criteria were quite satisfactory, but they fail in some aspects; in particular they were not able to furnish rational criteria for all concords; for instance they could not justify the concord 8:3.

Ptolemy first solved this particular problem, by observing that the interval 8:3 is the conjunction of an octave and a fourth (8:3 = 2:1 × 4:3). Assume three notes, A, B and C. A and B are separated by an octave, B and C by a fourth; consequently A and C are separated by an interval of 8:3. Based on hearing, A and B are nearly the same. Thus if there is concordance between B and C (and this is the case), there also should be concordance between A and C (Ptolemy, 2000: p. 20) .

Then Ptolemy put forward his hypotheses about ratios for which the notes that are not concordant are at least melodic. Ptolemy's interest is centered on intervals lower than a fourth, those in a tetrachord. His position is quite articulated. However it is not my goal here to fully grapple with his ideas, to this purpose reference can be made to (Barker, 2006) . In essence he proposed the following hypotheses that that may be classified as rational as they can be derived by reason according to principles of symmetry and simplicity. They are summed up in this quotation:

In a word, the homophones would come from both the first multiple ratio and those measured from it. Consonances are the first two superparticular ratios and those composed from them plus the homophones, while the emmelic are those super-particular ratios after the sesquitertian [4:3](Ptolemy, 2000: p. 24) .

Where the homophones are separated by octave intervals, the consonances are the first concords, the fourth and fifth (the two super-particular ratios); then the simply melodic intervals follow which are super-particular ratios smaller than a fourth. Ptolemy also suggested a criterion according to which the super-particular ratios (n + 1)/n with the smaller values of n are more melodic.

For the same reason, those ratios making their divisions nearly in half are more emmelic, as are those differences limiting greater, simpler parts than their excesses, since these, too, are nearer equality just as is primarily the half, and then the third and each of the succeeding ratios (Ptolemy, 2000: pp. 23-24) [9].

To these fully rational criteria Ptolemy added another of an empiricist nature that holds for the arrangement of the two moving notes of the tetrachord. According to this they should be organized so that of the three intervals the lowest is smaller than the other two:

Agreeing with our sense perceptions, we assume as an equally common element in the three genera that the following intervals of the three magnitudes are smaller than each of the remaining intervals; and it is particular to the genera which have pykna\footnote{Singular pyknon; two notes in the tetrachord very close each other. that the two bottommost intervals be smaller than the topmost interval, while it is particular to the apyknon genera that no interval be greater in magnitude than that of the two remaining intervals (Ptolemy, 2000: p. 48) .

Ptolemy was not troubled about mixing rational and empiricist principles; his epistemology permitted this, organized as it was in two distinct phases. The first inductive phase forwards the hypotheses. Here Ptolemy is completely free in his choice, that could even be made by chance -though this would be quite a modern approach. So any source could be assumed for the hypotheses made. In the second deductive phase theorems are derived from the hypotheses, assumed as principles of a more geometrico argument. Theorems which consist in propositions to be verified by hearing. If consent is gained the hypotheses are validated independently of how they were obtained.

Ptolemy believed that his criteria of melody had absolute value. For him it is part of human nature to distinguish which sounds are melodic and which are not. Today it is rather generally accepted that musical taste is largely influenced by culture. Specifically what Ptolemy and the Greeks considered pleasing melodies, if the reconstruction of Greek sounds are reliable, are not such for us. The following anecdote, I think, will clarify the idea.

Some years ago, for example, Sir Robert Stewart delivered a lecture in Trinity College, Dublin, on the Music of Distant Times and Places; and illustrated it by specimens from various nationalities and periods, an ancient Greek hymn being included in the number. It was the unanimous verdict of all the musicians present that, while the music of the less civilized nations was often crude, barbarous, and monotonous in the highest degree, the Greek hymn stood quite alone in its absolute lack of meaning and its unredeemed ugliness; and much surprise was expressed that a nation which had delighted all succeeding generations by its achievements in the other arts should have failed so completely in the art which it prized and practiced most (Aristoxenus, 1902: Introduction).

4.2. The Harmonic Canon

The fundamental assumption of harmonics is the representation of pitches through numeric values and relations between pitches through ratios. But since numbers cannot be found in sounds they are to be found in their causes. More precisely these are found in observable and measurable aspects of musical instruments.

Because circumstances are similar regarding in the case of sounds and hearing, there is a need, as there is for sight, of some rational criterion for them from the appropriate instruments, just as the straight edge has, so to speak, need of a ruler, and the circle and the measurement of its parts have need of a compass. In the same way, the hearing, which is, along with sight, for the most part a messenger for the theoretical and rational aspect of the soul, needs some reasoned approach, too, for those things its nature cannot judge accurately, which it will not contradict but confirm as correct. [...] The tool of this type of approach is called the harmonic canon [emphasis added] (Ptolemy, 2000: p. 6) .

After some fairly convincing arguments Ptolemy concluded that the most appropriate instrument to identify the ratios between tones is the monochord; it is simple to construct and is controllable, more than the aulos for instance[10], where the height of the sound also depends on the power of the player, and this is not observable. The best observable magnitude in the monochord-canon is its length and the numbers that describes it serve as a measure of the height of the tones.

Ptolemy devoted much thought to the description of the monochord. In essence it consists of a string of a suitable material, for instance the same material as the chords of the lyra, fixed at two ends. Plucking or drawing a bow, the string produces a sound whose pitch depends on various parameters. Among them Ptolemy specified; the length of the string, the diameter of the section, the density of the matter. Only subsequently did Ptolemy mention tension (notice that here, and in the following, tension means force of traction, not to be confused with stress, that is force per unit of surface).

There are three reasons for the difference between high and low sounds; the density, the diameter, and the lengths of the strings. The higher sound results from a denser and thinner string and a shorter length; tension here takes the place of density, for it creates greater tension and hardens also what would be similar if the string had a shorter length (Ptolemy, 2000: pp. 37-38) .

Here Ptolemy said that the greater the density?a term to which he associated, in an intuitive form, the present meaning ? the higher the sound. Actually the contrary is true. The reason for this error can be explained by the fact that when Ptolemy compared sounds produced by percussion of bronze and wood, the former for him produces higher sounds because it is denser (Ptolemy, 2000: p. 12) [11]. With some imprecision he added that tension (force) and diameter could compensate each other. That is strings with different diameters sound the same if their tension is proportional to the diameter. Actually the proposition is true assuming the area and not the diameter of the cross section of the string[12].

The monochord described by Ptolemy is shown in Figure 5(a) and described as follows:

Imagine a canon along the straight line ABCD. At its ends on each side are bridges equal and alike forming as much as possible spherical surfaces under the strings. F marks the center of the aforementioned surface on the line BE, and H similarly marks the center on CG, while the signs E and G are understood to mark the bisections of the two circular surfaces (Ptolemy, 2000: p. 26) .

How the monochord is presented and commented on is somewhat difficult to understand. Moreover, the use of two cylindrical solids to define the length of the string is unconvincing; surely no luthier would adopt this method. The presentation made by Ptolemy can only be justified by his need to formulate the problem from a formal point of view, to justify theoretically that the contact between the rope and medium terms precisely at the tangent points E and G.

To measure the length of the string Ptolemy referred to a calibrator (a graduated rule), that fits "to the string transferring the length EG on it, so that we can make our measurement more easy"(Ptolemy, 2000: p. 27) . Then Ptolemy spoke of a bridge to be put in an intermediate position, similar to these at the end E, G, but slightly higher. This ensured a good contact between the string and the bridge, effectively separating the string, as illustrated in Figure 5(b).

I must confess that I do not understand what Ptolemy's purpose was when he proposed to test the bridge in different positions. His goal should have been to check the construction of the monochord; in particular, to check if the string is homogeneous. Dividing the string in half and by vibrating the two halves separately if the string is uniform along its length one should hear the same sound, because the two halves are of the same length and the same tension. In addition to this type of verification other can be done. For example, by dividing the rope in a ratio of 4:3, the shorter string produces a sound that is a fourth interval far from the longest and the two strings can be made to vibrate at the same time so that the concordance can be perceived. Similarly, by dividing the rope in the ratio 3:2 one can play a fifth, in the ratio 8:3 an octave and a fourth (2:1 × 4:3).

Once the correct construction of the monochord is verified, it can be used to conduct musical experiments. Ptolemy expressed no doubts about this, even though the modern reader may remain somewhat skeptical about whether it makes sense to verify by way of experiment mathematical hypotheses that do not seem to admit alternatives. Today, but probably even then, the monochord is mostly used to demonstrate to students the goodness of the theory of harmony based on ratio, for particular values of ratios; for example 2:1, 4:3, 3:2. If for some reason hearing suggests a lack of concordance, we immediately believe that the canon has serious construction defects. Ptolemy seemed genuinely convinced, at least in some situations, that he was carrying out true experiments and even went so far as to show the reader how to do these. However, even though Ptolemy would never have conducted musical experiments himself, his epistemological proposal was extremely clear and was considered applicable, for example, to astronomy where the possibility that the experiment (observation) falsifies the theory is more plausible.

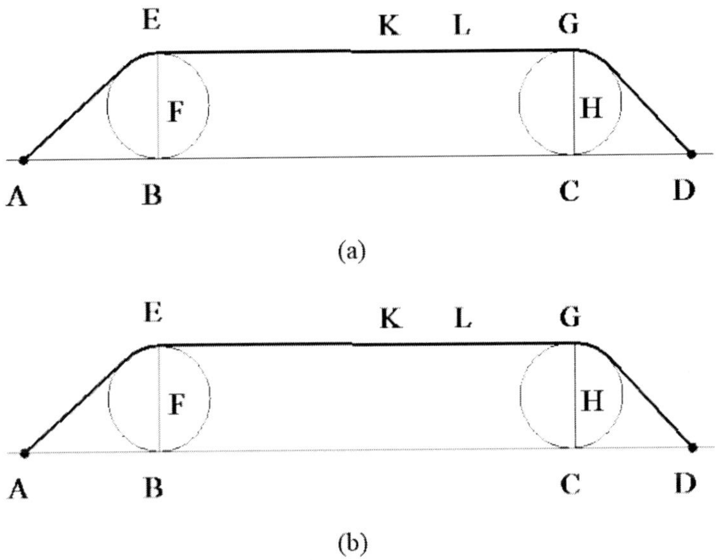

Figure 5. (a) Basic monochord[13]; (b) monochord with and intermediate bridge.

Ptolemy proposed tests to show that certain assumptions by Aristoxenus on the composition of musical intervals were wrong. In particular, he claimed that the hypotheses for which a fourth interval is constituted by two and a half tones, and one octave by six tones are false. Here I will not argue whether Ptolemy is falsifying the theory of Aristoxenus, as the two theories are essentially incommensurable; I will limit myself to showing that Aristoxenus's theory as reported by Ptolemy is falsified. My only comment is on the proof that the octave is not the sum of six tones, because in this case there is no need to define the semitone, and thus the proof is simpler. If the definition of the interval of a tone of Ptolemy is accepted, represented by the ratio 9:8, his conclusion is straightforward and was already found in Euclid's Sectio canonis (see previous sections). In fact it is $(9:8)^6 \simeq 2.002:1$ and thus slightly greater than the octave ($2:1$).\footnote {Note that for the definition of an interval Ptolemy has him state that the disjoint union of two tetrachords provides one octave, even if it is not the sum of six intervals. In fact the two tetrachords each provide an interval of 4: 3; the tone between them defines an interval of 9:8, the union of the three intervals gives (4:3) × (4:3) × (9:8) = 2:1[14].

Ptolemy argued that this difference could be observed experimentally. In order to do this he suggested building a tool of eight strings (octachord canon), the first seven separated by an interval of one tone, the last equal in length to half of the first string (then separated from it by an octave). The purpose of the experiment, I believe, is the empiric verification of the difference between the union of six tones and the octave. The use of multiple strings should have the function of being able to achieve a better control of the division into six tones. Ptolemy also discussed, albeit very briefly, the possibility of tuning the six strings separately, adjusting the force to which they are subject.

Together with monochord and octachord canons, Ptolemy proposed another instrument "which was devised by theoreticians" (Ptolemy, 2000: p. 66) [15] the helicon, which is shown in Figure 6 and described in the following quotation:

They propose a rectangle[16] ABCD, halve AB and BD at E and F, connect AF and BGC, and parallel to AC they lead EHK through E and LGM through G. In such a configuration AC will be double each of BF and FD, as will each of these be double EH since also AB is double AE, so that also AC will be quadruple EH and sesquitertian of the remainder HK[17] (Ptolemy, 2000: p. 66) .

The four strings AC, BK, LM, BD, all similar and equally taut, are divided by the inclined bridge AF in portions with various ratios to each other; thus many consonances and the whole tone could be revealed: The fourth (ratio 4:3) is formed by AC:HK, as well as GM:FD and LG:EH. The octave by AC:BF (or FD) and GM:LG. The octave plus a fourth (3:1) by AC:EH, the whole tone (9:8) by HK:GM.

The advantage of this instrument with respect to the monochord is that there is no need to move the bridge to reveal the various concordances.

A variant to the helicon is suggested in Figure 7. Here, there are still four strings, AC, LG, NH, BD, and a bridge AE, where E is such that AE = 2AB; there is no need that ABCD be square. The intermediate chords are positioned by measuring their distances from E. If P is a generic point on EC, the ratio EC: EP defines the ratio of the length of the strings with respect to AC. For instance, if P ≡ D, the chord DB is divided in F, so that AC: DF = EC:ED = 2:1; for P ≡ H, the chord HN is divided in M, so that AC: HM = EC:EH (for instance 3:2). The bridge AE can be moved to EO, with AC:AO a proper ratio. Thus different sets of ratios of pitches can be easily obtained.

The advantage of the first form of helicon (that shown in Figure 6), is that without moving the strings a large set of pitches can be achieved, this is the same advantage both helicons have with respect to the canon. Ptolemy stresses that moving a string creates problems from a practical point of views, needing tuning each time. The advantage of the second form of the helicon is that thanks to the particular arrangement of the bridge internal chords can be located and all sorts of ratios can be obtained.

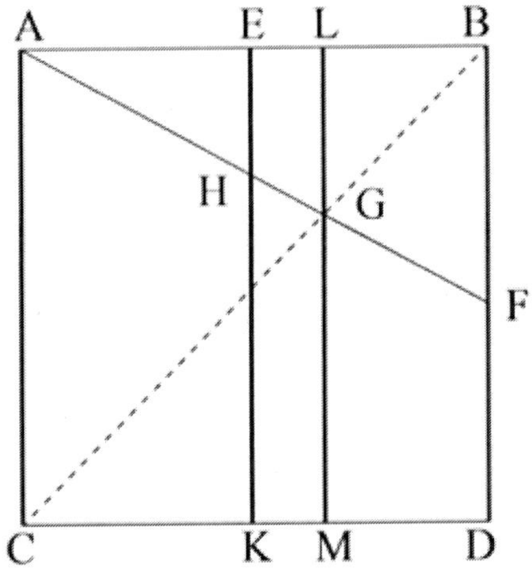

Figure 6. The helicon[18].

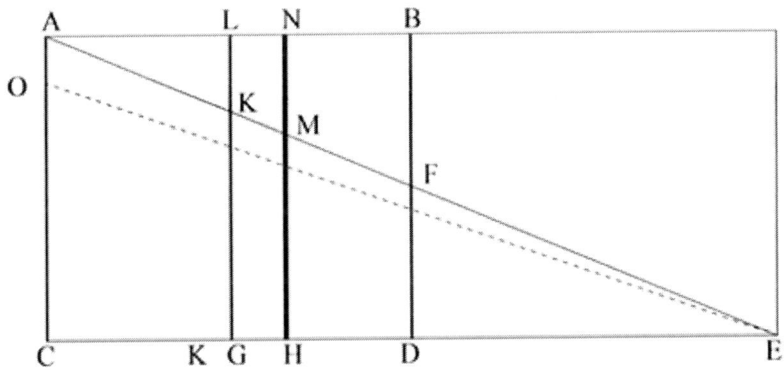

Figure 7. A variant of the helicon[19].

4.3. Extended Harmonics

In the third book of the Harmonica Ptolemy enlarged the concept of harmonics, from music to psychology and astronomy. He reasserted the definition of harmonics, or better harmonic function. This was descrkibed as a function of reason that regards the symmetry of movements, a knowledge of which is expressed by mathematics, that studies the ratio of audible differences (Feke, 2009) . This function of reason can be extended to any kind of regular motion:

It must be said in addition that it might be necessary for such a function to be present in all things which contain the beginning of motion, even to a small extent, just as the others," particularly and most extensively in those which share a more perfect and rational nature on account of the individual conditions of their creation. In these alone can it be shown universally and clearly to preserve as much as possible the similarity of those ratios which make what is suitable and harmonious in the different species. For generally, whatever is governed by nature shares in reason to a certain extent both in motions and the basic subject matter. And where reason is able to be guarded with due proportion, here can be found creation, nurturing, preservation, and everything which is spoken of as superior (Ptolemy, 2000: p. 66) .

I have omitted details on the parallel of harmonics in music and in the human soul and heavenly bodies. Only a list of the titles of the chapters of Book III which gives a clear idea of Ptolemy conceptions are provided.

1) How through the whole system we might both use and judge the ratios by means of the fifteen-stringed canon.

2) Methods for dividing up to the double diapason with only the eight notes.

3) In which genus must we place harmonic function and our knowledge of it?

4) That the function of the harmonic exists in the natures of all the more perfect things and that it appears particularly in both human souls and heavenly motions.

5) How the consonances are in accord with the first differences of the soul and their species.

6) A comparison between the harmonic genera and those of the first virtues.

7) How harmonic modulations resemble the circumstantial modulations of souls.

8) On the similarity of the perfect system and the circle through the middlc of thc zodiaco

9) How the harmonic consonances and dissonances resemble those in the zodiac.

10) That succession in notes resembles the longitudinal movement of the stars.

11) How the stellar movement in altitude compares with the harmonic genera.

12) That modulations of tonoi are like stellar crossings in latitude.

13) On the similarity of the tetrachords and the aspects of the sun.

14) By what first numbers might the fixed notes of the perfect system be compared to the first spheres in the cosmos.

15) How the ratios of their motions are calculated by the numbers.

16) How the combinations of the planets should be compared to those of the notes (Ptolemy, 2000: p. 127) .

5. CONCLUSION

Harmonics had been an important mixed mathematics from Classical and Hellenistic Greece, throughout the Middle Ages and the Renaissance. Despite this, it has been largely ignored by historians of science. This paper shows that harmonics merits a careful study and that its methods are the same as other sciences, such as astronomy and mechanics, to name but a few. Harmonics, perhaps, was the only scientific discipline where experience was subject to real experiments, very similar to those carried out in modern laboratories.

Two different approaches to music are compared, which are also found in the other mixed mathematics. Aristoxenus and his school followed the approach of mixed mathematics as practiced by most mathematicians and theorized by Aristotle. The starting point was of an experimental nature and was considered as true, even though it was not framed in metaphysics where it could somehow also be regarded as evident. Experimental results served as a starting point for mathematical reasoning, for example, to demonstrate relationships among intervals. The use of mathematics by Aristoxenus could perhaps have been a legacy of his mathematical education and could have been ignored, focusing on demonstrations based on logical relationships only. However, Aristoxenus did not do this.

Danilo Capecchi (the monochord-canon, the octochord-canon, and the helicon) whose length can be adjusted. The tests were to be repeated many times in order to eliminate possible errors of observation. Probably the epistemological position of Ptolemy in his harmonic science was merely rhetorical and in fact expressed his general ideas on science. However, it may also be envisioned that he had deliberately undertaken a series of actual musical experiments. In any case, the Harmonica, probably a late work, provided him with the opportunity to clarify his ideas about science. It was not the scope of this paper to show if these ideas were received by modern scientists. I just say that this is probably true for Galileo (Drake, 1970; Drake, 1978; De Pace, 1993) .

REFERENCES

1. Aristoxenus (1902). The Harmonics of Aristoxenus. Edited and Translated into English by H. S. Macran Oxford: Clarendon Press.
2. Aristoxenus (1954). Aristoxeni Elementa harmonica. Edited and Translated into Italian by R. Da Rios. Rome: Publicae Officinae polygraphicae
3. Barbera, C. A. (1977). Arithmetic and Geometric Divisions of the Tetrachord. Journal of Music Theory, 21, 294-323.
4. Barker, A. (1978). Music and Perception: A Study in Aristoxenus. The Journal of Hellenic Studies, 98, 9-16.
5. Barker, A. (1989). Greek Musical Writings (Vol. 2). Cambridge: Cambridge University Press.
6. Barker, A. (1994). Ptolemy's Phytagoreans, Architas, and Plato's Conception of Mathematics. Phronesis, 39, 113-135.
7. Barker, A. (2006). Scientific Method in Ptolemy's Harmonics. Cambridge: Cambridge University Press.
8. Comotti, G. (1991). La musica nella cultura greca e romana. Turin: E.D.T.
9. De Pace, A. (1993). Le matematiche e il mondo. Milan: Francoangeli.
10. Drake, S. (1970). Renaissance Music and Experimental Science. Journal of the History of Ideas, 31, 483-500.
11. Drake, S. (1978). Ptolemy, Galileo, and Scientific Method. Studies in History and Philosophy of science, 9, 99-115.
12. Düring, I. (1930). Die Harmonielehre des Klaudios Ptolemaios. Porphyrios Kommentar Zur Harmonielehre Des Ptolemaios. PhD Dissertation, Göthenburg: University of Göthenburg.
13. Düring, I. (1956) Greek Music: Its Fundamental Features and Its Significance. Journal of World History, 3, 302-329.
14. Feke, J. (2009). Ptolemy in Philosophical Context. Ph.D. Dissertation, Toronto: University of Toronto.
15. Laloy, L. (1904) Aristoxéne the Tarente et la musique de l'antiquité. Paris: Societé Français d'Imprimerie et de Libraire.
16. Litchfield, M. (1988). Aristoxenus and Empiricism: A Reevaluation Based on His Theories. Journal of Music Theory, 32, 51-73.
17. Manolova, D. (2014). Discourse of Science and Philosophy in the Letters of Nikephoros Gregoras. Ph.D. Dissertation, Budapest: Central European University.
18. Palisca, C. V. (1961). Scientific Empiricism in Musical Thought. In: H. H. Rhys (Ed.), Seventeen Science and the Arts (pp. 91-137). Princeton, NJ: Princeton University Press.
19. Palisca, C. V. (1985). Humanism in Italian Renaissance. Musical Thought. New Haven, CT: Yale University Press.
20. Ptolemy, C. (2000). Ptolemy Harmonics. Translated and commented by J. Solomon. Leiden: Brill.

NOTES

[1]Barker translates "has no ratio" (Barker, 1989: p. 284) . But I prefer irrational, though the meaning probably refers in both cases to mathematics.

[2]The use of modern concepts could be useful in recovering the association of the idea of spatial interval to ratios, using the logarithmic function. If we measure intervals not by the ratio of the pitches but rather by the logarithms of the ratio, we can say that intervals adds, as implied in the Aristoxenian musical theory.

[3]Super-particular, or epimoric, ratios are those in the form (n+1)/n.

[4]Redrawn from (Barker, 1989: p. 199) .

[5]Aristoxenus did not consider the 'experimental' construction of a ditone interval to be a problem, as shown in Figure 4.

[6]I report here, using an algebraic language, Aristoxenus's reasoning. Let ω_A, ω_B, ω_C, ω_D, ω_E, ω_F,the pitches associated with A, B, C, D, E, F respectively (measured for instance by the logarithm of their frequencies expressed in Hz), T is the interval of a tone and F the interval of a fourth. By construction it is: $\omega_C = \omega_B - 2T$; $\omega_D = \omega_A + 2T$; $\omega_E = \omega_D - F$; $\omega_F = \omega_C + F$. Simple calculation gives: $\omega_F - \omega_E = 3F - 4T$. Assuming ω_F ? $\omega_E = F + T$, that is a fifth, the previous relation gives: $F + T = 3F - 4T$, or $F = 2.5$ T, thus Aristoxenus thesis.

[7]The Greek version was edited in 1930 by Ingemar Düring (Düring, 1930: p. 319) its history and translation in the late Middle Ages and in the Renaissance are documented respectively in(Manolova, 2014) and (Palisca, 1985) . To my knowledge there are two modern English translations, one by Andrew Barker (Barker, 1989) , the other by Jon Solomon (Ptolemy, 2000) . Here Solomon's translation and critique is largely referred to, with some reference to Barker's that is more accessible.

[8]Probably Ptolemy is referring to Architas of Tarentum (Barker, 1994: p. 13) .

[9]Barker translates this piece as follows: "Of these ratios too, those that make divisions most nearly into halves must be more melodic, for the same reason as are all those whose differences contain larger simple parts of the things that are exceeded; for these, too, are nearer to the equal, just as the half is nearest of all, then the third, and then each of the others in succession" (Barker, 1989: p. 290) .

[10]The aulos is a wind instrument with double reed.

[11]Actually the difference in sound mainly depends on the greater stiffness of bronze.

[12]The frequency of a vibrating string is inversely proportional to length and diameter, directly proportional to the square root of ratio between tension and density, or alternatively inversely proportional to length and directly proportional to the square root of the ratio between tension and the product of area by density.

[13]Redrawn from (Ptolemy, 2000: p. 26) .

[14]Note that for the definition of an interval Ptolemy has him state that the disjoint union of two tetrachords provides one octave, even if it is not the sum of six intervals. In fact the two tetrachords each provide an interval of 4:3; the tone between them defines an interval of 9:8, the union of the three intervals gives 4:3 × 4:3 × 9:8 = 2:1.

[14]No reference is made at all to who those theoreticians were.

[15]No reference is made at all to who those theoreticians were.

[16]Barker translates squares instead of rectangle: “they construct a square, ABCD” (Barker, 1989: p. 319) . Indeed a little below Ptolemy said that AB = 2AE, and thus the rectangle is in fact a square.

[17]In the ratio 4: 3.

[18]Redrawn from (Ptolemy, 2000: p. 67) .

[19]Redrawn from (Ptolemy, 2000: p. 67) .

CHAPTER 4

Contribution of A. Jakštas-Dambrauskas (1860-1938) to Mathematics in Lithuania

Juozas Banionis

Vytautas Magnus University, Lithuanian University of Educational Sciences, Vilnius, Lithuania

ABSTRACT

The article, dedicated to prelate Aleksandras Dambrauskas (pseudonym: Adomas Jakštas, 1860-1938), a Lithuanian encyclopaedist, Honorary Doctor of Mathematics, summarizes the researcher's mathematical works and highlights his merits to the development of the science of mathematics in Lithuania. Most frequently his activities are associated with theology, philosophy, literature and politics. However, it is rarely mentioned that it was actually mathematics which led A. Jakštas' search to be true and beautiful in philosophy, literature and theology. Firstly, he stands out as a reviewer of the fundamentals of geometry and trigonometry in the development of the science of mathematics in Lithuania. The mathematical works that were written by the prelate fall into three groups. The first group includes the original works covering the field of fundamentals of trigonometry and geometry. The second group embraces popularising and informative articles and booklets, and the third group unites school textbooks on mathematics. A. Jakštas manifested himself as a supporter of mathematics. The prelate was interested in problems which could be practically applied in everyday life or were based on all the generally accepted laws of the Universe. The major part of the mathematical works by the prelate was devoted to school mathematics. Observing the dawn of

Lithuanian sciences, A. Dambrauskas' significant contribution to mathematics in Lithuania is acknowledged. Reading the mathematical works by the prelate, his ability to fill them with humanist philosophy is admired. A. Dambrauskas is a perfect example for all the mathematicians, and his mathematical works occupy an honourable place in the history of mathematics in Lithuania. This article is dedicated to A. Jakštas-Dambrauskas, an outstanding Lithuanian personality, who left indelible marks in the Lithuanian culture and education. The latter merits are usually enthroned, whereas the prelate's mathematical works are often neglected. In this article, the author returns to the mathematical contribution of A. Dambrauskas, supplementing his presentation delivered in Lithuania in 2011 with new sources, and aims to familiarise readers with this exceptional personality in English. To achieve the established goal not only already published but also archived materials stored in the Office of the Central State Archive of Lithuania and the Manuscripts Department of Vilnius University Library will be used. The research employs the methods characteristic of research work in humanities: text analysis, interpretation and historical description. Moreover, the criterion of objectivity and the principle of historicism are followed to revive the atmosphere of the end of the 19th century to the beginning of the 20th century.

Keywords: Philosophy, Theology, Fundamentals of Geometry and Trigonometry, Mathematical Terms, Popularisation of Mathematics, History of Mathematics in Lithuania

1. INTRODUCTION

The name of Lithuanian encyclopaedist and prelate Aleksandras Dambrauskas (pseudonym: Adomas Jakštas) is most usually associated with theology, philosophy, literature, politics, and, finally, mathematics. However, it is rarely acknowledged that only due to mathematics, the traces of A. Jakštas' search to be true and beautiful extend to philosophy, literature and theology. Firstly, in the evolution of the science of mathematics in Lithuania, he is perceived as a reviewer and "examiner" of the fundamentals of geometry and trigonometry.

The mathematical works of A. Dambrauskas were most widely discussed in the article of Otonas Stanaitis written in the period of Independent Lithuania (Stanaitis, 1938) . At the same time Petras Katilius (Katilius, 1938) classified the works of the prelate into three groups and characterised the main ones in general terms. Česlovas Masaitis (Masaitis,

1938) highlighted the philosophical aspect of A. Jakštas' geometry fundamentals and new systems of trigonometry.

A large number of authors of the aforementioned period such as L. Dambrauskas, J. Keliuotis, A. Maceina, P. Samulionis, J. Senkus, S. Šalkauskis, M. Vaitkus, P. Venckus and others presented not only memoir material in their works. They also characterised the prelate as a well-rounded personality, who was committed to church and science. His extraordinary interest in the philosophy of mathematics was emphasised and exceptional liking for mathematics was also revealed on fragmentary basis.

In our times Algirdas Ažubalis and Aleksandras Baltrūnas (Ažubalis, 1990; Baltrūnas, 1996)analysed his main mathematical works. The merits of the prelate to Lithuanian terminology were highlighted by Petras Rumšas (Rumšas, 1979) , whereas Nijolė Kalinauskaitė (Kalinauskaitė, 2011)revealed his innovations in development of trigonometric systems.

The article aims to summarize mathematical works of prelate A. Dambrauskas—A. Jakštas particularly emphasising his exceptional merits in enhancing research on the science of mathematics in Lithuania. To achieve the established goal not only published but also archived materials stored in the Office of the Central State Archive of Lithuania and the Manuscripts Department of Vilnius University Library will be used. The research employs the methods characteristic of research work in humanities: text analysis, interpretation and historical description. Moreover, the criterion of objectivity and the principle of historicism are followed to revive the atmosphere of the end of the 19th century to the beginning of the 20th century.

2. THE ROAD TO MATHEMATICS

A. Jakštas chose the following slogan in his life: "to be logical, i.e. to rely on the divine Logos that conducts the world affairs" (Jakštas, 1938: p. 349) . Therefore a certain precision-mathematical thinking is felt in all A. Dambrauskas' works from the above-mentioned areas (Figure 1).

Even more, employing mathematics he formulated his apologetic arguments. "I particularly liked mathematics in that respect that it employs unvarying, eternal truths; if a human soul is fed with eternal truths, it has to be eternal as well"-A. Dambrauskas wrote" (Keliuotis, 1931: p. 898) . While learning in Šiauliai gymnasium (1872), the mathematical abilities of the learner became visible. However, as he confessed himself, his interest in mathematics was rather accidental and conditioned by highly strict character of his mathematics teacher Horbacevičius (Samulionis, 1938: p. 304; Jakšto, A. fondas, p. 9) . Being in the fifth form, Aleksandras undertook independent studies of theories of geometry. At that time he even managed to demonstrate a "certain" new theorem but later it turned out that its proof had already been known in the science of mathematics (Senkus, 1938: p. 189) . Therefore, possessing an obvious talent for mathematics, A. Dambrauskas chose the Faculty of Mathematics and Nature of Saint Petersburg University in 1880.

Figure 1. A. Dambrauskas-Jakštas (1860-1938).

Such a choice was motivated in the following way: "I chose the Department of Physics-Mathematics because I found the study subjects taught there perfect. I was particularly fascinated by the truths of mathematics. They seemed absolute to me with two divine qualities: presence everywhere and permanence. There is no place without them. What is true in Saint Petersburg is true in Paris, in London, on the Moon and on Mars. In a similar way they are eternal and immutable: they were the same in the very beginning and they will stay the same forever. Thus, I consider it a great honour to serve mathematics. Mathematics has always seemed to me the queen of all sciences". On the other hand, mathematics for A. Jakštas was "an ideal world, where an individual is freest and where he or she is able express oneself best as a creator". Even more, he argued that "If a human creator may remind of God Creator, this is possible only in mathematical creation. The nicest worlds of opportunities are created there namely from nothing" (Maceina, 1938: p. 312-313) .

However, disappointment with some university professors as well as riots, which started in his university at that time, forced A. Dambrauskas to change his choice. It is no coincidence that a twenty year old young man, who failed to adapt to life in the capital of empire, wrote to his parent 22 February 1881: "I am just a mere Lithuanian, an individual from bones and blood, who will never be able to learn various customs of the big world or to give up own faith and religion"(Dambrauskas, 1939: p. 250) . Therefore, after less than one year of studies in mathematics, he left the university, returned to Lithuania and entered the Samogitian Theological Seminary. There, as the irony of fate may have it, after three years of studies, the archdiocese sent him to Petersburg again and this time-to the Theological Seminary. In 1888 he successfully graduated from the Seminary and was conferred the Master's degree in Theology. Unfortunately, A. Dambrauskas did not have a chance to enjoy the return to Lithuania: after service as a chaplain in Panevėžys gymnasium and occurring disagreements with tsarist officials, he was exiled to Ustyuzhna in Novgorod Governorate (Russia) in 1889. He got acquainted with Lithuanian writer Vincas Pietaris (1850-1902). The latter provided mathematics textbooks and priest A. Dambrauskas' passion for independent mathematical studies arose again and he engaged in the first mathematical research (Samulionis, 1938: p. 307) . After five years in exile, he got the right to return to Lithuania in 1894. Staying in Kaunas, A. Dambrauskas spent almost all his life there with exception of the period from 1902 to 1905, when he worked as professor in the Petersburg Theological Seminary and during the first two years of World War I (1914-1915), when he lived in Vilnius.

In 1905 St. Casimir Society, which was committed to publishing books in Lithuanian, was founded. Feeling the lack of scientific knowledge in the Lithuanian language and possessing an inclination for mathematics, A. Dambrauskas undertook writing articles for Lithuanian society and mathematics popularisation. Under the initiative of the priest, the literary, scientific and political journal "Draugija" (Society) was established in 1907. Next to numerous articles on various themes written by A. Jakštas, the journal also published his mathematical texts. After becoming a prelate in 1914, A. Dambrauskas further penetrated into the problems of mathematics ignoring disapproval and contempt of some clergymen. Such efforts did not get unnoticed and 14 December 1928 A. Dambrauskas-Jakštas was conferred the honorary title of Doctor of Mathematics for his merits to Lithuanian mathematics by Faculty of Mathematics and Nature (Lithuanian university in Kaunas)(VDU MGF fondas.(b), p. 232) .

3. MATHEMATICAL WORKS AND THEIR IMPORTANCE

The mathematical works written by the prelate may be divided into three groups. The first group would include original mathematical works. The second group would comprise popularising and informative articles and booklets, and the third group would embrace school textbooks on mathematics. Each group will be analysed separately.

The first group consists of original mathematical works covering the field of fundamentals of trigonometry and geometry. As early as 1893 and still being in exile in Ustyuzhna, A. Dambrauskas got interested in the problem of parallel lines. He investigated Euclid's Fifth Postulate, which had previously attracted enormous efforts of mathematicians from various epochs, who had tried hard to provide own proofs. Fully emerged into the problem, A. Jakštas introduced a certain curve (conditionally called "postuloida"), dependent on Euclid's Postulates, into geometry. The very idea was published in the newspaper "Wiadomości matematyczne" (Mathematical News) in 1898. Later corresponding with Prof. Boudier and following Prof. Meray, A. Jakštas continued to improve the aforementioned remarks about Euclid's Fifth Postulate. As a result, the article was written in Esperanto in 1902 and published in the French newspaper "Revue Bourguignonne···" (Bourguignonne Review). Moreover, it should be emphasised that A. Jakštas is one of the establishers of Lithuanian Esperanto Association (1919) and a publisher of the newspaper "Litova stelo" (Lithuanian Star) published in this international language. The aforementioned article was also published as a brochure in Berlin in 1906 (Dombrowski , 1906: p.14-16) (Figure 2).

Finally, having taken into account the recommendations of Prof. O.Voeros, in 1917 the prelate improved his article and published it under the title "Euklido V postulato kreivė ir metageometrijos pagrindų kritika" (The

Curve of Euclid's Fifth Postulate and Criticism of Meta-geometry Fundamentals). Its last variant was published in issue three of the journal "Logos" in 1926(Dambrauskas, 1926a) . The article consisted of two parts. The first part contained the translation of the article published in Esperanto. It characterised the kinds of new curves (pastulatoids). The second one contained criticism of fundamentals of meta-geometry, i.e., non-Euclidean geometries.

Bearing in mind that "from its very beginning geometry searched for the main concepts for its further creation", A. Jakštas discussed "the perfect straightness" of the line in the same article. Further, at the end of it the author made the following conclusion "Non-Euclidean geometries cannot be equalled to Euclidian geometry. These are absolutely different branches of geometry: the first one is linear geometry whereas the rest are ascribed to curvilinear geometries" (Dambrauskas, 1926a: p. 281) . Thus, assigning priority to Euclidian geometry, A. Jakštas critically assessed fundamentals of other geometries by this article. He endeavoured to find and to draw an explicit distinction between Euclidian and non-Euclidian geometries. The first one was ascribed to the science of mathematics, whereas the remaining ones, following the great Einstein, were thought to belong to branches of science of physics. In fact, these geometries are "purely physical and supplemented by material" (Jakštas, 1931) . Meanwhile, the prelate, following his own loyalty to conservatism, believed that he had managed to clear the foundations of geometry from "old paradoxes" and "new sophisms".

Inviting the return to the "true" geometry created by Euclidius, to Euclidian "real lines" and to Euclidian "real planes", A. Jakštas wrote two more works related to the aforementioned problems. One of them, i.e., "Kas yra tiesioji linija?" (What is the Straight Line?), was written on the basis the lecture delivered in the recently established University of Lithuania 28 November 1922. Later, after recommendations of Otto T. Volk (1892-1989), a German professor working in Kaunas, the article appeared in volume eight of Matematikos gamtos fakulteto darbai (The Works of Faculty of Mathematics and Nature) (Dambrauskas, 1926b) . Endeavouring to reveal the essence of the line, A. Dambrauskas grounded his work on three issues. Firstly he chose Lithuanian literary classicist, bishop Antanas Baranauskas (1835-1902) considerations about the line as a point of departure. Secondly he took into consideration that the definition of the line is the key problem to the whole geometry. Thirdly, in his work he indicated that the problem was not fully solved in mathematics. On the basis of the aforementioned arguments, the author raised the main problem, i.e., searched for the result of infinite prolonging of the line. He found three solutions, which namely meant the geometrical systems of Lobachevsky-Bolyai, Euclidus and Riemann. A. Jakštas also referred to them as to three hypotheses about the line. Having presented a broad historical overview of the line (from

Archimedes to C. Vöros), A. Jakštas formulated a conclusion once again emphasising that the true line is the Euclidian one, whereas the other two are non-Euclidian and, therefore, only “quasi-lines”. Thus, he undoubtedly prioritised Euclidian geometry over non-Euclidian one and certified own approach towards mathematics full of “certain a priori attitudes arising from intuition or being of metaphysical nature” (Masaitis, 1938: p. 292-293) .

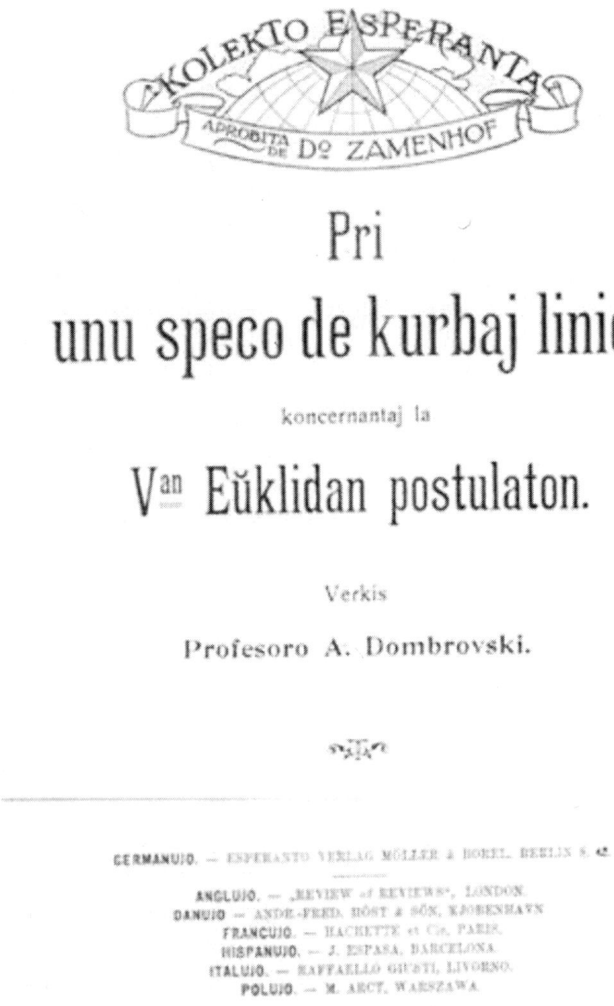
KOLEKTO ESPERANTA

APROBITA DE D-RO ZAMENHOF

Pri

unu speco de kurbaj linioj,

koncernantaj la

V^an Eŭklidan postulaton.

Verkis

Profesoro A. Dombrovski.

GERMANUJO. — ESPERANTO VERLAG MÖLLER & BOREL, BERLIN S. 42

ANGLUJO. — „REVIEW of REVIEWS“, LONDON
DANUJO — ANDR.-FRED. HÖST & SÖN, KJOBENHAVN
FRANCUJO. — HACHETTE et Cie, PARIS
HISPANUJO. — J. ESPASA, BARCELONA
ITALUJO. — RAFFAELLO GIUSTI, LIVORNO
POLUJO. — M. ARCT, WARSZAWA

Figure 2. About Euclid’s fifth postulate in Esperanto, Berlin, 1906.

It is understood that the essence of the analysed problem was predetermined by direct dependence of line and geometries. "Seeking to deeper penetrate into the essence of all the geometries, it is necessary to better cognize the line"-the author stated. Hence, the axioms analysed on the basis of geometry lead to related geometries. Analysing the line he noticed the concept of infinity, which resulted in the emerging paradoxes. Therefore, generalising the work A. Jakštas emphasised that the line itself approximates closer to the object of sciences of philosophy and theology.

It is necessary to emphasise that the author, who perfectly perceived the problem of the article, succeeded in an elaboration on the essence of reasoning about the concept of line. Further, the words of A. Jakštas at the end of the article should be understood as an encouragement for young people to engage in scientific research: "Thank God, we have already lived to see our university, but it consists not of walls but of people, who are particularly devoted lovers of truth and its diligent seekers" (Dambrauskas, 1926b: p. 391-392) . Writing the above-mentioned article, he followed the works by famous mathematicians of that time, such as P. Engriques, E. Isenkrhe, R. Sudre and others.

The second article, which specified the "advantage" of Euclidian geometry, was "Matematiškos sąlygos kreiviems paviršiams nuo plokščių ir kreivėms nuo tiesiųjų atskirti" (Mathematical Conditions for Distinguishing Between Curved Surfaces and Between Planes and Curves)(Dambrauskas, 1930) . This was as if a supplement to the previously published article "Kas yra tiesioji linija?". Following M. Geiger, A. Jakštas pointed out the conditions for compliance of planes and surfaces.

Another original work of mathematics, which is broad in its scope, was written penetrating into trigonometry. In his article A. Dambrauskas considered whether it is the absolute and the sole presently "applied science of triangles". Further he also expressed his opinion that if next to Euclidian geometry system there exist non-Euclidian ones, there should be alternative systems of trigonometry. Own search in that direction was concretised by A. Jakštas in 1902, when he described the first six elementary new trigonometric systems out of twenty eight ones. In the course of time, he succeeded in identifying not only the so-called "elementary" trigonometric systems but also in establishing how to construct a big number of complex ("sukrautinės") systems. A. Dambrauskas acknowledged a long-term search for a systemic delivery of a course in trigonometric systems. The article about geometric groups written by professor Bourlet in the French scientific newspaper "Internacia Scienca Revuo" (International Science Review) gave a new impetus for such search.

Soon A. Jakštas prepared his article and in October 1905 he sent it to the aforementioned professor Bourlet. The later author approved of the work and in 1906 the article was published as a separate publication in Esperanto in Berlin. Two years later (1908) the article "Naujos trigonometriškos sistemos" (New Trigonometric Systems) reached the readers in French, which was translated by professor E. Lefevre from Belgian Military School, who was interested in its theme. A. Jakštas returned to this work once again in 1921, when prepared it for publication in the Lithuanian language.

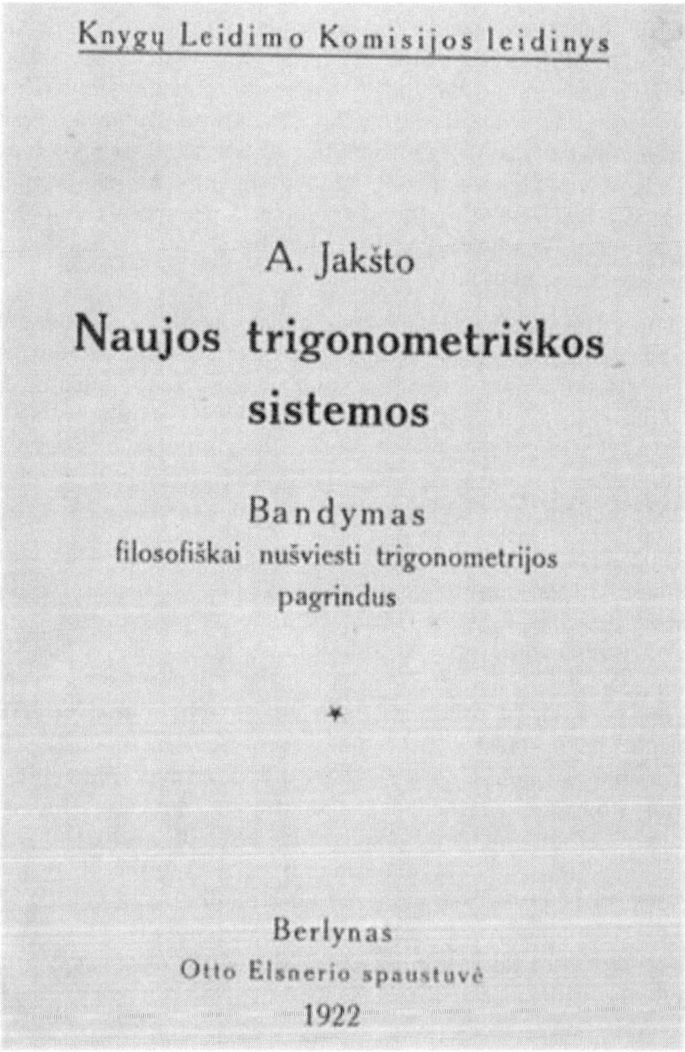

Knygų Leidimo Komisijos leidinys

A. Jakšto

Naujos trigonometriškos sistemos

Bandymas
filosofiškai nušviesti trigonometrijos pagrindus

*

Berlynas
Otto Elsnerio spaustuvė
1922

Figure 3. New trigonometric systems, Berlin, 1922.

He reorganised the article and added several chapters. In this way, through intermediation of Lithuanian writer, professor Vincas Krėvė-Mickevičius (1882-1954), in 1922 "Naujos trigonometriškos sistemos" were published in the Lithuanian language in Berlin. The publication was dedicated to the established Faculty of Mathematics and Nature at the University of Lithuania (Jakštas, 1922) (Figure 3).

In his "Naujos trigonometriškos sistemos" the prelate noticed that "the trigonometric system, which is applied at present, is one of the indefinite cases of one more general system" (Jakštas, 1922: p. 71) . Taking this into account, the author associated two more variables (a radius and an angle) to the six trigonometric lines already established in the classical trigonometry. Thus, eight values, two of which are regarded as constants, as well as trigonometric functions constructed like in classical trigonometry but differentiating from the ordinary ones, were received. In such a way the aforementioned twenty eight new systems, including the already used one, appeared. They were called elementary. Further, assuming that only one value is considered constant and linking the other values through algebraic equation that corresponds the plane curve, an infinity of trigonometric systems, which the author referred to as "sukrautinės", i.e., complex. Having introduced new trigonometric systems, in the conclusions A. Dambrauskas pointed out that the systems are characterised by functions, which may be geometrically demonstrated in curves. Moreover, a transition from functions of one system to those of other systems is also possible. He provided numerous examples of trigonometric line systems but did not present their complete classification (Kalinauskaitė, 2011: p. 159) (Figure 4).

Encouraged by mathematics professor O. T. Volk, the prelate directed his research on systems towards complex rectangular trigonometric systems. In 1926 and 1930 A. Dambrauskas published a two-part study "Apie sukrautines stačiakampes trigonometrines sistemas" (On the Complex Rectangular Trigonometric Systems) following the conclusions of the previous work. On the basis of the research on trigonometric functions of oblique triangles conducted by A. Braunmühl and Biehringer, in the first part of his study A. Jakštas introduced characterised curves from the aforementioned systems. In the second part he analysed a reverse problem- any curve described algebraically in the first part requires establishment of the corresponding rectangular trigonometric system. After finishing his study "Apie sukrautines stačiakampes trigonometrijos sistemas" and having devoted several years of his life to the problem, A. Dambrauskas made attempts to energise trigonometry and suggested new fields for research in the science of triangles. However, the new trigonometric systems did not find any practical application and were not further investigated.

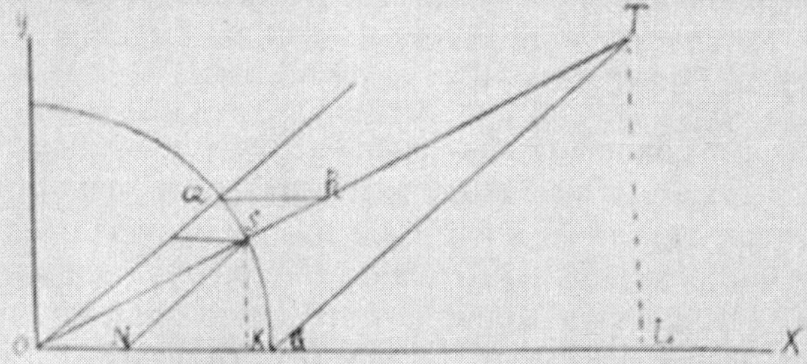

mums 8 brėžinys, kame bus SN = s, ON = c, OS = OG = ϱ, TG = t, OT = z, QR = k ir t. t.

8 bréž.

Lieka sužinoti, ką ištikrųjų reiškia kitas ieškomoje sistemoje nekintamas dydis a.

Tuo tikslu kreipkimės į 8 bréz. Iš trikampio SNK betarpiškai turime:

$$(136) \; . \; . \; \begin{cases} SK = y = s \cdot \sin \vartheta \\ NK = x - c = s \cdot \cos \vartheta . \end{cases}$$

Padauginę antrąjį lyginį (136) daugikliu sin ϑ, o pirmąjį — cos ϑ ir atėmę pirmąjį padaugą iš antrojo, gausime:

$$(137) \; . \; . \; x \sin \vartheta - y \cos \vartheta = c \cdot \sin \vartheta .$$

Pakeitę lyginyje (135) y ir x sin ϑ — y cos ϑ naujais reiškiniais iš (136) ir (137) ir patėmiję, kad šiame žygyje $\sqrt{x^2 + y^2} = \varrho$, gausime:

$$a \cdot c \cdot \sin \vartheta = s \cdot \sin \vartheta \cdot \varrho ,$$

iš kur turime:

$$a = \frac{\varrho s}{c} = t.$$

Tuo būdu ieškomoji sistema bus $S_{\varphi = \vartheta}^{t = a}$.

2. Teesie tūlos trigonometriškosios sistemos geometriškoji punktų S vieta kreivoji išreikšta lyginiu:

$$(138) \; . \; . \; \frac{(x - a)^2 + y^2}{x - a} = \frac{x^2 + y^2}{x \sin \vartheta - y \cos \vartheta}$$

5

65

Figure 4. Fragment of "New Trigonometric Systems".

A. Dambrauskas is the first author, who made attempts to evaluate the merits of bishop A. Baranauskas, as a Lithuanian mathematician. This was the article dedicated to the collection of A. Baranauskas' works prepared by Lithuanian writer Juozapas A. Herbačiauskas (1876-1944) and published in 1906. However, the publication of the book was delayed and the article was published in issue four of the journal "Draugija" in 1907. The fact that the author referred to unpublished works (personal letters written by the bishop to him in 1890-1900) increased the significance of that work. They allowed to learn that A. Baranauskas continued his acquaintance with mathematics, which started at school, during the studies in the Academy. There in Petersburg, he adopted knowledge of algebra and also engaged into analysis of the theory of numbers. Having mentioned the works in the area of primes, the prepared dissertation in the theory of numbers, which remained in manuscripts, A. Dambrauskas further discussed geometrical articles written by the bishop that penetrated into the problem of circle quadrature. Evaluating the merits of A. Baranauskas to mathematics in Lithuania, A. Jakštas, wrote that the bishop had not turned "into a mathematician specialist with narrow views. Mathematics was never a goal in itself but only a tool. He trained his own mind with its help. He used the exceptional God-given mathematical talent"(Dambrauskas, 1907) .

One more original work of mathematics, assigned to the history of mathematics, was prepared by A. Jakštas on the basis of the lecture delivered accepting the diploma of Honorary Doctor of Mathematics. The lecture was about Polish mathematician and philosopher J.M. Hoene-Wro?ski and the article titled "A.J.M. Hoenė-Wronski matematikas" (A.J.M. Hoene-Wro?ski as a Mathematician) appeared in the book "Užgęsę žiburiai" (Extinct Lights) in 1930. Here the prelate presented the most important biographic features of the scientist and emphasised his great merits in creation of foundations of algorithmic philosophy. Moreover, A. Jakštas was fascinated by Wronski's search for supreme and most common laws of mathematics and exalted his works in applied mathematics. Feeling the similarity between own attitude and that of the man of science from Poland, A. Dambrauskas stated: "Mathematics was not the goal in itself and only a means to attain higher goals. Due to this reason, he went towards mathematics following the philosophical path"(Dambrauskas, 1975: p. 401) .

The second group of A. Dambrauskas' mathematical works includes the articles that aimed to familiarise Lithuanian society with the achievements in that science. According to mathematician P. Katilius "Mathematics is accessible to the society to the degree the society is educated itself"(Katilius, 1938: p. 148) . Assuming this, A. Jakštas undertook dissemination of mathematical knowledge in the society, though popularisation of this science is most challenging.

"Trys garsieji matematikos klausimai" (The Three Famous Mathematical Problems) published in 1924 should be firstly distinguished from the mathematical legacy. In the aforementioned work A. Dambrauskas aimed at clarification of three main problems deriving from the ancient times: circle quadrature, doubling of cube and angle trisection (division into three equal parts). These problems only at first sight are clear in terms of their formulation and easily perceivable. However, their solution using a pair of compasses and a ruler was confusing and required thousands of years. The written work was dedicated to the 20th anniversary of the death of bishop A. Baranauskas. The problem of circle quadrature occupies the main place in the article because "this assignment surpasses the other two in terms of its renown and importance" (Jakštas, 1924: p. 5) . The essence of the later problem is to draw a square using a pair of compasses and a ruler, whose area is equal to that of the given circle. All the story of solving the problem continued more than four thousand years. The author divided such a long period of time into three parts: "elementary geometry, higher analysis and algebra". Having overviewed each of them, A. Jakštas highlighted the merits of A. Baranauskas ventilating the aforementioned problem. At the end the author acknowledge that he had got "contaminated" by the problem as early as gymnasium and arrived at a conclusion that "arithmetics provides π value, whereas analysis leads to its genesis" (Jakštas, 1924: p. 71) .

The second problem had a huge influence on the development of methodology of mathematics and originated from the Delos Island. It is well known that this classical Greek problem asks to draw a cube that is twice as big as the volume of the given cube. The author provided a detailed solution of the problem.

The third objective included angle trisection with the help of a pair of compasses and a ruler. A. Jakštas criticized the works written by some Russian authors (D. Viskovatov, V. Grigorjev and others) on the aforementioned theme. Strictly and theoretically substantiating he analysed the problem and proved that the latter drawings are only approximate. Moreover, the publication also introduced the three best known problems in cases of non-Euclidian geometries. Writing his own work, the author referred to the works by French, German and Hungarian scientists on the related issues. It is important to emphasise that he succeeded in persuading the majority of amateur mathematicians in Lithuania to give up attempts to complete the three well-known ancient problems using a pair of compasses and a ruler.

The latter group also embraces a bunch of less significant science popularisation works. Some of them are related to arithmetic, the science of numbers, for example, "Apytikris π nubrėžimas"(Approximate Drawing of π), "Naujas būdas nuošimčiam surasti apskaitant terminius mokesčius" (A New Way to Calculate Percentage Keeping Record of Terminated Taxes), "Apokalipsiškasai skaičius 666 matematikos žvilgsniu ir simboliškoji jo reikšmė" (Apocalyptical Number 666 from the Perspective of Mathematics and its Symbolical Meaning), "Sis tas iš humbugo bei hasardo srities" (Something on Gambling and Ardour), "Matematiškas vieno gešefto tyrinėjimas" (Mathematical Analysis of One Business). Some were linked with geometry, such as, "Bendras būdas taisyklingiems daugiakampiams ratilan įbrėžti" (A Universal Way of Inscribing a Regular Polygon into a Circle), "Matematiškas bičių instinktas ir vabalas matematikas" (Mathematical Instinct of Bees and a Beetle Mathematician), "Dieviškoji proporcija" (The Divine Proportion), "Matematiškas dailės dėsnis" (Mathematical Law on Fine Arts).

In terms of history of arithmetic there is an interesting article about magic number 666. This work was published as a separate publication and familiarised the reader with the attitudes of Talmud, ancient Greeks and the Bible towards that number. A year later, in his publication of amusing mathematics written in 1921 the author tells about the form of bee comb, grounding the rationale of the choice of hexagon (Jakštas, 1921b) . In the second part of the brochure, A. Jakštas referring to "beržo suklys" (Rhynchites betulae, "a birch leaf roller"), called it a beetle mathematician and presented mathematical generalisations of leaf rolling.

In the same year he published a book about the golden ration. In addition, the beginning of this work dates back to 1907, when "Draugija" published his article about "mathematical criterion of poetry". Later the article was expanded, showing how the golden ration reveals itself in music, architecture and living nature (Jakštas, 1921a) .

Filling in the gap of geometry textbooks, A. Jakštas announced a way to inscribe all the regular polygons into a circle in his "Švietimo darbas" (Educational Work) in 1929. The article is particularly significant as it contains explanation how to inscribe septangles, enneagons, tetradecagons and octadecagons (Dambrauskas, 1929) .

A few works by A. Dambrauskas, ascribed to that group, remained in manuscripts. They were as follows: "Vysk. A. Baranausko transcendentalinės progresijos matematiškoji teorija" (Mathematical Theory of Transcendental Progression of Bishop A. Baranauskas), "Apie sukrautinio periodiškumo kreives" (On Curves of Complex Periodicity), "Keletas žodžių apie menamąjį kintamąjį" (A Few Words on Apparent Variable) and "Plokščių ir tiesiųjų linijų tiesumo sąlygos" (Conditions for Straightness of Planes and Straight Lines). Some of the aforementioned manuscripts were used by the prelate writing other articles (VDU MGF fondas.(a), p. 94) . A. Jakštas also acted as a critic of works of mathematics. He reviewed several newly published textbooks of mathematics in the journal "Draugija". In such a way A. Dambrauskas contributed to publication one of the first arithmetic textbook by Lithuanian publisher Petras Mikolainis (1868-1934), published in Tilsit in 1909, and recommended it not only for children but also for "adults who do not know the taste of the science".

In the first years of World War I A. Jakštas became a member of Commission for Terminology. He actively involved in coining of Lithuanian mathematical terms. Therefore, later reviews of textbooks of geometry and trigonometry contained numerous corrections of terms. The prelate entered into polemics with Lithuanian educator Pranas Mašiotas (1863-1940) regarding suggested terms of "galia-laipsnis" (degree), "lyginis-lygtis" (equation), "logaritmų knygos-logaritmų lentelės" (logarithmic tables) and others. He argued with mathematics educator Marcelinas Šikšnys (1874-1970) criticising the coined neologisms and suggesting international equivalents: "diagonalė-įstrižainė" (diagonal), "diametras-skersmuo" (diameter), "perpendikuliaras-statmuo" (perpendicular) and others.

Despite the aforementioned contradictions, about 40 mathematical terms coined by A. Jakštas have been acknowledge and used up to now (Rumšas, 1979: p. 55) . For example, the following terms were naturalised and have been in use even now: "atstumas" (distance), "aukštis" (height), "atitinkamieji kampai" (corresponding angles), "brėžinys" (drawing), "daugiakampis" (polygon), "dvisienis" (dihedral), "erdvė" (space), "ilgis" (length), "išklotinė" (evolvement), "įžambinė" (hypotenuse), "kūgis" (cone), "kampas" (angle), "laužtinė" (broken line), "matlankis" (protractor), "trikampis" (triangle) and others.

However, a big number of terms suggested by A. Dambrauskas were not naturalised and were replaced by others: "šonas (now-kraštinė)" (side), "ratilas (skritulys)" (circle), "ratlankis (apskritimas)" (compass), "rutulys (ritinys)" (sphere), "stipinas (spindulys)" (radius), "skritulys (rutulys)" (circle) and others.

The prelate was constantly sincerely concerned about terms and even as early 1919 wrote with certain grievances: "There is a trouble with terms: they are coined by anybody and imposed on everybody in a form of immature discoveries" (Jakštas, 1919b: p. 359) .

Thus, all the works of mathematics popularisation and information by A. Jakštas were undoubtedly useful to the Lithuanian society. They disseminated new knowledge or supplemented the possessed one of classical mathematics as well as raised the problems addressed in the science and indicated possibilities for their solutions. Moreover, the works increased the interest of young people in mathematics.

A. Dambrauskas wrote two textbooks of mathematics. The Commission of Textbooks under the Lithuanian Scientific Society in 1915 committed a task of writing a geometry textbook to A. Jakštas. To complete the obligation, he chose and translated P. Mironov's "Parengiamsis geometrijos kursas" (Preparatory Course of Geometry). The author acknowledged that it was "work of starvation" because unavailability of textbooks necessitated their prompt preparation. However, the prelate was creative and supplemented the Lithuanian material with another chapter on geometry written by I. Yurevich. Moreover, at the end of the book he devised a glossary, where the terms were compared to the terms available in the geometry book published by Lithuanian public man and engineer Petras Vileišis (1851-1926) in 1900, and provided their equivalents in four languages. It should be pointed out that "Geometrijos vadovėlis dviklasėms mokykloms" (Geometry Textbook for Two-Class Schools) was published twice (1916 and 1920) (Jakštas, 1916) .

Encouraged by successful results deriving new systems of trigonometry, A. Dambrauskas showed his determination writing an original textbook "Plokštinės trigonometrijos" (Plane Trigonometries) in 1919. It was approached by the author as an introduction into the research on trigonometry systems. In addition, it was one of the first textbooks in Lithuanian. Therefore, A. Jakštas seriously studied the works in the related sphere written by foreign authors (F. Bendt, P. Cranz, H. Niewęglowski, A. Davidov, N. Rybkin and others). From the textbooks of the above-mentioned authors he chose "what was good, i.e. what was most appropriate from the perspective of science, methods and pedagogy" (Jakštas, 1920: p. 3) . The textbook based on theories and assignments consisted of three parts: "geometry, rectilineal trigonometry and geometrical and analytical expression of trigonometrical functions". Thus, it is necessary to point out that having familiarised with the latest achievements in the science, he made attempts to deliver a course of trigonometry in a modern way.

Following D. Mercier, G. Chelpanov, in 1918 A. Dambrauskas prepared the first Lithuanian textbook "Logika" (Logics). A year later the prelate wrote the book "Logikos evoliucija" (Evolution of Logics), which is of interest to the science of mathematics. Discussing the object of the aforementioned science, the author in his book singled out two aspects-mathematics of finite and constant numbers and mathematics of infinite and imaginary numbers. Later he explained that "the first one embraces the world of manifestations, which is only a fictitious image of our world. The second one includes the substantial and real world, which means the world the way it is" (Jakštas, 1919a: p. 12) . Referring to "real mathematics" A. Jakštas presented a multifaceted understanding of the concept of infinity.

A. Dambrauskas maintained contacts with famous Western European and Russian mathematicians (C. Vöros, E. Lefevre, F. Weinrich, S. Wiekstein, I. Pervushin and others) (A. Jakšto···VUB RS), was an active participant in activities of Mathematics Section of Lithuanian Society of Naturalists. The qualities of specification and criticism were ascribed to him. Consistently following the principle of "classical mathematics", he was also characterised by conservatism. The features of the true scientist were revealed in the personality of A. Dambrauskas. He would refuse to rely on the mathematical truths that he failed to understand himself or to completely clarify, even if they had already been claimed by authorities (Stanaitis, 1938: p. 392) .

4. THE ROLE OF A. JAKŠTAS IN LITHUANIAN MATHEMATICS (CONCLUSIONS)

Acting in the afore-mentioned spheres, A. Jakštas played a role of a supporter of applied mathematics. Therefore, "he was firstly concerned about specific, down-to-earth problems rather than about a priori transcendental statements" (Žg, 1938: p. 6) . The prelate was interested in the tasks, which could be applied in daily life or were grounded on commonly perceived laws of our world.

Jakštas' mathematical works are perceived not only as the very first ones in Lithuanian mathematics. This is confirmed by remembrances of engineer Narutavičius. The mathematics professors from Saint Petersburg University, who would find out about mathematical works of A. Dambrauskas from Lithuanian students, acknowledged them "as deserving attention". Later M. Vaitkus told about it to A. Jakštas and regretted that "he had ignored own vocation" and had not devoted himself fully to mathematics. At that time the prelate answered: "I wish at least one priest had talked to me like this in their time" (Vaitkus, 1938: p. 343) .

In fact, according to S. Šalkauskis, A. Jakštas' mathematical works show that he was "a zestful mathematician" (Šalkauskis, 1938: p. 1) . A. Jakštas-Dambrauskas relied on the thought: "Hence, theologists and philosophers should see mathematics as a friend and assistant of theology and philosophy rather than an enemy". On the other hand, being aware of the depth of the science of mathematics of those times and critically evaluating the completed work, the prelate, when called a mathematician, would say: "I am not much of a mathematician. I am, perhaps, a lover of mathematics" (Stanaitis, 1938: p. 495) . Thus, was A. Jakštas- Dambrauskas a mathematician? According to O. Stanaitis, he should be approached from two perspectives. If introducing the standard that a mathematician is the one, who has had an impact on at least one branch of mathematics, the prelate is considered as an amateur mathematician. On the other hand, if the ability to use mathematical truths was prioritised, A. Jakštas would be definitely qualified as a mathematician. At the same time, it should be pointed out that the bigger number of prelate's mathematical works is dedicated to school mathematics and, this is understandable as he lacks profound knowledge of higher mathematics.

Evaluating A. Jakštas' merits to Lithuanian mathematics, we have to consider the fact that the width of his works in this area was limited by the wide range of his activities. However, it is necessary to emphasise that developing the Lithuanian national education system at the end of the 19th century and establishing the modern Republic of Lithuania after 1918, the discussed works of the prelate undoubtedly disseminated the knowledge of mathematics and evoked interest in the science. The obvious truth was understood that mathematics is "accessible" to the society to the extent the society itself is mathematically educated. It is also significant that publishing his mathematical works, A. Dambrauskas had great hopes to encourage future Lithuanian generations, firstly young people, to penetrate into the science of mathematics and to seek discovery of new truths.

Thus, seeing the dawn of the Lithuanian science, we have to acknowledge undeniable merits of prelate A. Dambrauskas to our mathematics. Reading the mathematical works of the prelate, we admire his ability to transfuse philosophy of humanism to them. His research works implanted a continuous pursuit of truth seeking in mathematics, denied the existence of the eternal "ignoramus". This makes A. Dambrauskas a model to follow for present and future generations of mathematicians and his mathematical works occupy an honourable place in the history of mathematics in Lithuania.

ACKNOWLEDGEMENTS

Translated by Jovita Bagdonavičiūtė, Lithuanian University of Educational Sciences, Vilnius, Lithuania.

REFERENCES

1. Azubalis, A. (1990). Adomas Jakstas-matematikas. Mokslas ir Gyvenimas, 8, 30-31. [Paper reference null]
2. Baltrūnas, A. (1996). A Dambrauskas-Jakstas matematikos populiarintojas. Alfa ir Omega, 2, 43-48. [Paper reference null]
3. Dambrauskas, A. (1907). Vyskupas Antanas Baranauskas, kaipo matematikas. Draugija, 4, 332-341. [Paper reference null]
4. Dambrauskas, A. (1926a). Euklido V-ojo postulato kreive ir metageometrijos pagrindu kritika. Logos, 3, 257-294. [Paper reference null]
5. Dambrauskas, A. (1926b). Kas yra tiesioji linija? Lumgfd, 3, 373-394. [Paper reference null]
6. Dambrauskas, A. (1929). Bendras būdas taisyklingiems daugiakampiams ibrezti. Klaipeda. [Paper reference null]
7. Dambrauskas, A. (1930). Matematiskos salygos kreivems pavirsiams nuo ploksciu ir kreiviems nuo tiesiuju atskirti. Vdumgfd, 5, 124-176. [Paper reference null]

8. Dambrauskas, L. (1939). Keli seniausi Dambrausko-Jaksto laiskai. Draugija, 4, 249-251. [Paper reference null]
9. Dambrauskas, A. (1975). AJM Hoene-Wronskis Matematikas. In Uzgese ziburiai (p. 401). Roma. [Paper reference null]
10. Dombrowski, A. (1906). Pri unu speco de kurbaj linioj koncernantaj la V-an Euklidan postalaton. Berlin. [Paper reference 1]
11. Jakstas, A. (1916). Geometrijos vadovelis dviklasems mokykloms. Kaunas. [Paper reference 1]
12. Jakstas, A. (1919a). Logikos evoliucija. Kaunas. [Paper reference 1]
13. Jakstas, A. (1919b). Masiotas. Geometrijos uzdavinys. Draugija, 7-8, 359-360. [Paper reference 1]
14. Jakstas, A. (1920). Plokstine trigonometrija. Pilnas sistemiskas kursas. Kaunas. [Paper reference 1]
15. Jakstas, A. (1921a). Dieviskoji proporcija. Matematiskas dailes desnis. Kaunas. [Paper reference 1]
16. Jakstas, A. (1921b). Matematikos biciu instinktas ir Vabalas matematikas. Kaunas. [Paper reference 1]
17. Jakstas, A. (1922). Naujos trigonometriskos sistemos. Berlynas. [Paper reference 2]
18. Jakstas, A. (1924). Trys garsiausieji matematikos klausimai. Kaunas. [Paper reference 2]
19. Jakstas, A. (1931). Recenzijos. Filosofija. Vairas, 3, 388. [Paper reference 1]
20. Jakstas, A. (1938). Jakstas kalbaimus. Zidinys, 3, 347-353. [Paper reference 1]
21. Jaksto, A. fondas. VUB RS1 F. 1. E306-RS1279; 575. [Paper reference 1]
22. Kalinauskaite, N. (2011). Aleksandro Dambrausko naujosios trigonometrines sistemos. Lietuvos matematikos rinkinys. LMD Darbai, 52, 156-160. [Paper reference 2]
23. Katilius, P. (1938). Dambrauskas-Matematikas. Kosmos, 3-4, 148-151. [Paper reference 2]
24. Keliuotis, J. (1931). Pasikalbejimas su prof. A. Dambrausku-Jakstu. Naujoji Romuva, 38, 897-901. [Paper reference 1]
25. Maceina, A. (1938). Jaksto filosofija. Gamtos problema. Naujoji Romuva, 13, 312-314. [Paper reference 1]
26. Masaitis, C. (1938). A. Dambrauskas matematikoje. Naujoji Romuva, 12, 292-293. [Paper reference 2]
27. Rumsas, P. (1979). Prie matematikos terminijos istaku. In Zigmas Zemaitis (Collection of Articles, pp. 55-56). Vilnius. [Paper reference 2]
28. Salkauskis, S. (1938). A. Dambrauskas, mano mokytojas ir kritikas. Logos, 1, 1-53. [Paper reference 1]
29. Samulionis, P. (1938). A. Dambrausko-Jaksto gyvenimo bruozai. Zidinys, 3, 303-310. [Paper reference 2]
30. Senkus, J. (1938). Netekome mokslininko ir kultūros veikejo. Trimitas, 8, 189. [Paper reference 1]
31. Stanaitis, O. (1938). Jakstas ir matematika. Zidinys, 4, 491-496. [Paper reference 3]
32. Vaitkus, M. (1938). 30 metusu A. Jakstu. Zidinys, 4, 340-346. [Paper reference 1]
33. Zg, J. (1938). A. Jakstas-Matematikas. Lietuvos Aidas, 28, 6. [Paper reference 1]
34. 2LCVA—Central state archive of Lithuania. [Paper reference 1]
35. VDU MGF fondas.(a) LCVA2. F. 631, 1, 137, 94. [Paper reference 1]
36. VDU MGF fondas.(b) VUB RS. F. 96. VDU4, 232. [Paper reference null]

NOTES

[1]VUB RS—Vilnius university library, departament manuscrips.

[2]LCVA—Central state archive of Lithuania.

CHAPTER 5

The twenty-third ICMI study: primary mathematics study on whole numbers

Sybilla Beckmann

Department of Mathematics, University of Georgia, Athens, Georgia 30602, USA

ABSTRACT

A new study on the teaching and learning of whole number arithmetic in primary school and pre-school will be conducted by the International Commission on Mathematical Instruction (ICMI). The study is organized around themes concerning foundational aspects of whole number arithmetic (cultural, historical, epistemological, as well as (neuro)cognitive), the teaching and learning of whole number arithmetic, and the connection of whole number arithmetic with other mathematical topics. The International Program Committee invites papers for the new study.

Keywords: Whole numbers, Arithmetic ICMI study, Primary mathematics, Pre-school mathematics

1. BACKGROUND

The International Commission on Mathematical Instruction (ICMI) announces a new study on whole number arithmetic in the official discussion document posted at http://www.mathunion.org/ icmi/ conferences/icmi-studies/ongoing-studies/icmi-study-23/. This study, the twenty-third led by ICMI, addresses for the first time mathematics teaching and learning in primary school (and pre-school as well) for all, taking into account inclusive international perspectives including socio-cultural diversity and institutional constraints. One of the challenges of designing the first ICMI primary school study is the complex nature of primary mathematics. For this reason, a specific focus has been chosen, as the key and driving feature, with a number of questions connected to it. The broad areas of whole number arithmethic (WNA), including operations and relations and the solution of arithmetic word problems, are the kernel or core content of all primary mathematics curricula. The study of this key core content area is often regarded as foundational for later mathematics learning. However, the principles and main goals of instruction in the foundational concepts and skills in these aspects are far from universally agreed upon, and practice varies substantially from country to country. An ICMI study that provides a meta-level analysis and synthesis of what is known about this core area of primary mathematics would provide a useful base from which to gauge gaps and silences and an opportunity to learn from the practice of different countries and contexts.

Whole numbers are part of everyday language in each culture, but there are different views on the most appropriate age at which to introduce whole numbers in school. Whole numbers, in some countries, are approached in pre-school, with nearly all the children before the age of 6 attending pre-school. The OECD has reported that, in general, participation in pre-school produces better learning outcomes for 15-year-old students (OECD PISA FOCUS 2011). In some countries, primary school includes grades 1 to 6; in others, it includes grades 1 to 5. Also the entrance age of students for primary school may vary from country to country. For all these reasons, this study addresses teaching and learning whole numbers from the early grades, i.e., the periods in which whole numbers are systematically approached in the formal school, hence, when it is the case, also in pre-school.

2. ICMI STUDY 23

In Berlin, in January 2014, the International Program Committee (IPC) (see 'Appendix') for ICMI Study 23 met and agreed upon four principles.

First, it was decided that *cultural diversity* and how this diversity impinges on the early introduction of whole numbers would be one major focus. The study will seek contributions from authors from as many countries as possible, especially those in which cultural characteristics are less known and yet they influence what is taught and learned. In order to foster the understanding of the different contexts in which authors have developed their studies, each applicant for the conference will be required to prepare background information (on a specific form) about this context.

Second, it was decided to find better ways to involve *policy makers* (who have the duty to offer to every child the opportunity to go to school and to learn arithmetic) and, in order to take care of this specific aim, to solicit also contributions in the form of commented and annotated video clips about practical examples with a (potentially) strong impact.

Third, it was decided to collect experiences about teaching and learning *for all*, including students with special needs, considering that in some countries, they have special classrooms and teachers and even special schools while in others, they are enrolled in mainstream classes.

Fourth, it was decided to focus also on *teacher education and professional development*, considering that in order to teach elementary mathematics, there is a need for sound professional knowledge, both in mathematics and in pedagogy.

In order to meet this complex set of principles, the IPC delineated a set of *themes* to serve as the organizing framework for the study conference.

This document presents the background of the study, together with its challenges and aims. These sections lead to the description of the five organizing themes of the study. Because the study conference will be organized around discussion within each theme (with some overarching sessions), each proposed contribution to the study should be addressed to the theme into which it will fit best (with a first and a second choice, according to possible multiple foci). Finally, this document briefly outlines the organization, timing, and location of the study conference and the timetable of the milestones leading up to the conference and to ICMI publication.

3. BACKGROUND OF THE STUDY

Primary schooling is compulsory in all countries, although with different facilities and opportunities for children to take advantage of it. Mathematics is a central feature of all primary education. The content and quality delivery of that curriculum is important in all countries for the kinds of citizens each seeks to produce.

In the international literature, there are many contributions on primary school mathematics. In many cases, especially in the West, early processes of mathematical thinking, usually observed in early childhood (i.e., 3- to 8-

year-old children), are also investigated by cognitive and developmental psychologists. These researchers sometimes study the emergence of early processes of mathematical thinking in laboratory settings, where children are stimulated by suitable displays (to observe the emergence of one-to-one correspondences, counting, measuring, and so on). In several countries, Piaget's theory is very influential despite its critics. Also, neuroscientists have been studying for some years the emergence of 'number sense,' but it has been observed (UNESCO 2013) that what is still missing is a serious and deep interdisciplinary work with experts in mathematics education.

4. DISCUSSION

A recent document prepared by ICMI's Past President Michéle Artigue and commissioned by UNESCO (2012) discusses, from a political perspective, the main challenges in basic mathematics education. It reads, in part:

We live in a world profoundly shaped by science and technology. Scientific and technological development has never been faster, has never had an impact as important and as immediate on our societies, whatever their level of development. The major challenges that the world has to face today, health, environment, energy, development, are both scientific and human challenges. In order to take up these challenges, the world needs scientists able to imagine futures that we barely see and able to make these possible, but it also needs that the understanding of these challenges, the debate on the proposed changes, are not reserved for a necessarily limited scientific elite, but are very widely shared. Nobody can now doubt that positive, sustainable and equitable evolutions cannot be achieved without the support and contribution of the great majority of the population.

Drawing on this idea, ICMI has acknowledged that it is timely to launch, for the first time in its history, an international study that especially focuses on early mathematics education that is both basic and fundamental mathematically. Primary school mathematics education has been present in other ICMI studies, but in most cases, secondary school mathematics education has been predominant. When foundational processes are concerned, a strong epistemological basis is needed. This might be the added value of ICMI involvement with respect to the analysis carried out in other fields. Such epistemological analysis was part of classical works of professional mathematicians (e.g., Klein, Smith, Freudenthal) who played a big role in the history of ICMI (ICMI 2008) and considered mathematics teaching as a whole. It is worthwhile to mention here a comment by Felix Klein, the first president of ICMI, used as an epigraph in the website on the history of ICMI (ICMI 2008).

I believe that the whole sector of Mathematics teaching, from its very beginnings at elementary school right through to the most advanced level research, should be organised as an organic whole. It grew ever clearer to me that, without this general perspective, even the purest scientific research would suffer, inasmuch as, by alienating itself from the various and lively cultural developments going on, it would be condemned to the dryness which afflicts a plant shut up in a cellar without sunlight (Felix Klein, 1923).

One cannot study school mathematics teaching without focusing also on the teacher's role and responsibility. The attention towards mathematics teacher education and professional development has been a constant preoccupation of ICMI. The case of primary school and (more generally) the case of early education deserve special attention. The complex nature of arithmetic and its foundational value for mathematics are well known by mathematicians and mathematics educators. However, primary school teachers work within systems which may or may not support a rigorous professional environment in which they are knowledgeable and respected professionals who are experts on both the mathematics and the pedagogy of what they teach. In some systems, teaching WNA may be treated as something that virtually any educated adult can do with little specific training; WNA may be viewed by some as straightforward and intuitive and involving no more than showing children how to cope with everyday life and to carry out algorithms.

There are systems where primary mathematics teachers are specialists and others where they are generalists. It is not within the aims of this study to enter deeply into the pedagogical debate about specialist vs generalist teachers in early education, as both models show advantages and disadvantages. What is important to highlight is that much is already known from research about productive ways to teach WNA, yet this knowledge cannot be enacted in systems in which teachers are not proficient in elementary mathematics and the particular pedagogical approaches. Effective teacher education may require a backdrop of a culture in which teachers are expected to be highly educated professionals.

5. AIMS OF THE STUDY

This study aims to produce and share knowledge on the sustainable ways of realizing teaching and learning WNA for all, taking into account the large body of theory and research already existent as well as socio-cultural diversity and institutional constraints. In particular, the following specific aims were acknowledged by the IPC, for the early teaching and learning WNA:

- Bring together communities of international scholars representative of ICMI's diverse membership across regions and nationalities in addressing the theme of WNA for the production of a study volume.
- Provide a state-of-the-art expert reference on the theme of WNA.
- Contribute to knowledge, better understanding, and resolution of the challenges that WNA faces in diverse contexts.
- Collectively represent the great variety of concerns in the field of WNA and reflect upon it.
- Facilitate multi and interdisciplinary approaches (including cooperation with other bodies and scientific communities) to advance research and development in WNA.
- Disseminate scholarship in mathematics education—research; methodologies, theories, finding and results, practices, curricula—in the theme of WNA.
- Pave the way towards the future by identifying and anticipating new research and development needs of WNA.
- Be of interest and a resource to researchers, teacher educators, policy, and curriculum developers and analysts and the broad range of practitioners in mathematics and education.
- Promote and assist discussion and action at the international, regional, or institutional level.

6. THE THEMES OF THE ICMI STUDY 23

The ICMI Study will be organized around five themes that provide complementary perspectives on the early approach to whole numbers in mathematics teaching and learning. Contributions to the separate themes will be distinguished by the theme's specific foci and questions, although it is expected that interconnections between themes will emerge and merit attention.

The five themes are as follows:

1) *The why and what of whole number arithmetic*
2) *Whole number thinking, learning, and development*
3) *Aspects that affect whole number learning*
4) *How to teach and assess whole number arithmetic*

5) *Whole numbers and connections with other parts of mathematics.*

Themes 1 and 2 address foundational aspects from the cultural-historic-epistemological perspective and from the (neuro)cognitive perspective. What is especially needed are reports about the impact that foundational aspects have on practices (both at the micro-level of students and classrooms and at the macro-level of curricular choices).

Themes 3 and 4 address learning and teaching, respectively, although it is quite hard, sometimes, to separate the two aspects, as suggested by the fact that in some languages and cultures (e.g., Chinese, Japanese, Russian) the two words collapse into only one.

Theme 5 addresses the usefulness (or the need) to consider WNA in connection with (or as the needed basis for) the transition to other kinds of numbers (e.g., rational numbers) or with other areas of mathematics, traditionally separated from arithmetics (e.g., algebra, geometry, modelling).

Each theme is briefly outlined and followed by exemplary questions that could be addressed in the submitted contributions. An overarching question which cuts across all the themes concerns teacher education and development:

How can each of the themes be effectively addressed in teacher education and professional development?

7. THE WHY AND WHAT OF WHOLE NUMBER ARITHMETIC

This theme will address cultural-historic-epistemological issues in WNA and their relation to traditional, present, and possible future practices.

The sense of numbers is constructed through everyday experience, where culture and language play a major role; hence, ethnomathematics has paid attention to the different grammatical constructions used in everyday talk (e.g., Maori numbers as actions; Aboriginal Australians' spatial approach to numbers). Ways of representing whole numbers and making simple calculations (e.g., with fingers or other body parts; with words; with tools, including mechanical and electronic calculators; with written algorithms) have enriched the meaning of whole numbers through the ages.

The base-ten system is critical for our current sophisticated understanding of WNA. The long and difficult development of place value systems is well documented in the history of mathematics (the introduction of place value in China and India; the migration to Europe through the Arabic culture; the invention of zero; the strategies for mental calculation) and indicates the need to study place value and the base-ten system deeply for understanding.

The above issues (and others) have been considered in different ways by different cultures throughout history. Beside the use of numbers in practical activities, there is evidence (in the history and in educational research) that the exploration of the properties of whole numbers, relations, and operations paves the way towards the introduction, with young students too, of typical mathematical processes, such as generalizing, defining, arguing, and proving.

Some references may be found in the ICMI Studies 10, 13, 16, and 19.

The following possible questions will help to illuminate this theme further:

- *What goals underlie the teaching and learning of WNA?*
- *Taking a mathematical perspective (as practiced by the current community of mathematicians) combined with an educational perspective, what are core mathematical ideas in paths to developing WNA?*
- *What are distinctive features concerning whole number representation and arithmetic in your culture? What is the grammar of numbers? In what ways does language or ways of representing and using numbers influence approaches to calculation or problem solving? How do these features interact with the decimal place value system?*
- *What is the role of mathematical practices and habits of mind in teaching and learning WNA? How can teaching and learning WNA support the development of mathematical practices and habits of mind?*
- *How much is the base-ten place value emphasized in your curriculum?*
- *How much computational facility is important for later mathematics learning and learning in other areas? What about mental calculation? What about speed of calculation?*
- *How do policies and the educational environment and system support or not support a culture in which teaching WNA is seen as requiring detailed, specific professional knowledge?*
- *What were the main historic features and their origins of WNA in (ancient) west/east? What were some factors that led to such historic features? What were the effects on the development of mathematics curriculum?*
- *How does your curriculum develop understanding of the structural features of whole number arithmetic and its extensions?*

8. WHOLE NUMBER THINKING, LEARNING, AND DEVELOPMENT

This theme will address the relationships between cognitive and neurocognitive issues and traditional, present, and possible future practices in the early teaching and learning of WNA.

The idea of number sense was in use for decades in the literature on mathematics education before entering into the cognitive and neurocognitive literature, with some similarities and differences. (Neuro)cognitive scientists have focused the children's spontaneous tendency to focus on numerosity in their environment, the development of rapid and accurate perception of small numerosities (subitizing) in connection with visualization and structuring processes, the ability to compare numerical magnitudes, and the ability to locate numbers on a (mental) number line. There are models for children's informal knowledge of counting principles and informal counting strategies and their development into more formal and abstract arithmetic notions and procedures.

A recent focus concerns developmental dyscalculia, as a difficulty in mathematical performance resulting from impairment to those parts of the brain that are involved in arithmetical processing, without a concurrent impairment in general mental function.

Recent debates concern the embodied cognition thesis resulting in the evidence, shared by many researchers that, although mathematics may be socially constructed, this construction is rooted in, and shaped by, the body and bodily experiences.

Some references may be found in OECD EDUCERI 2010 and UNESCO 2013.

The following possible questions will help to illuminate this theme further:

- *To what extent is basic number sense inborn and to what extent is it affected by socio-cultural and educational influences? How is the relationship between these precursors/foundations of WNA, on the one hand, and children's whole number arithmetic development?*
- *What can we learn from the (neuro)cognitive studies in WNA? Do their findings essentially confirm insights that are present (and were already present for a long time) in the mathematics education community or do they point to truly new insights and recommendations about the kind of tasks and instructional approaches children need? How do we integrate different perspectives about the foundations and development of whole number arithmetic concepts and skills?*

- *What are specific effects of the structure of the individual finger counting system on mental and linguistic quantity representation and arithmetic abilities in children and even in older learners and adults?*
- *How can an embodiment framework can be used to analyze and/or design educational approaches based on suitable representations (* e.g., *through the number line) or on manipulatives and modern technological devices (touch screens)?*
- *What are appropriate ways of analyzing the multimodal nature of mathematical thinking (* e.g., *the role of bodily motion and gesture)?*
- *What is the relationship between the embodied cognitive approach and older approaches, for example, Montessori and Piagetian, which had a strong influence on elementary school mathematics worldwide?*
- *How can the tools of the embodiment framework/analysis be integrated/combined with socio-cultural perspectives to compare/contrast approaches where embodiment is exploited or hindered?*
- *How can teachers be educated in order to exploit the (neuro)cognitive foundations for WNA?*

9. ASPECTS THAT AFFECT WHOLE NUMBER LEARNING

This theme will address some aspects affecting learning of WNA in both positive and negative ways.

Socio-cultural aspects influence enumeration practices, algorithms, and representations as well as metaphors or models (e.g., the number line). Hence, students' language and culture may help or hinder the construction of WNA not only in schools but also in informal settings. On the one hand, the recourse to tools from the history of mathematics (e.g., counting sticks; different kinds of abaci; reproduction of ancient mechanical calculators) may be effective to foster learning of WNA with explicit reference to the local culture. On the other hand, intentionally designed tools may address the effective learning processes evidenced in the literature (e.g., technological tools including the multitouch ones).

Low achievement in WNA is a major focus in debates at all levels, from school practice to international studies. Literature shows that it may depend on very different aspects: context variables (e.g., marginalized students; migrant and refugee students; education in fragile democracies), institutional variables (e.g., different languages in school and out of school context), learning disabilities (dyscalculia; sensual impairment for deaf and blind students), on affect factors (e.g., self-beliefs, anxiety, motivation, gender issues), on didactical obstacles (e.g., a too-limited approach as in the case of teaching addition separate from subtraction or multiplication as a repeated addition only), and on epistemological obstacles (related to the historical process of constructing WNA by mankind).

Some references may be found in the ICMI Studies 17 and 22 and, for general issues concerning the contexts, UNESCO 2010.

The following possible questions will help to illuminate this theme further:

- *What are the features of your language related to whole numbers, operations, and word problems that could affect learning in a positive or negative way? How these features are mirrored in formal, informal, or not formal settings?*
- *What main challenges for learning WNA are faced by marginalized students or, in general, in difficult contexts?*
- *What main challenges are faced for learning WNA by students with sensual impairments (blind and deaf)?*
- *What main challenges are faced for learning WNA by dyscalculic students?*
- *In your country, are students with special needs enrolled in mainstream classes (inclusive systems) or in special education classes? To what extent may the strategies especially developed for students with special needs be useful for all students in WNA?*
- *In your country, are there evidence that the literature on either didactical or epistemological obstacles had impact on classroom practice?*
- *Which tools (from the ancient or new technologies) are useful to enrich the classroom activity for all or to help low achievers for WNA? Are there evidence on effective use of traditional manipulatives (including the ones rooted in local cultures), virtual manipulatives, and technologies (including the recently developed multi touch technologies)? Are there classroom studies on the comparison of different kinds of tools?*
- *What strategies may be implemented by teachers in relations with the above issues?*

10. HOW TO TEACH AND ASSESS WHOLE NUMBER ARITHMETIC

This theme will address general and specific approaches to teach and assess the learning of WNA. WNA appears in the standards for mathematics of every country (see http://www.mathunion.org/icmi/other-activities/database-project/introduction/), in specific international studies (e.g., the Learner's Perspective Study, with 16 country teams). In some countries, also, independent research communities have developed projects on teaching and assessing WNA, which in some cases, are internationally acknowledged (e.g., Realistic Mathematics Education in the Netherlands; NCTM Curriculum and Evaluation Standards in the USA; Davydov's math curriculum in Russia; the Theory of Didactical Situations in France). In the ethnomathematics trend, projects sensitive to the local cultures and traditions have been developed (e.g., in Australia, Latin America, the USA, and Canada). A specific Symposium on Elementary Mathematics Teaching (SEMT) is held every second year in Prague since 1991.

Some issues to be focused may be the following: textbooks and future teaching aids (e.g., multimedia; e-books) for WNA, tools to approach specific elements of WNA (e.g., manipulatives, technologies), specific strategies for some fields (e.g., for word problems, the Chinese tradition of problems with variation, Singapore's model method, the extended literature on word problems and relations with real life situation), examples of practices rooted in local culture, and metacognitive aspects in national curricula (e.g., early approach to mathematical thinking processes).

In recent years, the assessment debate at the local and school level has been very much biased by the results of international studies (e.g., OECD PISA, TIMSS), which are likely to produce assessment-driven curricula. An ICMI study on assessment was produced in the early 1990s (ICMI Study 6), but updating might be necessary for the relevance and the media-wide appeal of the international studies.

Some references for this theme may be found in the proceedings of ICMI Congresses and Regional Conferences (http://www.mathunion.org/ icmi/Conferences/introduction/).

The following possible questions will help to illuminate this theme further:

- *What are the consequences of policy decision making related to WNA teaching based on evidence in comparison with policy decision making based on opinions?*
- *How is the intended curriculum reflected in textbook and other teaching aids?*

- *What are the changes (if any) that have resulted from the use of technology to teach WNA?*
- *How complete is understanding of the place value system developed, and at what points in the/your curriculum are key features of place value explored in greater depth?*
- *How does the/your curriculum foster the transition from a counting or additive view of number to a ratio/multiplicative/measurement view of number?*
- *How do children acquire WNA concepts and procedures outside of school? How can teachers built up on the knowledge children acquire outside school?*
- *What are the approaches that have proven to be effective in your school setting to teach elements of WNA, for example, number sense, cardinality, ordering, operations (subtraction with re-grouping, etc.), problem solving, estimation, representing, mental computation...?*
- *Problem solving context, should it be realistic? Should it be authentic? Always? What is the place (if any) of traditional word problems? What is the role of (real world) context in WNA? Always necessary?*
- *How to develop positive attitudes towards mathematics while teaching WNA?*
- *How teachers promote the development of student's metacognitive strategies during the learning of WNA?*
- *What main challenges are faced by teachers when teaching and assessing WNA?*
- *What innovative assessment approaches are used to evaluate the learning outcomes of WNA? What are the changes (if any) in assessment WNA that have resulted from the media appeal of international studies like PISA or TIMSS?*

11. WHOLE NUMBERS AND CONNECTIONS WITH OTHER PARTS OF MATHEMATICS

This theme will address WNA in terms of its interrelationships with the broader field of mathematics.

Table 1 Members of the International Program Committee

Member	Division	Institution	Email address	Position
Maria G. (Mariolina) Bartolini Bussi	Department of Education and Human Science	University of Modena and Reggio Emilia, Italy	bartolini@unimore.it	Co-chair
Xuhua Sun	Faculty of Education	University of Macau, China	xhsun@umac.mo	Co-chair
Berinderjeet Kaur	Centre for International Comparative Studies	National Institute for Education, Singapore	berinderjeet.kaur@nie.edu.sg	
Hamsa Venkatakrishnan	Mathematics Education Unit	University of the Witwatersrand, Johannesburg, South Africa	hamsa.venkatakrishnan@wits.ac.za	
Joanne Mulligan	School of Education	Macaquarie University, Sydney, Australia	joanne.mulligan@mq.edu.au	
Jarmila Novotna	Faculty of Education	Charles University, Praha, Czech Republic	jarmila.novotna@pedf.cuni.cz	
Lieven Verschaffel	Faculty of Psychology and Educational Sciences	KU Leuven University, Belgium	lieven.verschaffel@ppw.kuleuven.be	
Maitree Inprasitha	Faculty of Education	Khon Kaen University, Thailand	inprasitha@hotmail.com	
Sybilla Beckmann	Department of Mathematics	University of Georgia, Athens, GA US	sybilla@math.uga.edu	
Sarah González de Lora		Pontificia Universidad Católica Madre y Maestra, República Dominicana	sarahgonzalez@pucmmsti.edu.do	
Roger E. Howe	Department of Mathematics	Yale University, New Haven, CT US	howe@math.yale.edu	ICMI liason
Ferdinando Arzarello	Department of Mathematics	Univeristy of Turin, Italy	ferdinando.arzarello@unito.it	ICMI President
Abraham Arcavi	Department of Science Teaching	The Weizmann Institute of Science, Rehovot, Israel	abraham.arcavi@weizmann.ac.il	ICMI Secretary General

Some connections concern the following: pre-algebra and algebraic thinking (e.g., looking for patterns; schemes for the solution of world problems), geometry or spatial thinking (e.g., triangular or square numbers and similar; number lines), rational numbers and measurement (e.g., Davydov's curriculum for arithmetic), and statistical literacy (e.g., mean, median and mode, interval, scale, and graphical representation).

Evidence suggests that the earliest formation of WNA can support the learning of mathematics as a connected network of concepts and, vice versa, embedding WNA in the broad field of mathematics can foster a better understanding. Some references for this theme may be found in the ICMI Studies 9, 12, 14, and 18.

- *How can WNA teaching and learning contribute to understand other interconnected mathematical ideas and build on one another to make students view mathematics as a coherent body of knowledge?*
- *In your country, to what extent are connections between WNA and other mathematics topics pointed out in the curriculum syllabus and textbooks, and how are they approached?* i.e., *WNA and measurement, WNA and elementary statistics? Pre-algebra patterns, WNA and algebra?*
- *In your system/country, are symbolic and non-symbolic approaches to word problems compared? To what extent are connections made? Is the analogy of between base-ten arithmetic and polynomial arithmetic emphasized? To what extent are the rules of arithmetic/properties of operations used as a guide in learning manipulation of algebraic expressions?*
- *In your country/system, to what extent are connections between WNA and other mathematics topics stressed in the teachers' education programs?*
- *In what ways does the connection between WNA and specific themes in other areas of mathematics contribute to students' understanding of these themes?*
- *What learning conditions enable students to make connections between WNA and other mathematics topics?*
- *In which ways does the practice of connecting WNA to other areas of mathematics contribute to the development of mathematical thinking?*
- *How can the connection of WNA with other areas of mathematics improve communication of mathematical ideas?*
- *How can technology be used to make connections between WNA and other mathematics topics?*

- *How does the use of representations in WNA teaching and learning contribute to build connections with other mathematical areas? For example, to what extent is the number line used to exhibit the connections between WNA and arithmetic of fractions?*

The following possible questions will help to illuminate this theme further:

12. CONCLUSION

As in every ICMI study, the ICMI Study 23 is built around an international conference and directed towards the preparation of a published volume. The study conference will take place in Macau, China, and will be hosted by the University of Macau (*June 3* to *7, 2015*).

As is the usual practice for ICMI studies, participation in the study conference will be by invitation only for the authors of submitted contributions which are accepted. Proposed contributions will be reviewed and a selection will be made according to the quality of the work, the potential to contribute to the advancement of the study, with explicit links to the themes contained in the discussion document and the need to ensure diversity among the perspectives. The number of invited participants will be limited to approximately 100 people. Deadlines for contributions are published in the discussion document:

http://www.mathunion.org/icmi/conferences/icmi-studies/ongoing-studies/icmi-study-23/

The *first product* of the ICMI Study 23 is an electronic volume of proceedings, to be made available first on the conference website and later in the ICMI website: it will contain all the accepted papers as reviewed papers in a conference proceedings (with ISBN number).

The *second product* is a gallery of commented video clips about practices in WNA, to be hosted in the conference website and, possibly, later, in the ICMI website.

The *third product* is the ICMI study volume. The volume will be informed by the papers, the video clips and the discussions at the study conference as well as its outcomes.

The IPC for ICMI Study 23 invites submissions of contributions of several kinds: theoretical or cultural-historic-epistemological essays (with deep connection with classroom practice, curricula, or teacher education programs), position papers discussing policy and practice issues, discussion papers related to curriculum issues, reports on empirical studies, and video clips on explicit classroom or teacher education practice. To ensure a rich and varied discussion, participation from countries with different economic capacity or with different cultural heritage and practices is encouraged.

The ICMI Study 23 website can be found at the address http://www.umac.mo/fed/ICMI23/.

The website contains the official discussion document for ICMI Study 23 and will be regularly updated with information about submission procedure and the study conference and will be used for sharing the contributions of those invited to the conference in the form of conference pre-proceedings. Further information may be asked at the following address:

REFERENCES

1. ICMI: *The First Century of the International Commission on Mathematical Instruction. History of ICMI, by Fulvia Furinghetti & Livia Giacardi*. 2008. (last visited March 23, 2014) http://www.icmihistory.unito.it (last visited March 23, 2014)
2. Klein F 1923.http://www.icmihistory.unito.it Retrieved from (last visited on August 2, 2014)
3. OECD EDUCERI: *Dyscalculia Primer and Resource Guide by Anna J. Wilson*. 2010. retrieved from (last visited on March 23, 2014) http://www.oecd.org/edu/ceri/ dyscalculiaprimerandresourceguide.htm retrieved from (last visited on March 23, 2014)
4. OECD PISA FOCUS: *Does Participation in Pre-Primary Education Translate into Better Learning Outcomes at School?*. 2011. retrieved from (last visited on March 23, 2014) http://www.oecd-ilibrary.org/education/does-participation-in-pre-primary-education-translate-into-better-learning-outcomes-at-school_5k9h362tpvxp-en;jsessionid=eos7dj2qr9bgo.x-oecd-live-02retrieved from (last visited on March 23, 2014)
5. UNESCO: *Challenges in Basic Mathematics Education, by Michèle Artigue*. 2012. retrieved. (last visited March 23, 2014) http://unesdoc.unesco.org/images/0019/001917/191776e.pdf retrieved. (last visited March 23, 2014)
6. UNESCO: *Educational Neuroscience: More Problems than Promise? By Ilkka Tuomi*. 2013. (last visited March 23, 2014) http://www.unescobkk.org/resources/e-library/publications/article/educational-neurosciences-more-problems-than-promise-education-policy-research-series-discussion/ (last visited March 23, 2014)

CHAPTER 6

The classical origin of modern mathematics

Floriana Gargiulo1* , Auguste Caen2 , Renaud Lambiotte1 and Timoteo Carletti1

1 NaXys, University of Namur, 8 rempart de la Vierge, Namur, Belgium.

2 DICE, Inria, 15 parvis René Descartes, Lyon, France

ABSTRACT

This paper introduces a data-driven methodology to study the historical evolution of mathematical thinking and its spatial spreading. To do so, we have collected and integrated data from different online academic datasets. In its final form, the database includes a large number (N~200KN~200K) of advisor-student relationships, with affiliations and keywords on their research topic, over several centuries, from the 14th century until today. We focus on two different issues, the evolving importance of countries and of the research disciplines over time. Moreover we study the database at three levels, its global statistics, the mesoscale networks connecting countries and disciplines, and the genealogical level.

Keywords: academic genealogies, history of mathematics, directed acyclic graphs

1. INTRODUCTION

The statistical analysis of scientific databases, including those of the American Physical Society, Scopus, the arXiv and ISI web of Knowledge, has become increasingly popular in the complex systems community in recent years. Important contributions include the development of appropriate scientometric measures to evaluate the scientific impact of scholars, journals and academic institutions [1, 2, 3, 4, 5] and to predict the future success of authors [6, 7] and papers [8]. In parallel, the structure of collaboration has attracted much attention, and collaboration networks have become a central example for the study of complex networks, thanks to the high quality and availability of the datasets [9]. From a dynamical point of view, different papers [10, 11] studied the mobility of researchers during their academic career, showing that the statistical properties of their mobility patterns are mainly determined by simple features, such as geographical distance, university rankings and cultural similarity.

Limitations of the aforementioned datasets include their relatively narrow time window extension, at best, over 100 years and the difficulty to disambiguate author names, and thus to correctly distinguish career paths across time. The original motivation of this paper was to address these issues by performing an extended study of *The Mathematics Genealogy Project*, a very large, curated genealogical academic corpus [12]. The dataset, whose basic statistics have been already analysed elsewhere [13, 14], extends over several centuries and contains pieces of information allowing us to retrieve the direct genealogical mentor-student links, but also university affiliations at different points of a career and the research domains. Data from the same website have already been used to assess the role of mentorship on scientific productivity [15] and to study the prestige of university departments [16].

Our main goal is to analyse the history of modern mathematics, through the processes of birth, death, fusion and fission of research fields across time and space. In particular, we focus on the temporal evolution of the roles and importance of countries and of disciplines, on the structure of 'scientific families' and on the impact of genealogy on the development of scientific paradigms. As it is often the case when performing a data-driven analysis of historical facts [17, 18], the data set is expected to be incomplete and to present biases, mainly for the more ancient data. In the present case, the website collects the data in two ways: a participative method, based on the spontaneous registration of scholars (who can also register their students and their mentors), and a curated method, based on historical facts and performed by the creators of the web site.

The presence of biases calls for the use of appropriate statistical measures, in preference based on ranking instead of absolute measures. In this work, we have also introduced data-mining methods to correct and enrich the data structure. A first contribution of this work is thus methodological, with the design of a methodological setup that could be applied to other systems. We have then performed an analysis of the system at three levels of granularity. First, a global one investigates the fully aggregated 'demography' (population in terms of countries and disciplines) of the database, with the aim to classify countries and disciplines according to their normalised activity behaviour. Tracking the evolution of the rankings helps identify transition points in the mathematical history, associated to emerging fields of research. Second, we have constructed directed weighted networks where nodes are scholars endowed with a set of attributes (thesis defence date, thesis defence location, thesis disciplines) and linked to other nodes using the genealogy associated to the mentor-student relation. This 'mesoscale' network allows us to investigate the relationships between the attributes and to identify a strong hierarchical structure in the scientific production in terms of countries as well as its evolution in the course of time. Finally, using an approach typic of kinship networks studies [19, 20], we focus on the statistical properties of the tree structure of the genealogy in terms of family structures. We conclude by showing the presence of strong memory effects in the network morphogenesis.

To summarise this paper has a twofold goal: first to propose, in the framework of data science, new tools to collect and analyse historical databases, second and in complement with the former, to provide a narrative on the history of mathematics as extracted from data. In both cases, this work opens interesting perspectives. Because of their generality, the presented tools could clearly be used to study different databases with a genealogical structure, for instance in the case of bibliometrics or Wikipedia studies. In addition, our results provide a first glimpse of the potential use of data and algorithms in the study of the history of science. An important future step would consist in complementing and interpreting this data-driven view with that of epistemologists and historians of science, as briefly outlined in the conclusions.

2. DATASET AND ASSOCIATED NETWORKS

The core of our dataset has been extracted from the website 'Mathematical Genealogy Project'. It is one of the largest academic genealogy available on the web, consisting of approximatively 200K not-isolated scientists (186,505) with information on their mentors and students. The data cover a period between the 14th century until nowadays. For a majority of mathematicians, we have detailed information about his/her PhD, including the title (for 88% of the scholars), the classification according to the 93 classes proposed by the American Mathematical Society [21] (for 43% of the scholars), the University delivering the degree as well as the year of its defence. However, because a large part of the database is spontaneously filled by the scientists, the data is imperfect and attributes may be wrong or missing. A first step has thus consisted in comparing the database with additional data form Wikipedia [22]. In particular, we first downloaded, when available, the Wikipedia pages of all the scholars present in the Mathematics Genealogy Project database. Disambiguation of the names is assured by the fact that Wikipedia pages have a direct link to the Mathematics Genealogy Project site. In the text of the Wikipedia pages, we then searched the keywords associated to the AMS classification in order to expand the information about authors. This external dataset allowed to assign a discipline to the 54% of the mathematicians.

For more recent entries, we retrieved the affiliations with the Scopus profiles of scientists [23]. Notice that we only extracted the information required for our needs and that additional information, e.g. about their scientific impact or on their geographical links, could be collected in order to address other research questions.

It is worth noting that our analysis are biased by the actual scientific and socio-political environment. First, the countries' borders changed in time. In the Mathematics Genealogy Project, the location of the PhD defence is determined according to the position of the university in the current geo-political setting. We kept in our analyses this county classification, but it would be interesting in the future to consider, for example, the resilience of the system to borders shifts. Similarly, the concept of discipline is also very delicate to define on such a long time scale [24], and we decided here to use the current classification from the AMS for all authors.

After this preliminary phase, we have enriched the information available for the authors, by developing algorithms aimed at correcting the dates and assigning to each thesis a discipline. The algorithm for fixing errors in temporal entries is based on the topological structure of the genealogical network and uses the available statistics on the age difference mentor-student to identify and suitably correct wrong time sequences (e.g. the cases where the mentor has completed its PhD after its student, or where the time distance between mentor's and the student's PhD is too large). The missing disciplines (not previously extracted from Wikipedia) have been learned based on the thesis title using a Bayesian supervised dictionary learning technique.

As previously stated, these algorithms, summarised in Additional file 1 (Sections I.B-I.D), are general and could be applied in other contexts. After the enrichment, all the scholars of the database have a corrected date, 88%

of them have an associated discipline and 94% an associated country. As a next step, we have exploited the enriched database in order to study the geographical and temporal evolution of mathematics. Different data representations, described below, have been adopted to mine different typologies of information from the dataset.

2.1 The mesoscale networks

As a first step, we built a multi-partite network, where the different kinds of nodes are scholars and their attributes extracted from our database, namely universities, cities, countries and disciplines. Hence we considered several possible 'projections'; for instance considering the case of universities, we add a direct link between two universities-nodes, say *A* and *B*, if there exists in the database, and hence in the multi-partite network, a couple of scientists such that the one with attribute 'university *A*' has been the PhD supervisor of the second one having attribute 'university *B*'. Observe that our data have time stamps, corresponding to the time of the defence of the PhD thesis, then under the assumption that a supervisor is in the same university as his/her PhD student, directed links are therefore a proxy for the mobility in time of a scholar, but also of the flow of knowledge between different places. In the case of disciplines, directed links correspond to a transfers of knowledge from one scientific discipline (the one of the mentor) to another one (the one of the student), from one generation to another, and how disciplines at a certain time may inherit, in terms of ideas and methods, from research fields at previous times. Let us note that the network has a large number of self-loops associated to the frequent situation when the supervisor and the student got their PhD in the same university or in the same discipline. Moreover the links are weighted by the number of researchers connecting attributes.

The procedure is illustrated, for the case of flows between countries, in Figure 1. In the following, we will perform a longitudinal study of the system, by considering the evolution of networks observed in different time windows. Note here that the data are not uniformly distributed across time, with a strong bias towards recent times.

2.2 The genealogical tree and its partitions into families

The genealogical graph is the most obvious representation of our dataset, consisting in an oriented acyclic graph [25] linking a mentor to her/his students. This defines automatically the structure of hierarchical generations. Notice however that the structure of our data is not simply a tree due to the several cases where a student has two advisors.

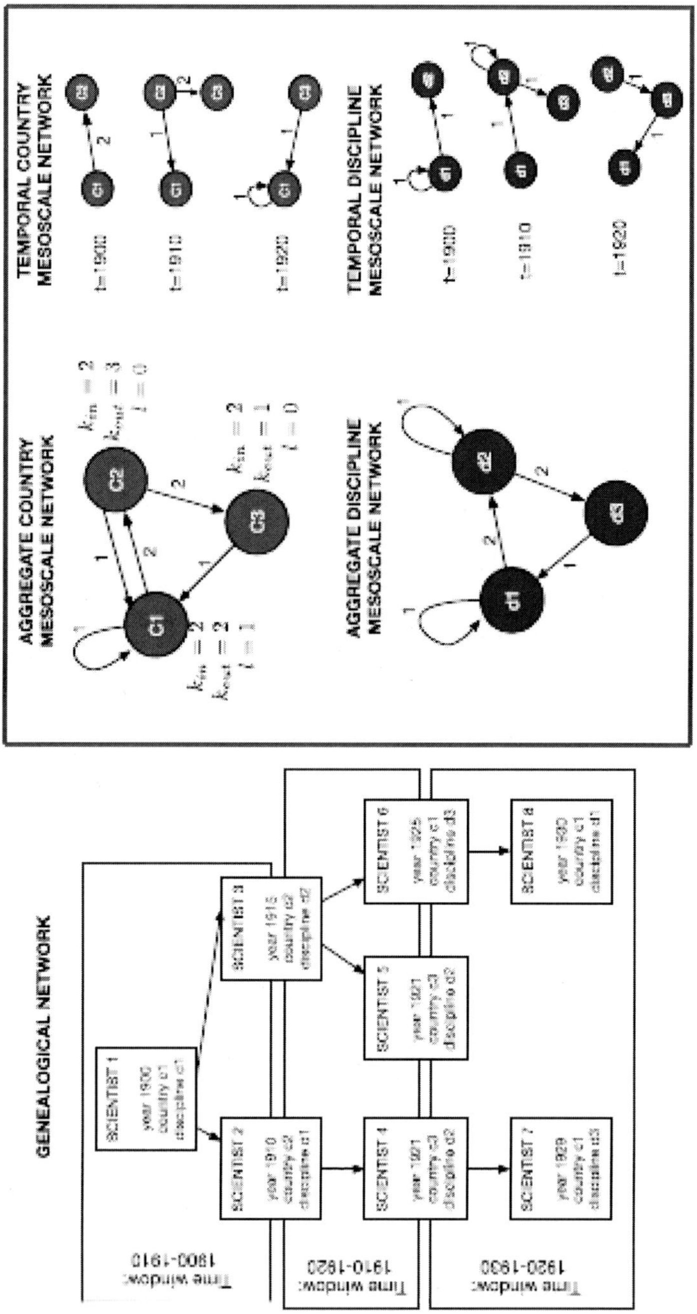

Figure 1. Mesoscale networks. An example of the procedure to derive the mesoscale network from the genealogical data.

A very common process in kinship is to cut the genealogical directed acyclic graphs into linear trees (alliances) where each individual has a single progenitor (the mother for representing the uterine links and the father for the agnatic ones). In this representation, the links between alliances represent the matrimonial structures between the different alliances in the society. In our context, when a scientist has more than one advisor, it is not clear which links should be cut to retrieve the original ancestors (our dataset prevents us from identifying the principal supervisor from the secondary one, if any). We thus propose a method to reproduce the optimal ancestry lines and to identify the important families in the genealogy. The method, fully described in Additional file 1 (Section I.D), is based on the decomposition of the network into pure linear trees, and their statistical clustering based on probabilistic arguments; roughly speaking given two nodes *A* and *B* that can be linked in more than one way, thus implying the presence of non-trivial loops, we assign to every link in such paths the probability that *A* and *B* will be disconnected if the link is removed. We thus select links to be removed by maximising the probability that *A* and *B*are still linked. The resulting partition of the graph into families identifies 84 families; remarkably, the 24 most populated families cover the 65% of the scientific population in the database. Let us observe that alternative methods for family identification do exist, see for instance [26, 27].

3. RESULTS

3.1 Global statistics

Let us define the *relative abundance profile* of each country in different periods $fI(t)=NI(t)/N(t)fI(t)=NI(t)/N(t)$, where $N(t)N(t)$is the total number of scientists whose country is known in the database at time t and $NI(t)NI(t)$ is the number of scientists in country I at time t. Notice that, at this stage, the genealogical information is not used. From such profiles we can asses the evolution of the importance of the countries had in the history of mathematics due to the their different historical dynamics. In order to compare the profiles of different countries independently from their total human capital, we normalise each of these profiles by their 'volume' ($f{\sim}I(t)=fI(t)/\sum tfI(t)f{\sim}I(t)=fI(t)/\sum tfI(t)$) and then we classify them based on their Kolmogorov-Smirnov distance (see Figure 2 where the results are reported using a dendrogram, and the SI where we reported the profiles for the top 10 countries in the database). We observe different prototypical behaviours: countries with a central role in the ancient history whose centrality has decreased in the last centuries (for instance Italy, France and Greece), countries with a central role before the world wars (e.g. central

Europe countries), countries emerging after the world wars (such as Japan and India), countries recently emerging (among which China and Brazil). Because of the normalisation procedure we used, we obtain a cluster where USA is linked with ex-USSR countries and show a similar decreasing behaviour in the latest decades (impossible to observed in a non-normalised context).

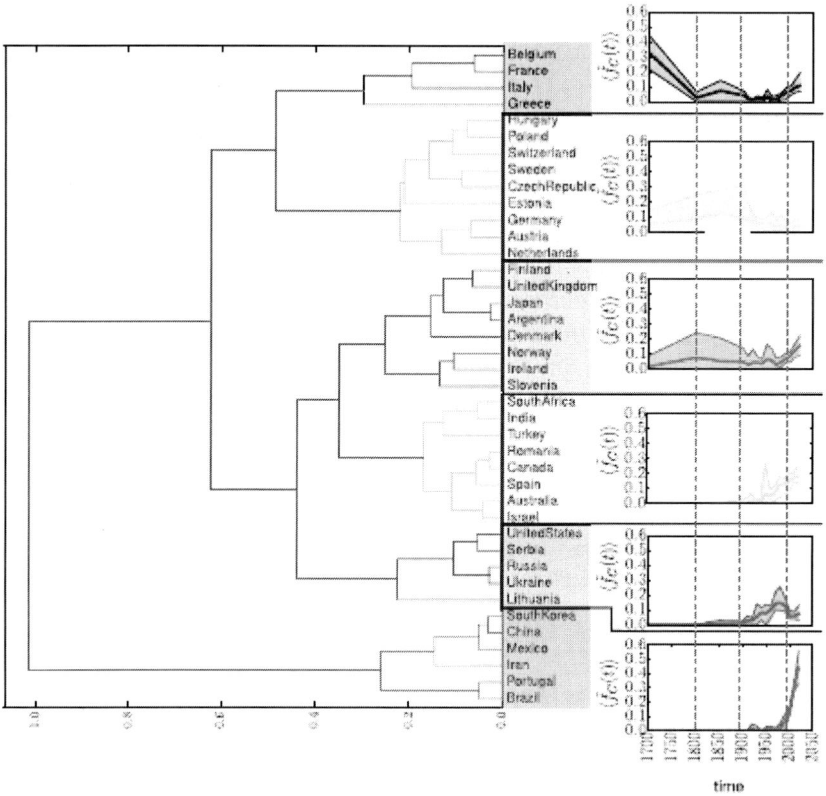

Figure 2. Country prevalence dendrogram. Clustering of the countries according to their time prevalence profile f~I(t)f~I(t). The lines in the plots on the right describe the average prevalence profile for the cluster $\langle \tilde{f}_C(t) \rangle = \sum_{l \in C} \tilde{f}_l(t) / (\sum_{l \in C} 1)$. The shadowed area is included between the minimum value and the maximum value of the prevalence profile on the cluster $(\min_{l \in C} \tilde{f}_l(t), \max_{l \in C} \tilde{f}_l(t))$.

Additional information on how the total number of mathematicians compares with that of scientists would make these results more significative, but this type of information is difficult to be retrieved in electronic archives.

The same procedure is applied to disciplines and results reported in Figure 3 allow to identify three main blocks of disciplines: the disciplines that were more central during the industrial revolution (before 1900) are associated to physical applications (such as thermodynamics, mechanics and electromagnetism). The disciplines reaching their maximum of expansion around the 1950 are more abstract, even if several links exist to applied topics, such as telecommunication and quantum physics. Finally, the last decades have witnessed the emerging dominance of applied mathematics (e.g. statistics, probability) and computer science. This last point shows the considerable impact of the computer revolution on the evolution of mathematics: the magenta subfield in Figure 3 (operations research; systems theory; category theory; computer science) has emerged at the expense of many other fields, but the yellow one, which may have also been helped by the introduction of computers.

To capture the rise and fall of countries or disciplines, we have compared the rankings of the top 10 countries and disciplines in different time periods. Standard indicators for rank comparison, such as the Kendall-Tau index, cannot be applied here since the elements in the top-*k* lists are not conserved in time [28]. For this reason, we have used a distance measure based on a modified version of the Jaccard index allowing to compare ranked sets, J(rank1,rank2)J(rank1,rank2) (more information is provided in Additional file 1, Section A). As for the original Jaccard index the modified version is such that J(rank1,rank2)=1J(rank1,rank2)=1 when the rankings rank1rank1 and rank2rank2 are completely equivalent, and gives a value 0 when these latter are not correlated at all. This information is then transformed in a distance by taking dJ=1−JdJ=1−J. Increases in the distance measure, dJdJ, indicate major reshaping of the rankings.

As we can observe in the upper plot of Figure 4, corresponding to countries, we observe several transition points; for example a transition can be observed during the First World War, with the decreasing centrality of Austria and Hungary due to the end of the Austro-Hungarian emperor and the entering in the ranking of Russia. Another transition is connected with the European political reshaping during the Second World War and with the massive migration of Jewish and dissident scientists to US due to fascism in Europe. This is the period at which for the first time, USA surpasses Germany in the ranking. A third transition, around the 1960s, shows the increase of centrality of the Soviet Union (testified by the presence of several East-European countries in the ranking). Finally, more recently, we observe the decline of Russia and the emergence of new countries such as Brazil.

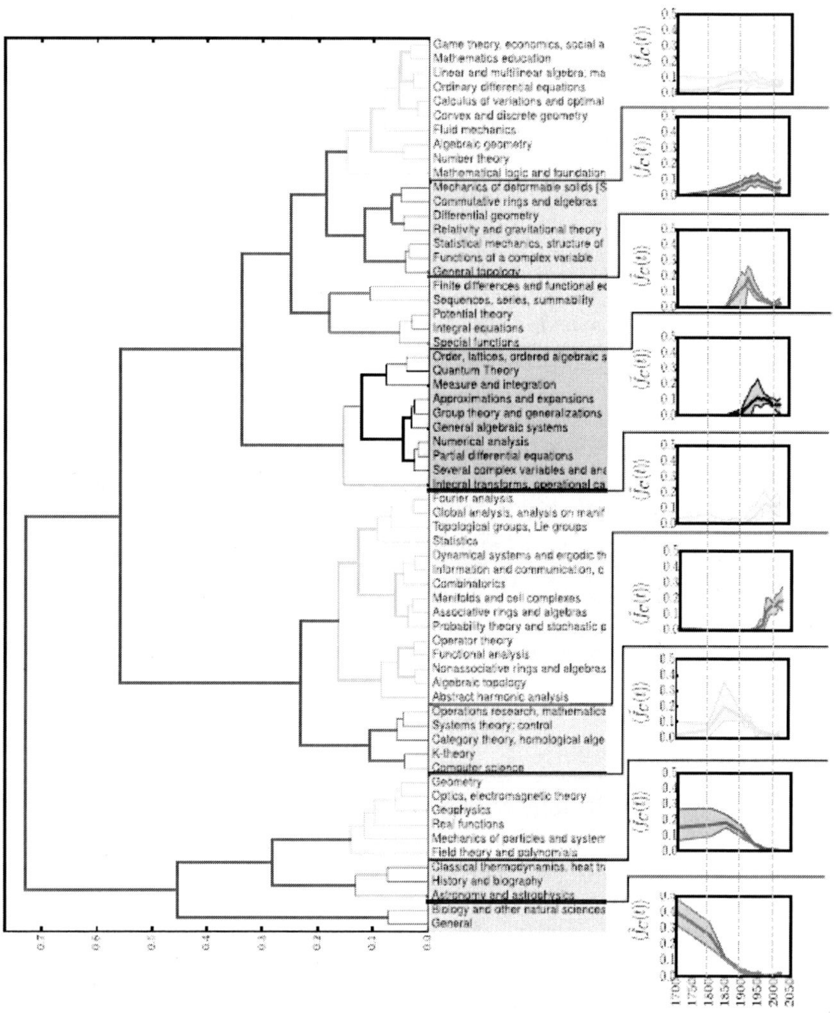

Figure 3. Discipline prevalence dendrogram. Clustering of the disciplines according to their time prevalence profile f~I(t)f~I(t). The lines in the plots on the right describe the average prevalence profile for the cluster $\langle \tilde{f}_C(t) \rangle = \sum_{l \in C} \tilde{f}_l(t) / (\sum_{l \in C} 1)$. The shadowed area is included between the minimum value and the maximum value of the prevalence profile on the cluster $(\min_{l \in C} \tilde{f}_l(t), \max_{l \in C} \tilde{f}_l(t))$.

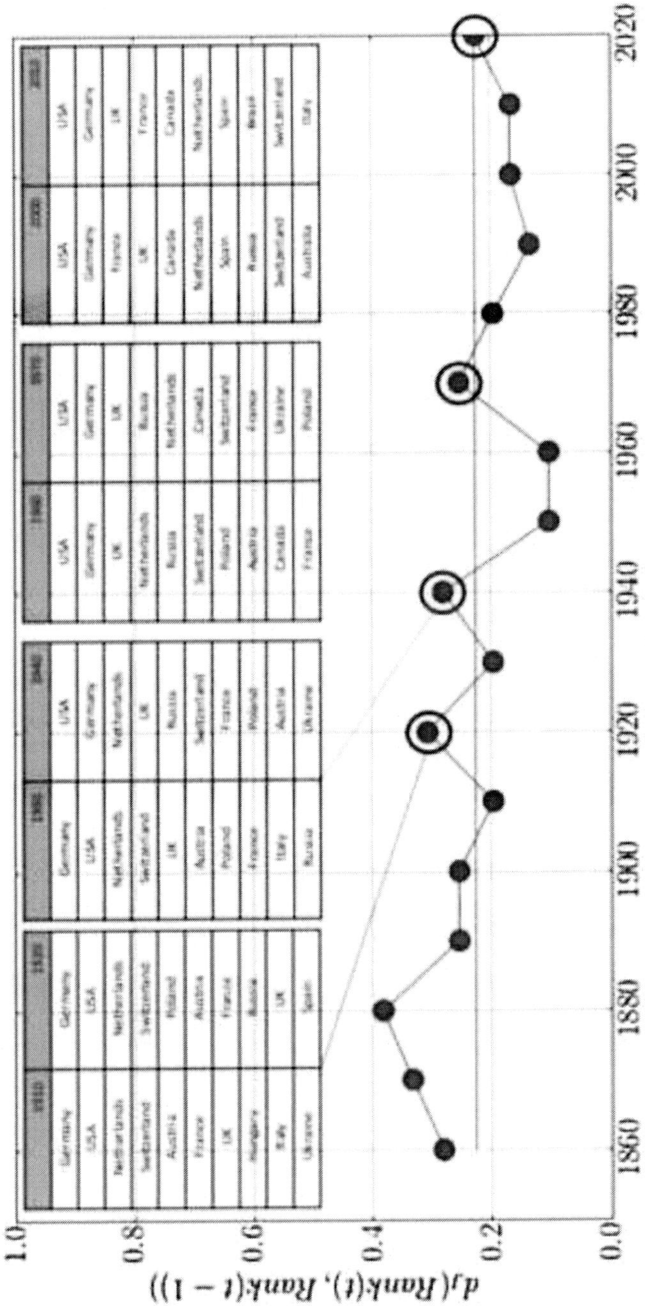

Figure 4. Countries' centrality tipping points**.** Modified Kendall-Tau index comparing the countries' rankings in different periods.

A similar analysis can be performed for disciplines. Results reported in Figure 5 show the presence, among the others, of three significative tipping points. The first one is connected to industrial revolution and to the emergence of disciplines related to the physics of machines (such as thermodynamics and electromagnetism).

The second one is connected to the emergence of fields linked to telecommunication and cryptography (e.g. number theory, spectral functions) during the Second World War period.

Finally the third one, in the 1980s, concerns the emergence of computer science and statistics.

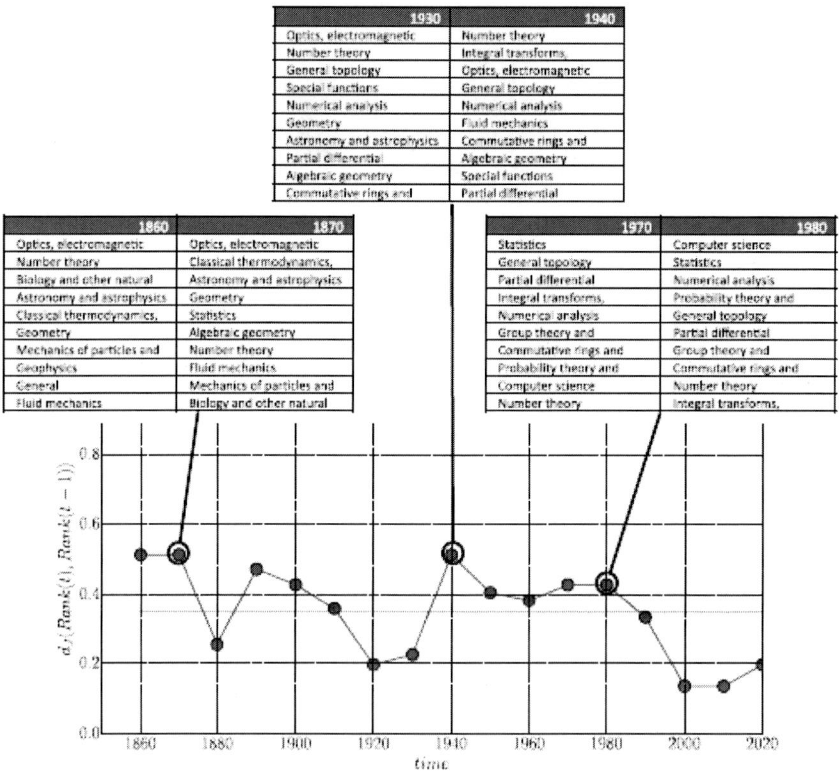

Figure 5. Disciplines' centrality tipping points. Modified Kendall-Tau index comparing the disciplines' rankings in different periods.

3.2 Mesoscale networks

3.2.1 Network of countries

The countries network can be used to represent the *knowledge flows* from one country to another one, associated to the transition of a student in a country, becoming a professor and PhD supervisor in another country. The network presents few important hubs, that are the gravity centres of the scientific research (USA, Germany, Russia, UK). Each of these hubs tends to be surrounded by a community of countries. These communities can be associated to historical divisions, for instance a large block connected to USA scientific production, the Commonwealth nations, the ex-Soviet block, the central European countries. The betweenness of countries allows to detect countries at the interface between different communities, such as France connecting the central European countries with the USA-centred community or Poland connecting European research and the ex-Soviet area.

Another important index of the countries network is the weighted *in* (*out*)-degree of the nodes and the number of self loops. The *in*-degree of a country represents the number of scientists obtaining their PhD elsewhere and mentoring a PhD student in that country. Therefore a high *in*-degree is associated to a country with a strong capacity of attracting scholars and absorbing knowledge from abroad. On the contrary, a high *out*-degree represents a country producing scholars and exporting knowledge elsewhere. Each country can be therefore characterised by three normalised quantities, the fraction of scientist formed inside the country and remaining there, the fraction of scientist formed inside and leaving and the fraction formed abroad and absorbed by the country. In Figure 6A we display the different positions of the most important countries with respect to these indexes, the closer a country is to one of the triangle vertices the larger is the associated index. The size of the dot is a measure of the production of a country, i.e. estimated by its number of PhDs. The most productive countries tend to be the most *scientifically autarchic* ones, with a large fraction of self-loops. The most important *exporters* are Russia and the UK. Countries with small scientific production show a tendency for *importing* scholars. Observe that these indexes evolve in time (see Figure 3 of the SI). An important inversion point between kin(t)kin(t) and the kout(t)kout(t) is often observed around the Second World War. Moreover, a key signature of emerging scientific countries seems to be the presence of kin(t)>kout(t)kin(t)>kout(t).

To characterize the mobility of scientists across countries, we show in Figure 6B the fraction of the total production and the total absorption of *migrant* scientists for the first *r* countries respectively in the *in* and *out*-degree ranking. The distribution is highly skewed, as one observe from the fact that the top 7 countries produce 80% of the total international scholars.

On the contrary, the curve concerning the total absorption has a lower slope, depicting a larger worldwide spreading around the world. This shows a strong hierarchical structure in academic research where few countries ensure a large share of the worldwide diffusion of scientific knowledge. This scenario obviously evolved in time, as we observe in Figure 7. Remark that scientific leaders changed at different points in time, but also that the scientific leadership group (countries producing the 80% of the whole scientists production) is more restricted in recent times. It is interesting to notice that the minimal size of the scientific *elite* has been reached in the sixties during the world bi-polarisation resulting from the cold war. Since then, the size increased again with the emergence of globalisation.

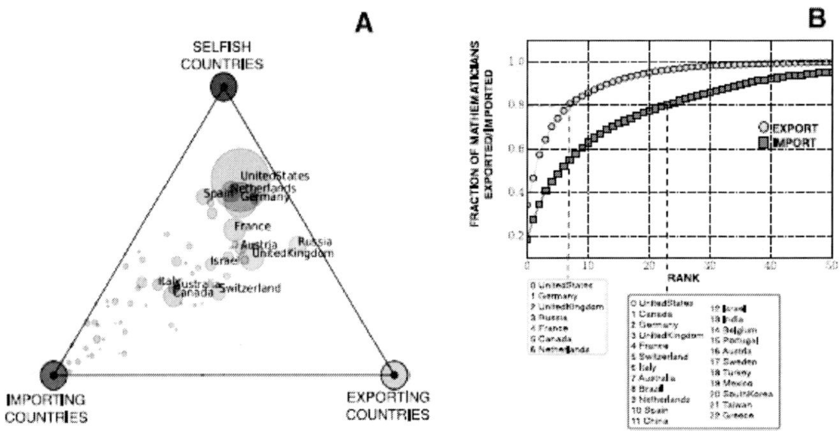

Figure 6. Autharchic, exporting and absorbing behaviors in the countries network. Panel **A**: relative position of the countries between scientifically autarchic/exporting and absorbing behaviours. Panel **B**: fraction of scientists produced and absorbed from the first countries in the rankings respectively of kinkin and koutkout. In the boxes are displayed the countries producing and absorbing the 80% of scientists. Both the panels concern the temporal aggregate of the network.

More information about this network, in particular the properties of the aggregated transition networks concerning the whole historical period, can be found in Additional file 1 (Section III).

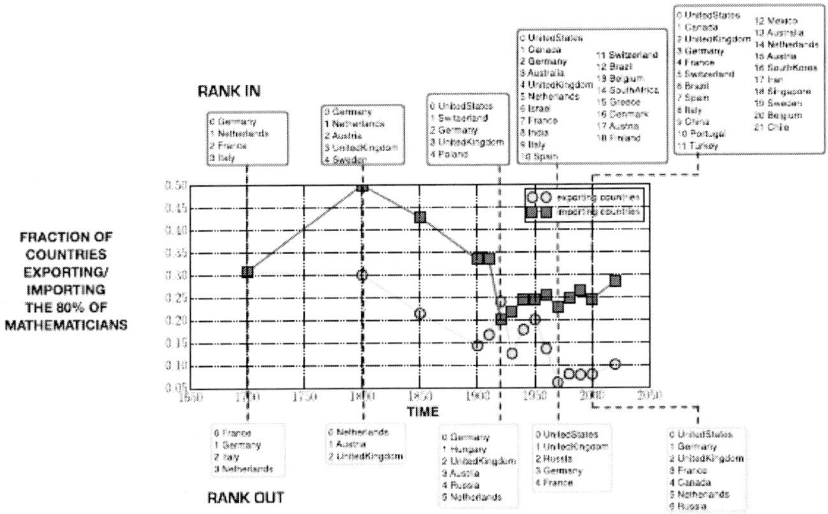

Figure 7. Temporal network. Fraction of countries producing (and absorbing) the 80% of the scientists in different historical periods.

3.2.2 The transition network of disciplines

The transition network of disciplines represents transfers of knowledge from one scientific discipline (the one of the mentor) to another one (the one of the student). The structure of this graph is quite homogeneous in terms of degree and four major topological communities can be identified using standard community detection algorithms working on the topological structure of the weighted network [29]: computer science, geometry, analysis and physics. Each community represents the disciplines exchanging more knowledge between them than with other research fields, and therefore can be interpreted as the scientific paradigms (according to Thomas Kuhn definition) at a certain period.

In Figure 8 we show the normalised mutual information (NMI) between the community structures obtained from different temporal slices of the network. The NMI index varies between one, when the two partitions are equal, and zero, when the two classifications are completely disjoined. A low value of the NMI indicates a 'revolution' in the sense of a strong reorganisation of the knowledge structures, previously non interacting research fields start to exchange knowledge. The figure shows two important points where the NMI is low. The first transition, observed between 1930 and 1940, can be associated to the period when Statistics and Probability merged together, attracting then more applied disciplines like information theory, game theory and statistical mechanics, and leading to the emergence of the field of applied mathematics. The second transition is between 1970 and 1980, where computer science and statistics form one community, together with dynamical systems and applications in other fields of science. The latest transition is expected to be a spurious effect, due to a lack of data in recent years (the last time window starts in 2010 and therefore can contain data only for 5 years). Another potential approach, alternative to the measure of the NMI, to identify the structural changes in these structure could be the one proposed in [30].

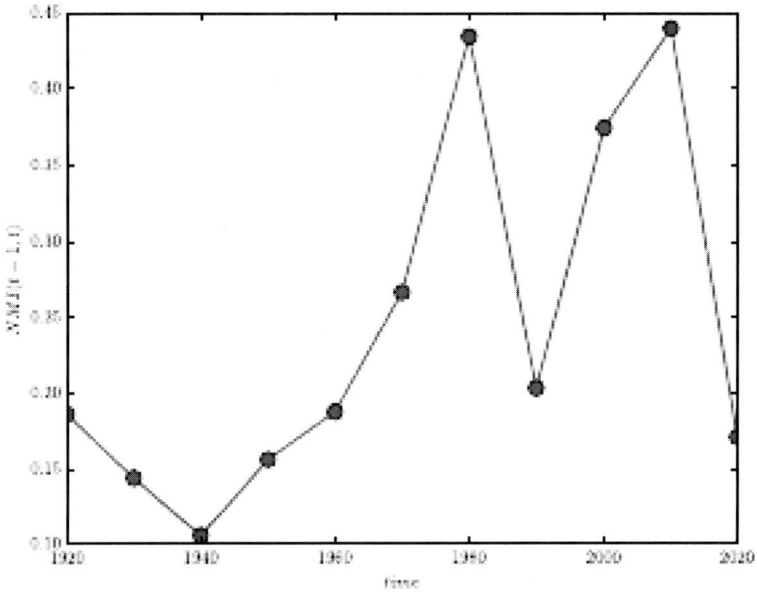

Figure 8. Mutual information between the communities structures of the discipline network in different historical periods.

3.3 The genealogical structure

This last section is devoted to the study of the genealogy tree reconstructed from our data and of its relevance in the evolution of the history of mathematical science.

The first result is the presence of a strong memory effects in the network morphogenesis, as students very often do research in the same discipline their mentor did. To quantify this idea we analysed the genealogical chains where the 'filiation' link connects a mentor with a student maintaining the same discipline of the mentor. We call these objects iso-discipline chains.

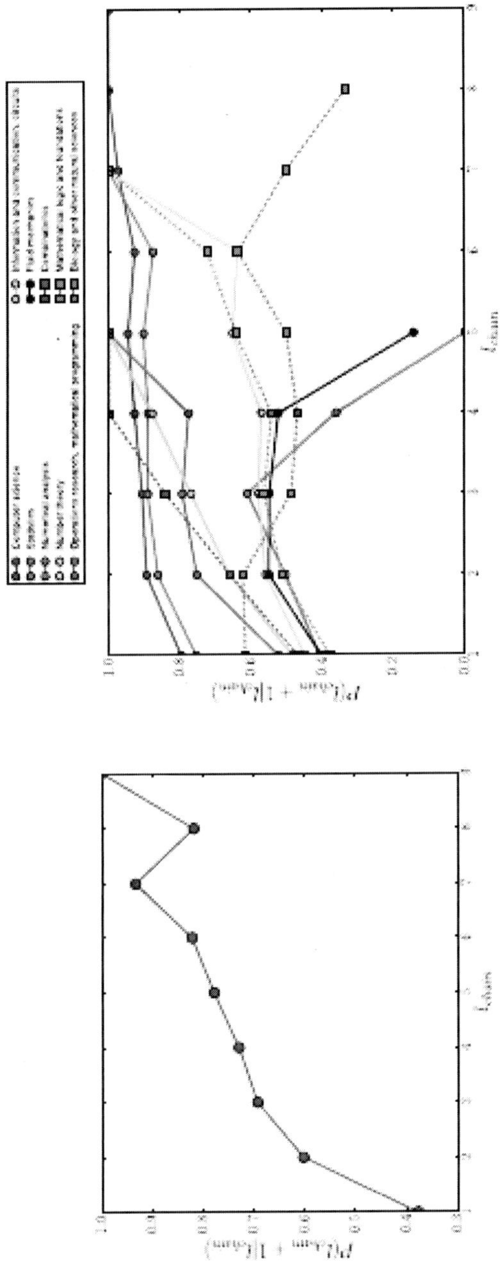

Figure 9. Conditional probability of having an iso-discipline chain of length n+1n+1n+1n+1 , **having a chain of length *n* .** Left panel: aggregated data all disciplines together; right panel: some selected disciplines.

Let us observe that our analysis is data driven and that we only have information about 'filiations' present in the Mathematics Genealogy Project. In Figure 9 (left panel) we show results concerning the conditional probability of having a chain of length n+1n+1 given a chain of length *n*, in other words the probability to have one more descendant working in the same research field of the whole chain, aggregating data over all the disciplines. The first point thus represents the probability for a student to have the same discipline of his/her mentor, one can clearly appreciate that as the chains get longer the probability to continue the same iso-discipline increases. This very marked memory effect in the network can be associated to the existence of 'schools' where a long tradition in a discipline exists such that new students are attracted and continue the tradition. Observe however (Figure 9, right plot) that this phenomenon strongly depend on the discipline.

As previously explained, we have partitioned the network into disjoint families of scholars. Figure 10A shows that the 65% of the scientists can be divided into 24 macroscopic families with size S>500S>500.

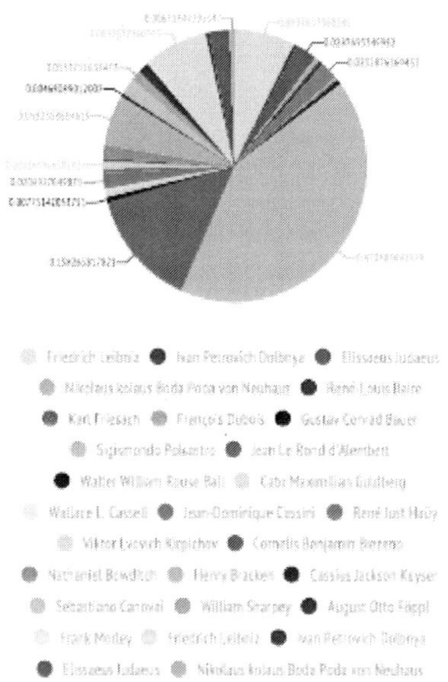

A

B

	ϵ_0	c_x	s_x
OBS	0.95	0.25	0.99
EXP	0.27	0.07	0.99

Figure 10. The families in the mathematical academic genealogy. Panel **A**: relative size of the different families and family's initiator name. Panel **B**: table with the values of the topological indicators for the real (observed) network in the first row and for the randomised model (expected) in the second row.

The largest family is the one originated in 1415 by the Italian medical doctor, Sigismondo Policastro. The second one, is the family originated by the Russian mathematician Ivan Petrovich Dolby, at the end of the 19th century. The large size of this family, born more recently than other families and geographically located mostly around Russia, is due to a high 'fecundity rate' in the Russian School of Mathematics.

The aggregated network between the families, reminiscent of kinship of the alliance networks defined in [19, 31], can be described using some typical topological indicators [20]: (1) the endogamy index, ϵ0ϵ0 describing the fraction of loops in the network (links between the same family); (2) the concentration index cxcx denoting the heterogeneity of the concentration of links between pairs of families (cx=1cx=1 when all links are concentrated on a single pair and cx=1/n2cx=1/n2 when links are homogeneously distributed among the *n* families); (3) the network symmetry index sxsxthat varies from 0 in case of total link unbalance, namely the outgoing flux and the ingoing one are very different each other, to 1 in case of perfect symmetry of fluxes. To asses the relevance of such indicators computed for our genealogy network, we compared them with the expected values for a random multinomial reshuffling - null model - (see Figure 10B), we can observe that, while the symmetry is a structural property, being unchanged by the reshuffling, the endogamy and the concentration are typical signatures of this network and moreover they are much higher than in traditional kinship networks [20]. These results imply that the obtained scientific families are structurally very distant between them and that their relationships are very hierarchical (being these mediated by the largest families).

This strong separation between the genealogical families can be a signature of the existence of tacit knowledge in mathematics [32]. It would be interesting to study the historical development of the kinship structure in order to better address this phenomenon.

Finally, we studied the distribution of families across countries and disciplines. As shown in Figure 11A, with the exception of few cases, the most important countries (in term of production) are present in all the families, while the remaining countries are represented in a very low number of families (from 1 to 3). This feature implies a strong correlation between the genealogical structures and the geography. A similar behaviour can be observed for disciplines (Figure 11B) even if, in this case, the curve describing the number of families with members working in a given discipline is smoother. We can therefore conclude that the genealogical families are strongly specialised in terms of geography and epistemic content.

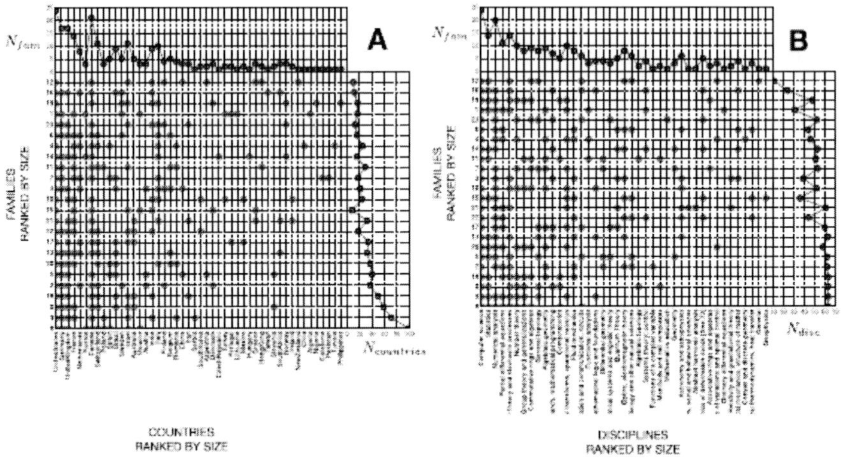

Figure 11. Country and disciplines distribution inside the families. Panel **A**: countries are set on the horizontal axis, ranked by the relative presence in the database, while in the vertical axis, we report the families ranked by their size. A point at the intersection of the country-column and family-row indicates that the country is present in this family. The upper plot, is the column-marginal of the matrix represented in the central plot, representing the number of families where each country appears. The right plot is the row-marginal of the matrix represented in the central plot, representing the number of countries present in each family. Panel **B**: on the horizontal axis we put the disciplines, ranked by the relative presence in the database, in the vertical axis, the families ranked by the size. A point at the intersection of the discipline-column and family-row indicates that the discipline is present in that family. The upper plot, is the column-marginal of the matrix represented in the central plot, representing the number of families where each discipline appears. The right plot is the row-marginal of the matrix represented in the central plot, representing the number of disciplines developed in each family.

4. CONCLUSIONS

In this paper, we have presented a data-driven study of the history of mathematical science, based on the Mathematical Genealogy Project. A first important aspect has been the cleaning and correction of the incomplete and sometimes inaccurate dataset. This operation was performed by means of machine-learning and by incorporating data from other sources, including Wikipedia.

We have then considered three different approaches to analyse the data: a demographic approach analysing the time evolution of the prevalence of certain attributes (i.e. country or disciplines); a mesoscale network approach focusing on the connections between these attributes; a 'kinship' approach based on the clustering of genealogical trees. Our analysis reveals important transition points in the history of mathematics and allows us to categorise countries according to their capacity to attract, export and self-maintain knowledge. Moreover, the community structures of the network of disciplines allows us to better describe the transformation of knowledge across time. Finally, we have also identified important scientific families, associating them to their founder, and described their geographical and disciplinary distribution.

Interesting lines of research for the future include the integration of additional datasets, based on different methodologies, to extend the scope of this work beyond the mathematical sciences.

Another research direction still connected to history of mathematics, could be to analyse how the scientific labor market reacts to exogenous events [33, 34] or to study the innovation dynamics due for instance to the impact of the computer age, of the Internet, of the peer review practices, etc. in the disciplinary prevalence.

Finally it would be worth also to build an abstract agent based models of innovation diffusion, that could be calibrated and implemented on this framework, in order to forecast future events and thus to add a predictive character to this dataset.

Other interesting research directions could include the analysis of gender roles in scientific production, using methods similar to the ones proposed in [35, 36].

ACKNOWLEDGEMENTS

The work of FG, TC and RL presents research results of the Belgian Network DYSCO (Dynamical Systems, Control, and Optimization), funded by the Interuniversity Attraction Poles Programme, initiated by the Belgian State, Science Policy Office.

COMPETING INTERESTS

The authors declare that they have no competing interests.

AUTHORS' CONTRIBUTIONS

FG collected the data and performed the analyses. AC developed the pretreatment algorithms for enriching the database. FG, TC and RL wrote the paper. All authors read and approved the final manuscript.

REFERENCES

1. Bergstrom CT, West JD, Wiseman MA (2008) The eigenfactor metrics. J Neurosci 28(45):11433-11434
2. Ramasco JJ, Dorogovtsev SN, Pastor-Satorras R (2004) Self-organization of collaboration networks. Phys Rev E 70:036106
3. Pan RK, Kaski K, Fortunato S (2012) World citation and collaboration networks: uncovering the role of geography in science. Sci Rep 2:902
4. Radicchi F, Fortunato S, Markines B, Vespignani A (2009) Diffusion of scientific credits and the ranking of scientists. Phys Rev E 80:056103
5. Radicchi F, Castellano C (2011) Rescaling citations of publications in physics. Phys Rev E 83:046116
6. Wang D, Song C, Barabási AL (2013) Quantifying long-term scientific impact. Science 342(6154):127-132
7. Acuna DE, Allesina S, Kording KP (2012) Future impact: predicting scientific success. Nature 489(7415):201-202
8. Shen HW, Wang D, Song C, Barabási AL (2014) Modeling and predicting popularity dynamics via reinforced Poisson processes. In: AAAI 2014, pp 291-297
9. Newman ME (2001) The structure of scientific collaboration networks. Proc Natl Acad Sci USA 98:404-409
10. Deville P, Wang D, Sinatra R, Song C, Blondel VD, Barabási AL (2014) Career on the move: geography, stratification, and scientific impact. Sci Rep 4:4770
11. Gargiulo F, Carletti T (2014) Driving forces of researchers mobility. Sci Rep 4:4860
12. The mathematical genealogy. http://genealogy.math.ndsu.nodak.edu/index.php
13. Narayan P (2011) Mathematics genealogy network. Thesis, University of Oxford. http://people.maths.ox.ac.uk/porterm/research/priya_thesis_final.pdf
14. Engin A, Gunes MH, Yuksel M (2011) Analysis of academic ties: a case study of mathematics genealogy. In: 2011 IEEE GLOBECOM Workshops (GC Wkshps). IEEE Press, New York
15. Dean MR, Ottino JM, Amaral LAN (2010) The role of mentorship in protégé performance. Nature 465(7298):622-626
16. Myers SA, Mucha PJ, Porter MA (2011) Mathematical genealogy and department prestige. Chaos 21(4):041104
17. Schich M, Song C, Ahn YY, Mirsky A, Martino M, Barabási AL, Helbing D (2014) A network framework of cultural history. Science 345(6196):558-562
18. Sinatra R, Deville P, Szell M, Wang D, Barabási AL (2015) A century of physics. Nat Phys 11(10):791-796
19. White D, Jorion P (1992) Representing and analyzing kinship: a new approach. Curr Anthropol 33:454-462
20. Roth C, Gargiulo F, Bringé A, Hamberger K (2013) Random alliance networks. Soc Netw 35(3):394-405
21. 2010 mathematics subject classification. http://www.ams.org/msc/msc2010.html
22. Wikipedia. http://www.wikipedia.org
23. Scopus. http://www.scopus.com
24. Sugimoto CR, Weingart S (2015) The kaleidoscope of disciplinarity. J Doc 71(4):775-794

25. Karrer B, Newman M (2009) Random acyclic networks. Phys Rev Lett 102(12):128701
26. Karloff H, Shirley KE (2013) Maximum entropy summary trees. Comput Graph Forum 32(3):71-80
27. Clough JR, Gollings J, Loach TV, Evans TS (2015) Transitive reduction of citation networks. J Complex Netw 3(2):189-203
28. Fagin R, Kumar R, Sivakumar D (2003) Comparing top *k* lists. SIAM J Discrete Math 17(1):134-160
29. Blondel VD, Guillaume JL, Lambiotte R, Lefebvre E (2008) Fast unfolding of communities in large networks. J Stat Mech Theory Exp 2008(10):P10008
30. Rosvall M, Bergstrom CT (2010) Mapping change in large networks. PLoS ONE 5(1):e8694
31. Hamberger K, Houseman M, White D (2011) Kinship network analysis. In: Carrington P, Scott J (eds) The SAGE handbook of social network analysis. Sage, London, pp 533-549
32. Polanyi M (1969) Personal knowledge: towards a post-critical philosophy. University of Chicago Press, Chicago
33. Borjas GJ, Doran KB (2012) The collapse of the Soviet Union and the productivity of American mathematicians. Q J Econ 127:1143-1203
34. Moser P, Voena A, Waldinger F (2014) German-Jewish emigres and US invention. Am Econ Rev 104:3222-3255
35. Larivière V, Ni C, Gingras Y, Cronin B, Sugimoto CR (2013) Bibliometrics: global gender disparities in science. Nature 504(7479):211-213
36. King MM, Bergstrom CT, Correll SJ, Jacquet J, West JD (2016) Men set their own cites high: gender and self-citation across fields and over time.

CHAPTER 7

The development of Euclidean axiomatics

Vincenzo De Risi[1]

MPIWG, Boltzmannstraße 22, 14195 Berlin, Germany

ABSTRACT

The paper lists several editions of Euclid's *Elements* in the Early Modern Age, giving for each of them the axioms and postulates employed to ground elementary mathematics.

In the following pages I give a list of the systems of principles adopted in the main editions of Euclid's *Elements* in the Early Modern Age, as well as some information about the axioms in Antiquity and in the Middle Ages. The aim of such a list is to provide the reader with a quick survey of the development of mathematical axiomatics, stressing the most important turning points in the foundations of geometry.

As is well known, the *Elements* of Euclid represented for centuries the very model of scientific and deductive reasoning, and their diffusion and influence in Europe were only matched by the Bible and by a few other writings of the Fathers of the Church. They were translated, edited and commented upon hundreds of times, and these editions and commentaries shaped the scientific tools, the methodological standards and the mathematical language of many centuries. Euclid's theorems were used to build further and more daring mathematical theories, or applied in the physical sciences, while the deductive structure of the proofs was studied by mathematicians, logicians and epistemologists as the ideal of reason itself. In this connection, a special historical role was played by the *principles* employed as grounds and foundations of the entire construction. Euclid had begun the *Elements* with a few unproven assumptions which had to be granted in order to establish his grand mathematical system. Since Antiquity, however, Euclid's system of principles has been repeatedly discussed and challenged: A few gaps in the proofs were found and the missing arguments provided with additional axioms; other Euclidean principles were in turn proven from more elementary statements and thus removed from the list of axioms; still others were worded in simpler ways, or changed to accommodate philosophical qualms or to satisfy architectonic considerations; and a few more were added in order to extend the geometrical results beyond the boundaries fixed by Euclid. These changes in the system of principles grounding elementary mathematics were among the most important outcomes of the foundational studies carried out during the Early Modern Age. From the Middle Ages up to the beginning of the nineteenth century, hundreds of different editions of the *Elements* provided their own principles for the same corpus of mathematical theorems; while the statements of the Euclidean propositions remained more or less the same, and the proofs themselves were seldom changed, we witness, in this period, an remarkable explosion of new principles aimed at rigorizing, systematizing and improving Euclid's work. The enormous creativity of these foundational attempts is attested by the invention of some 350 different axioms for elementary mathematics, with some works employing as many as fifty of them to ground the theorems of the *Elements* (Euclid himself probably only had ten principles). Many of these new axioms have great mathematical significance and introduced into the foundations of mathematics several important ideas that were to be given a full development only in the nineteenth and twentieth centuries. They concern continuity issues in the Euclidean plane, properties of incidence in two or three dimensions, Archimedean and non-Archimedean quantities, mereological relations, congruency and rigid motion, the Parallel Postulate and the Euclidean metric, or the first attempts at an axiomatization of arithmetic and algebra. In this connection, it is very important to stress that the best-known works on the foundations of mathematics produced in the modern age, such as Dedekind's, Frege's and Peano's works on the principles of numbers, or Pasch's, Pieri's and Hilbert's famous axiomatizations of geometry, by no means represent the sudden discovery of the need for rigorous foundations, but are rather the late and highly perfected outcomes of a long tradition that had been developing almost continuously from the time of Euclid onward. Even if it may be difficult to precisely ascertain a direct influence exerted by these early modern programs on, say, Hilbert's *Grundlagen der Geometrie*, there can be no doubt that these important mathematical works of the previous centuries did indirectly affect modern, formal research on axiomatics. I will not pursue here, however, the search for the sources of nineteenth-century foundational studies, as this would require an altogether different approach. In any case, while the development of the foundations of mathematics in the last two centuries has already been studied in some depth, the history of axiomatics in the Renaissance and the Early Modern Age is still largely neglected, and I hope that the present essay can represent a beginning on, and an invitation to, this rich field of research, as well as a useful tool for further investigations.

I take this opportunity to express my heartfelt gratitude to Fabio Acerbi and Bernard Vitrac for their accurate reading of a draft of this paper that largely contributed to improve the present version. I greatly benefited from further suggestions offered by Niccolò Guicciardini, Jens Høyrup, Alexander Jones, John Mumma and Victor Pambuccian. I thank Sonja Brentjes for her invaluable help with the Arabic sources and for having helped me with the pseudo- Tūsī text, and Philip Beeley, who informed me about Wallis's unpublished notes on Euclid and provided me with a copy of them. My special gratitude goes to the Library of the Max Planck Institute for the History of Science, for having provided me with several difficult-to-find editions of the *Elements*, and to Massimo Mugnai for his gift of a copy of Grandi's version of Euclid. Any further suggestion will be very welcome and will improve future editions of this essay.

1. INTRODUCTION

1.1 The Euclidean tradition

In the present essay I consider the axiomatic systems of about a hundred editions of the *Elements*, ranging from Byzantine manuscripts down to early nineteenth-century printed volumes. In drawing up this list, I did not strive for completeness but rather for perspicuity, and I have selected a few important works among the enormous mass of manuscripts and books dedicated to the recension, translation and interpretation of Euclid. More exhaustive inventories of the editions of the *Elements* may be found in the bibliographical catalogues drawn up by Riccardi (1974) and Steck (1981); these catalogues are, however, themselves quite incomplete (in fact, a few of the books that I consider here are not even quoted in these accounts). I have mainly taken into consideration the early modern editions, although I have mentioned a few medieval manuscripts which are especially important for the history and the reception of the *Elements*. I also concentrated on Western mathematics, even though a few important developments in the Euclidean tradition first appeared in the medieval Islamic world; these works are only mentioned for the echoes that they have produced in the Latin tradition. Although a complete recension of the Arabic editions of Euclid is still missing, it should be noted, in any case, that it does not seem that they offered a great diversity of axiomatic systems (see for instance Brentjes 1997). Among the early modern editions, I have normally only listed the printed and published works. In this respect, however, I have made a few exceptions, including in my list Maurolico's, Torricelli's, Wallis's, Pascal's, Roberval's and Newton's handwritten work on the principles of geometry, which seemed to me too important to be ignored. Several other similar essays (such as Leibniz's unpublished papers on the *analysis situs*) were not suitable to be easily included in the list and in any case did not exert any historical influence on the respective authors' own contemporaries.

As a general rule, I have tried to include all those editions which proposed new axioms and principles, as well as those editions produced by mathematicians important in their own right (even if these editions did not represent any advancement in the axiomatic field), and the editions that possessed some special historical significance (by reason of their broad diffusion, for instance, or because they were the first editions in a particular modern language). I have only quoted the first edition of each work, unless the editions that followed introduced important changes in the system of principles (as is the case, for instance, of the second edition of Clavius's commentary, or of Arnauld's *Nouveaux Élémens*).

A serious limitation of the field examined here is the fact that I have only considered editions of the *Elements* themselves, leaving aside all other material on the foundations of geometry. It should, however, be borne in mind that several important changes in the axiomatic systems were proposed in monographic essays dealing with special problems of the Euclidean text, as is the case, for instance, with several Arabic treatises, Wallis's or Saccheri's studies on the Parallel Postulate, or Galileo's new theory of proportions. A few more principles of mathematics were discussed in general mathematical works (such as certain treatises on algebra, or an essay on fluxions) which did not comment at all on Euclid. That development of the axiomatics, then, which is to be found in the editions of the *Elements* only gives a part of the whole story: and yet, this is (by far) the largest part, as the main discussions on the foundations of geometry in the Early Modern Age did indeed take place in the Euclidean commentaries. Some of these commentaries, moreover, introduced innovations and alterations with respect to the Greek text of the *Elements* to such an extent as to become almost new treatises on the foundations of geometry. There is, therefore, a certain degrees of arbitrariness in the decision to include certain works in the list of the Euclidean commentaries, even though they departed very far from the original structure of the *Elements*, while excluding certain others as moving really too far away from the original to be still considered discussions of Euclid's work itself. This is especially true as regards eighteenth-century manuals on elementary mathematics, which often offered a complete reworking of the content of the *Elements* and no longer had the form of a commentary at all. I have reproduced here the systems of axioms contained in a few of these manuals, choosing the most important representatives of the age (such as Simpson's or Kästner's books); the reader should, however, be aware that dozens of such volumes were produced in Europe at this time—and in eighteenth-century Germany in particular.

It may perhaps be useful to provide, at this point, a short synoptic sketch of the general development of the systems of axioms employed in the various editions of the *Elements* from Antiquity up to the modern era.

1.1.1 Antiquity and the Middle Ages

Euclid's *Elements* (in thirteen Books) were probably written in the third century BC, drawing together materials from a more ancient mathematical tradition, and surely underwent several changes and modifications already during the Hellenistic Age. Several centuries later, the mathematician Theon of Alexandria (fourth century AD) prepared an edition of the *Elements* which has had a great importance in the following tradition of the text. The oldest extant copies of the *Elements* are, however, Byzantine manuscripts dating back to the beginning of the ninth century; later Greek manuscripts were in circulation among scholars during the Renaissance and in fact provided the Greek text for the first editions of Euclid in the modern world. One further important manuscript was discovered only at the beginning of the nineteenth century by François Peyrard, and it apparently contained a version of the *Elements* older than any other, and possibly predating Theon's editing of Euclid. At least, it seems that its author had access to both Theonine and non-Theonine manuscripts (it should be remarked that this statement, endorsed by many scholars, has been challenged by some others; see Knorr 1996). It is still considered the best guess which we possess regarding the original form of Euclid's work, and has been employed by the Danish philologist Johan Ludvig Heiberg to prepare the only critical edition of the *Elements* (1883–1916) that we currently have. For a history of the Greek text of the *Elements*, see for instance Rommevaux et al. (2001).

All these Greek manuscripts (both those pertaining to the "Theonine" tradition, and Peyrard's text) show many definitions prefacing several books of the *Elements*, and a set of fourteen of fifteen other principles (depending on the manuscripts) at the beginning of Book One. The latter principles were considered to be the only statements on which the mathematical construction of Euclid was to be based, and are divided into a first list of *postulates*(ΑἸΤΉΜΑΤΑ)and a second list of *common notions*(ΚΟΙΝΑῚ ἜΝΝΟΙΑΙ) or, as they were later to come to be called, *axioms*. The rationale behind this division of the principles into 'postulates' on the one hand and 'common notions' on the other was a subject of debate and controversy already in ancient times, and several philosophers and mathematicians advanced their own opinions on the epistemological status of these two distinct kinds of assumptions. We will not discuss this difficult topic here, but will note that all the editions of the *Elements* in the Early Modern Age, with no exception, were to preserve this distinction (even if they interpreted it in different ways).

Some principles among the fourteen or fifteen contained in the oldest manuscripts, however, came to be considered spurious already in Late Antiquity, and Proclus (fifth century, the author of a crucially important commentary on the First Book of the *Elements*) informs us that a number of them were in fact later interpolations. Modern philology inclines to accept Proclus's opinion and attempts to prove (even if on different grounds) that some of the principles must have been added in the ancient transmission of the text. A largely (but not unanimously) endorsed opinion among scholars is that Euclid originally had only ten principles (five postulates and five common notions), while the other axioms are Hellenistic additions. Some further doubts are usually raised about two common notions (CN4 and CN5) that some scholars would like to regard as interpolations as well; if this latter view holds true, then Euclid only had eight principles. We know, in any case, that several ancient mathematicians discussed the Euclidean system of principles, and began to add, change or remove a number of axioms and postulates; this seems to have been the case, for instance, of Apollonius, Geminus, Ptolemy and Pappus. In particular, Heron of Alexandria (probably first century AD) wrote a commentary on the *Elements* in which he addressed a few gaps in the Euclidean proofs and tried to fill them introducing new assumptions (which he may, however, have regarded as provable). While Heron's commentary is now lost, it was still known in the Islamic Middle Ages (see for instance BRENTJES 1998, or the introductory essay to Acerbi and Vitrac 2014). A few fragments of it, moreover, circulated in the commentary to the *Elements* written by Proclus, whose text enjoyed only a very limited circulation in the Middle Ages, but was rediscovered in the Renaissance. Similarly, Simplicius's commentary to the *Elements* (sixth century) was extensively read in the Arabic-speaking world, but was later lost. A few of the axioms added in the Middle Ages and the Renaissance, in fact, derive from these Greek commentaries, and from Heron in particular, as the philosophers Proclus and Simplicius did not dare to add their own principles to the "most perfect" work of Euclid.

During the early Middle Ages knowledge of Euclid disappeared from the West (except for a few fragments preserved in a very elementary mathematical work attributed to Boethius), but flourished in the Islamic world. The *Elements* were translated into Arabic twice in the early ninth century by Al-Hajjaj, and again a few years later by Isḥaq ibn Ḥunain (a better version than Al-Hajjāj's, revised by the great mathematician Thābit ibn Qurra). These editions were translated from Greek manuscripts that were not very dissimilar to those which we still possess, and thus presented the extended system of principles (fourteen or fifteen of them) which were already common in Greek Late Antiquity. In the tenth century, the mathematician An-Nayrīzī wrote an important commentary on Euclid, drawing on Simplicius's and Heron's (and those by others) new additional assumptions. Even though this was not an edition of the *Elements*themselves, but rather a commentary thereon, it was clear that Euclid's system of principles could, and probably should, be enlarged to fill the gaps present in several proofs. An-Nayrīzī's commentary on Euclid survived as regards Books I–VI and X, with a very short fragment on Books VII and VIII and a longer fragment on Book IX.

The *Elements* of Euclid reappeared in the West in the first half of the twelfth century, when Adelard of Bath translated into Latin an Arabic manuscript containing (a modified version of) Al-Hajjāj's translation. Adelard's edition was followed by several other translations made from similar Arabic sources (and thus with similar principles), and by the different translation of the Isḥāq–Thābit's version made by Gerardo da Cremona. Gerardo, however, also translated An-Nayrīzī's commentary, and this paved the way for a reworking of the Euclidean axiomatics. Already in the edition of John of Tynemouth a few new principles were added at the beginning of Book VII of the *Elements*, which were partly drawn from An-Nayrīzī. In the thirteenth century, Campano da Novara edited a final version of the Euclid of medieval tradition, collating together several previous translations and even an independent work (the *Arithmetica* by Jordanus de Nemore) to fashion a highly successful edition of the *Elements*. It contained sixteen geometrical principles in Book I and fourteen arithmetical principles in Book VII. Campano's version was widely used in the following centuries, and it was the first edition of the *Elements* to be printed, in 1482.

1.1.2 The Renaissance

Campano's edition and its axiomatic system were used several times in the following century. Luca Pacioli produced an improved version of it, and Niccolò Tartaglia translated it into Italian in 1543 (altering a few axioms thereby). Tartaglia's edition inaugurated the translations of Euclid into modern languages and was followed by Xylander's translation into German (1562), Forcadel's into French (1564), Billingsley's into English (1570), Zamorano's into Spanish (1576) and Dou's into Dutch (1606). In the meantime, Greek manuscripts had begun to arrive in Europe, and Bartolomeo Zamberti had translated the *Elements* into Latin directly from the Greek (1505), thus opening a new era in the philological reception of the text. In 1533 Simon Grynaeus published the *editio princeps* of the

Greek text. The latter editions produced on the basis of the Greek text once again proposed the ancient system of principles which featured no arithmetical axioms nor any of the further corrections made by Campano. Grynaeus's edition, however, was based on a Greek manuscript that (probably following Proclus's hint) changed the division between postulates and axioms, and only admitted three postulates (P1, P2, P3, i.e., the clearly constructive ones), listing the others (P4, P5, P6) among the axioms. Grynaeus, in fact, even appended an edition of Proclus's commentary to Euclid's text: so that the system of principles contained in the *editio princeps* of the *Elements* was in fact a Neoplatonic philosophical construction. In any case, Grynaeus's edition of the Greek text remained the only one available for a very long time, and most of the "philological" editions of Euclid in Latin or in modern languages which were prepared in the following two centuries simply followed Grynaeus's altered system of axioms and postulates (e.g., Caiani, Camerarius, Scheybl, Gracilis, Xylander, Dasypodius, Forcadel, Candale, Dou, Henrion, Keill).

A debate on the superiority of Zamberti's and Grynaeus's Greek sources over Campano's Arabic ones then ensued; comparative editions were prepared; and even if, in the end, the Greek tradition prevailed in the hearts of the humanists and Campano's version was abandoned by the end of the sixteenth century, this latter's richer system of principles continued, nonetheless, to charm several geometers who prized mathematical perspicuity (and soundness) above philological exactitude.

A clearly recognizable line of French mathematicians dared to modify Euclid's system of principles to comply with mathematical correctness. Jacques Lefèvre d'Étaples published a new, enlarged system of arithmetical principles (drawing on Campano); Oronce Fine began to add a few new geometrical axioms; and Jacques Pelletier showed no hesitation in criticizing Euclid for his use of rigid motion in geometrical proofs, thus changing the system of principles in such a way to be able to dispose of these kind of demonstrations. A few years later, the mathematician Pedro de Monzón, who had translated Fine into Spanish, collected the fruits of this innovative tradition and published an edition of Euclid that comprised forty-three principles (both geometrical and arithmetical). Among the studies on the foundations of geometry in the sixteenth century, we should also mention the edition of Euclid produced by Herlinus and Dayspodius in 1566, which rewrote all the proofs of the *Elements*, molding them into syllogistic form; this important advancement in the logical understanding of geometrical reasoning and the deductive structure of the *Elements*, however, simply accepted Grynaeus's system of principles, and added nothing to the science of axiomatics in the strictest sense. The culmination of the sixteenth-century debates on the *Elements* was attained in the 1570s, when Commandino and Clavius printed their important editions of the *Elements*.

Federico Commandino published in 1572 his Latin (and Italian) version of Euclid, which was based on different Greek manuscripts and was in many respects more correct than Grynaeus's *editio princeps* (even though Commandino produced no edition of the Greek text itself). Even if Commandino clearly had important philological aims, he was also an outstanding mathematician and endorsed several changes in the axiomatic system which had been found necessary by the preceding tradition, adding a few further modifications himself; he added for the first time certain axioms in Book V and in Book X of the *Elements*. The very remarkable outcome is an edition of Euclid which can certainly be trusted for its philological accuracy but which nonetheless appeared much richer in foundational discussion (with a total of thirty-nine axioms and postulates), and many subsequent editions, both in Latin and in other, modern languages, in the course of the seventeenth and eighteenth Centuries, were simply to be based on Commandino's text.

In 1574 the Jesuit mathematician Christoph Clavius published another crucially important edition of Euclid in Latin. Clavius's edition was by far the longest and most complete edition of the *Elements* ever produced, and was accompanied by a very long commentary that collected almost everything that had been published about the Euclidean text in the previous centuries. For this reason it quickly became the reference work for any further discussion on the foundations of geometry in the following two hundred years, and easily the most important edition of Euclid that was ever published. Clavius believed that the original Greek text was important and had to be preserved but that mathematical perspicuity was a higher aim; his compromise went further than Commandino's in correcting Euclid. Clavius himself envisaged a few new axioms that were useful, in his view, in order to rigorize the proofs of the *Elements*; given the enormous success of this edition (and the fact that it immediately became the model for all the further Jesuit editions of Euclid), these new Clavian axioms were successively endorsed by almost the whole community of mathematicians. Another important aspect of Clavius's work on elementary geometry is that, in the 1580s, he came into contact with an Arabic edition of the *Elements* attributed to Nasīr ad-Dīn at-Tūsī (thirteenth century), and here he discovered the great difficulties that the Arab mathematicians had faced in attempting to prove the Parallel Postulate. In the 1589 edition of his commentary Clavius added Nasīr ad-Dīn's alleged proof of the Parallel Postulate, together with a few reflections of his own, and this episode marked the beginning of the rich epoch of the attempts to prove the Parallel Postulate in Europe—an epoch that culminated in the eighteenth century, with the subsequent discovery of non-Euclidean geometry. Clavius's complete system of principles, which was to remain a milestone in the following developments, contained four postulates and twenty axioms in Book I (elementary plane geometry), one axiom in Book V (the theory of proportions), three postulates and twelve axioms in Book VII (arithmetic), and one postulate and three axioms in Book X (irrational magnitudes).

1.1.3 The seventeenth century

The seventeenth century was the most creative age in the invention and development of new principles for geometry and arithmetic, and in many respects the highest point of research into the foundations of mathematics in the Early Modern Age. Following an age in which the main goal seemed to be that of the restitution of an accurate text of the ancient mathematical work after the corruptions of the Middle Ages, the new century decidedly aimed at the reform and improvement of the *Elements* beyond what Euclid himself had been able to do. To this effect, several completely new systems of axioms were devised in order to ground elementary mathematics. This innovative line of attack to the ancient text was especially popular in France and Italy.

French mathematicians often endorsed an epistemology of Ramist origins (that was later contaminated by similar stances emerging out of the new Cartesian tradition) which was intent on designating as an "axiom" any clear and evident mathematical proposition, and in any case any proposition the proof of which could be considered more complicated, and thus less evident, than the statement itself. Such an epistemology, which was stated in its most classical form in the *Logique de Port-Royal* (1662), led to a flourishing of "evident" principles in this century's editions of Euclid which was without precedent in previous ages. Several mathematicians simply wrote down dozens of axioms and postulates at the beginning of the *Elements* without even troubling themselves about whether they were actually needed for the proofs or whether they could be deduced one from the other. Into these highly pleonastic systems of principles, which were not inspired by any consideration of economy of means, a few good mathematicians also, indeed, inserted several important axiomatic principles which had been overlooked in the past and which were to play a central role in future axiomatizations. We may mention in this connection Herigone's edition of the *Elements* (1634), which contains forty-nine axioms and postulates, the greater part of them being extensions of the Euclidean common notions (i.e., principles on mereology and equality), and also the edition by Claude Richard (1645), with forty-six principles mostly dealing with geometry and introducing a few "topological" axioms to guarantee the continuity of the figures and the existence of their intersection points (the so-called Line–Circle or Circle–Circle continuity axioms). Richard was a Jesuit, and, like him, other French Jesuit masters continued Clavius's work on the foundations of mathematics, with the additional aim of producing pedagogically efficient handbooks of geometry; among their most remarkable outcomes, we may mention Tacquet's radical reform of the *Elements* and their principles, as well as the work by Dechales and Fabri. On the other side of the political–religious divide of the day, the Jansenist theologian Antoine Arnauld wrote an important new reworking of elementary mathematics (1667, 1683), which also included many new axioms (forty-two in total) and was followed by several editions loosely inspired by it (by Lamy, Malézieu, Sauveur, and others). The most important breakthrough was probably achieved, however, by Pascal and Roberval, who wrote two very important essays on the foundations of geometry, both of which clearly show their mathematical genius. Neither of them, however, was published. Roberval's work survived in its entirety, while we only have the introduction (and the axiomatic basis) of Pascal's essay on elementary mathematics. The latter text was copied by Leibniz in Paris in 1675, and largely inspired Leibniz's own work on the foundations of geometry (the *analysis situs*) which also offers a thoroughgoing discussion on axiomatics, but remained unpublished (and thus historically ineffectual) among Leibniz's papers in Hannover (on Leibniz's *analysis situs*, see De Risi 2007, 2015b).

Italian studies on Euclidean axiomatics were mostly carried on by the school of Galileo and concentrated especially on the reform of the theory of proportions, since Galileo himself had attempted such a construction (in the unpublished *Fifth Day* of the *Discorsi*) and the topic was central to his project of the mathematization of mechanics. It is to Torricelli (1647), Borelli (1658), Viviani (1674), Giordano (1680) and Marchetti (1709) that we mainly owe the progress made in this field. As compared to the French tradition, these Italian axiomatizations show less creativity in producing new axioms, but a greater attention to their systematic function and the economy of principles. From a purely logical point of view, some of these books may be considered to be the culmination of the logical reflection of the *Elements*, and Borelli's *Euclides restitutus*, in particular, marks a high point in researches on the foundations of mathematics.

1.1.4 The eighteenth century

In the eighteenth century, the foundational discussion on mathematics shifted for the most part outside the scope of the *Elements*. Most efforts were now dedicated to the many issues raised by the new Calculus, and reflection on the foundations of mathematics was often aimed rather at justifying the employment of infinite or infinitesimal quantities; these discussions, however, are seldom to be found in an edition of Euclid. Even elementary mathematics is from time to time grounded on different principles, and algebra (for instance) offered the possibility of a dissimilar foundational approach, grounded in symbolic manipulations rather than simple geometrical constructions. The *Elements* were no longer regarded as the proper place for advancing a scientific discussion, but rather just as the most classical handbook for a basic schooling in mathematics. The outcome of this modified approach was, on the one hand, a growing attention to pedagogical improvements in the teaching of elementary geometry, and thus a proliferation of simplified editions for students, and on the other hand a renovated search for a more reliable and philologically accurate edition of the *Elements*.

The former attitude was especially widespread in France and Germany, where a large number of Euclidean textbooks were produced, often largely departing from the original text of the *Elements*. These systematizations of elementary mathematics do not offer much, however, as regards the foundational domain, and their axiomatizations are often oversimplifications of older editions. Whereas the epistemology of the "clearness and evidence" of mathematical principles had resulted in France, in the previous century, in an overproduction of axioms, the same epistemology in the eighteenth century often yielded axiomless editions of the *Elements*, which merely made mention of the fact that several evident assumptions were implicitly employed in the development of the proofs (this is the case, for instance, of the celebrated book on elementary geometry written by Clairaut in 1741); a few other editions listed a small number of axioms, claiming that these were just examples of an infinite array of principles that cannot but remain implicit (see Deidier, 1739; or Legendre, 1794). In Germany the surviving Scholastic idea that all the axioms should ultimately be proven from the definitions also engendered several axiomless textbooks which enjoyed a wide diffusion (the most important representatives of these being Christian Wolff's, Andreas Segner's and Wenceslaus Karsten's German and Latin treatments of elementary mathematics). Nevertheless, there can be noted a few interesting developments in the area of axiomatization in this period, such as Samuel König's and Abraham Kästner's axiomatizations from 1758.

The philological attitude toward the *Elements*, on the other hand, was especially common in Britain, where David Gregory produced an emended edition of the Greek text (1703) which may be considered the first improvement on this latter since Grynaeus. Robert Simson, likewise, published a highly successful edition of the *Elements* (1756) which aimed at restoring the Euclidean text to the state in which it existed before the interventions of Theon of Alexandria; this was in fact later to be accomplished by Peyrard through his discovery of an actual pre-Theonine manuscript (or, at least, a manuscript in which non-Theonine elements are to be found); Simson's attempt was based merely on some lucky guesses. If these philological editions offer nothing in the way of an actual development of axiomatics, the commentaries by Thomas Simpson (1747) and, to a lesser extent, John Playfair (1795), nonetheless added several important principles, and attempted a new foundation of elementary mathematics. It may be mentioned that the axiomless attitude toward elementary geometry is well attested in Britain as well, with John Leslie's celebrated *Elements of Geometry* from 1809, and a famous *Geometry without axioms*, published by Perronet Thompson as late as 1833.

We end our survey with Peyrard's edition of Euclid, published in 1814. This book, in which the Greek text was established from several manuscripts, marked the most relevant advancement in textual analysis so far and opened the rich era of nineteenth-century Euclidean philology which eventually found its culmination in Heiberg's 1883 edition of Euclid (largely based on the manuscript found by Peyrard). The discussion of the system of principles in the nineteenth century mostly developed in the form of monographic essays or independent treatises (such as those by Pasch, Frege, Peano or Hilbert) which completely abandoned the literary form of commentary on the text of Euclid. In the same years in which Peyrard was searching for Greek manuscripts, and in the following decades, when his recension of the text begun to acquire authority and to be translated into several languages, Poncelet and Staudt were developing new foundations for geometry starting from the projective properties of space, and Lobachevsky and Bolyai were publishing their treatises on hyperbolic geometry. The gradual development of projective geometry and the non-Euclidean systems, as well as the subsequent birth of abstract algebra and set theory, shifted most of the foundational attention to these new fields. It became clear that the *Elements* of Euclid could no longer stand as the systematic ground of modern mathematics, since the ancient text could not bear the weight of these new axiomatizations. A completely different approach to the foundations of geometry was then in need.

1.2 Axioms and postulates

I cannot attempt to present here a whole history of the epistemology of mathematics; nor to discuss of the meaning of the notions of axiom and postulate from Antiquity down to modern times; nor to give a list of the many and diverse goals that mathematicians may have aimed at in reshaping the system of the Euclidean principles (for a few contemporary suggestions, see for instance Schlimm 2013). It should at least be noted, however, that many mathematicians in the Early Modern Age believed that axioms and postulates were in fact provable from definitions. They were not so much considered to be statements which *admittedof* no proof as statements which *required* none (since they were sufficiently evident already in themselves). In fact, however, a few of the most important editions of Euclid *did* explicitly prove the axioms and postulates; the commentary by Clavius, for instance, gave demonstrations for all mathematical principles except definitions. It then became a point of controversy whether or not it was useful to prove the axioms at all; many French authors during the seventeenth century, for instance, claimed that

Clavius's exercise had plainly been futile. A further disagreement concerned the criterion which was to be applied in distinguishing a principle from a theorem (as both of these latter are provable); the most common (and quite generic) answer given here was that, whereas an axiom or a postulate is *immediately* proven from the definitions, a theorem is only *mediately* proven from these latter (and from other theorems, and from other axioms); the notion of "immediacy" evoked here, however, remained quite vague. As the number of principles grew their demonstration was often omitted, and in many cases it is not easy to understand how such principles could have been proven or explained away. It has already been noted that many authors also claimed that the proper number of axioms should be regarded as infinite, and that the explicitly stated principles should be considered as merely examples and instances of the innumerable statements which immediately follow from the definitions themselves (or from the "essences" of the geometrical entities defined). I have found the first trace of this claim in a few medieval manuscripts of the *Elements* in the tradition of Robert of Chester (even though previous, more subtle traces of this attitude may be found in Ibn al-Haytham and the Arabic tradition), from whom the idea passed to Campano, and thus on into the mainstream epistemology of geometry. In any case, it should be clear that there is a sense in which *definitions alone* were considered the proper principles of mathematics, while axioms and postulates were often regarded as derivative statements.

All this notwithstanding, I have, in the following pages, only listed axioms and postulates, leaving aside the definitions. The omission, however, is not as serious as it may at first appear. It should be remarked that the system of definitions is much more stable than the system of axioms and postulates, and the definitions found in the early modern editions of the *Elements*are quite close to the list found in the Greek manuscripts. When the moderns add or modify a few definitions, moreover, this seldom has any effect on the further mathematical development of the treatise. Some definitions were only changed in order to dispel certain purely epistemological qualms (for instance, some authors define a circle as the trace left by the endpoint of a rotating segment rather than as a line equidistant to the center, with the aim of giving an explicit construction and thus a "*real definition*" of the geometrical entity in question); other definitions were added in order better to explain a few passages in Euclid, or so as to be able to add a few theorems. In most cases, the original Euclidean definition is simply placed side by side with the new definition, which is then regarded as equivalent to this latter, without any proof being given for this assumption. In short, then, the system of definitions was usually only modified in respect of foundational preoccupations which did not affect the further mathematical treatment. The axioms and postulates which are derived (or which should be derived) from these definitions are the statements which are properly employed in the proofs and which therefore affect the deductive structure of the *Elements*.

Nevertheless, it should be noted that a number of new definitions had real consequences for the axiomatics: This is especially true for the definition of a straight line (which troubled several mathematicians over a period of many centuries), for the definition of parallel lines (as several such definitions were conceived in order to prove the Parallel Postulate) and for the whole definitional system of Book V of the *Elements* on the theory of proportions. It is still impossible, however, to account for them in a synoptic survey like this one, and the reader is simply warned that definitions may play a supplementary role in the system of principles that does not always appear in an entirely evident manner in the list of axioms and postulates. In order, then, to follow the developments consisting in the important attempts to prove the Parallel Postulate or to reform the Eudoxian theory of proportions, further references may be required.

I have also omitted a small number of principles which are to be found in a few encyclopedic works on mathematics which also contain a section on the *Elements* inasmuch as these principles are irrelevant from a foundational point of view. Some of them are simply logical or metaphysical statements, and have no mathematical meaning. For instance, it is not uncommon in some Scholastic works to find stated, as an axiom, the Principle of Contradiction, or the Aristotelian claim that mathematics is a discipline abstracted from sensible magnitudes, or the claim that it is possible to consider the surface of a thing while abstracting from its body. Some other principles are properly mathematical, but have no foundational meaning. For instance, a few handbooks list, among the arithmetical axioms, certain principles explaining the working of the positional notation of numbers. These kinds of principle may be important for the history of notations and mathematical culture, but their relevance is one that lies far wide of the aim of the present essay. An example of both kinds of principles is to be found in the *Arithmetica perfecta* by Georg Henisch (1609, not included in the following list of mathematical editions), which contains sixteen axioms and seven postulates for arithmetic, none of which has anything to do with the theory of numbers in the modern sense; the first axiom states that numbers are potentially but not actually infinite, the third that a number cannot perform any action, the seventh that the natural order of numbers is from the left to the right, the ninth that zero means nothing in itself but adds much value to another number if it is placed at the end of it, while the second postulate states that it is possible to distinguish a whole number from its decimal part by placing a comma in the right place. Interesting as they are, I have excluded these kinds of principle from my list. It should always be remembered, in any case, that a mathematical treatise of the Renaissance or of the Early Modern Age (even an edition of

Euclid) may include an assortment of topics which look very odd from a contemporary perspective.

In listing the various principles, I have divided them into thirteen different classes, in accordance with their different mathematical meanings. It should be noted that this division was only carried out in order to facilitate the comparison between principles and has no roots in the real historical development: The actual classification and grouping of principles in the various Euclidean editions is given under each entry, and this classification and grouping is normally only concerned with a division in books and between postulates and axioms. My own arrangement is loosely based on Hilbert's work on the foundations of geometry (which offers a classification of principles that is still useful and fully perspicuous), but I was obliged to incorporate a few changes to Hilbert's list in order to accommodate the incredible variety of ancient axioms. In particular, it will be noted that Hilbert's "axioms of order" are absent from our list: This is an important feature of ancient and early modern mathematics which never thematized relations of order as such. These began to be studied in projective geometry and had to wait until a still later period to be incorporated into an axiomatic system. Hilbert himself attributed the introduction of this kind of axioms to Pasch (1882), and this event should be regarded as an important breakthrough in the history of the foundations of geometry; Pasch's axiom is nonetheless hinted at in a few early modern axiomatizations (see below). To facilitate a comparison with the modern discussion on axiomatics, I include Hilbert's system of axioms in the *Grundlagen der Geometrie* (1899, with a few revisions in the following editions) after the other lists of principles. For Hilbert's own historical sources, see Hallett and Majer (2004), and Volkert (2015) (which also offers a short history of geometrical axiomatics).

A few remarks are in order to explain some of the principles and their classification.

1.2.1 Ancient principles (Groups 1 and 2)

The Euclidean list of five Postulates and five Common Notions follows Heiberg's edition and finds its main foundation and justification in a number of comments made by ancient scholars, who had in their hands editions of the *Elements* with lengthier systems of principles but claimed that some of them had been added in the centuries following Euclid. The additional principles are here listed as P6* and CN6*–CN9*. Axiom CN8*, in particular, is attributed by Proclus (*In Euclidis* 197) to Pappus (along with M1 and M2, see below). Besides the principles which can be found in the Greek manuscripts of the *Elements*, we also mention a few other axioms and postulates which are explicitly stated in ancient commentaries and which clearly influenced the following editions, as well as a few important variants of these. Among these axioms and postulates, the extra assumptions EX1, EX2 and EX3 are attributed by Proclus (*In Euclidis* 198) to Pappus as well, and traces of them are evident in several further axioms, as well as in the definitions of a straight line and a plane; they are, however, nowhere formulated as axioms in a system of principles (even though Albert the Great mentioned them as possible additional axioms in his commentary to Euclid, see Tummers 2014). Other ancient assumptions which *did* find their way into medieval or early modern axiomatizations will be listed under the respectively appropriate category.

A few postulates changed their formulation in different editions of the *Elements*. In particular, Postulate 2 on the extendibility of a straight line states, in its original form, that the straight line may be produced *continuously*(ΚΑΤὰ ΤὸΣΥΝΕΧἶω). Already in the translation of Adelard of Bath (from the Arabic), however, the postulate is altered so as to make it into a statement about the extendibility of a straight line to *an arbitrary length* ("*assignatam lineam rectam quantolibet spacio directe protrahere*"), dropping the aspect of continuity and stressing rather the length of the extension. In Gerardo da Cremona we also find a version stating the *indefinite*extendibility of the line, the original meaning of which may have been the same as that conveyed by Adelard's translation. Both versions of the axiom were to be much used in the Renaissance and the Early Modern Age. The modified statement was employed by Campano, and it is common in the tradition stemming from him (in Tartaglia, for instance, or in some Jesuit textbooks), while the Greek formulation is to be found in Zamberti, Grynaeus, Commandino, Clavius and other authors more attentive to philological qualms. This notwithstanding, the two traditions mixed very soon, and it is not uncommon to find the statement about indefinite extension also in editions translated from the Greek; early instances of this attitude are the *Elements* edited by Xylander, Forcadel and Grienberger (who was normally copying his principles from Clavius), but many more may be found in the seventeenth and eighteenth Centuries. I have not signaled these oscillations, as the various spelling of the principle were clearly considered as equivalent, and in a few volumes their equivalence is in fact explicitly stated. It should be remarked that the version of P2 stating the *indefinite* extendibility of the line was sometimes transformed into a claim regarding the *infinite* extendibility of it (instances are Herlin & Dasypodius, Richard, Arnauld and others). Even though the latter spelling may have played a role in establishing the modern non-constructivist theory (i.e., that straight lines are just infinite), the early modern authors that employed the word "infinite" in this connection always added that it was a potential infinite (or a "syncategorematic infinite," following Richard's Scholastic wording), and therefore it could be equated with the standard notion of an "indefinite" extension.

The most important geometrical principle added in Antiquity, however, was the one stating that two straight lines do not enclose a space (P6*). This axiom is often stated in this form, even though it is also found from time to time (and more frequently in the Early Modern Age) as the simpler statement that two straight lines have at most one point in common (I17; the first instance of it seems to be in Arnauld, 1663), or a variation of the latter (e.g., I31, I32). While principles P6* and I17 are not equivalent in general (in elliptic geometry I17 obtains and P6* does not), the counterexamples were too abstract to be made specific note of by early modern mathematicians, who regarded the two axioms as interchangeable.

1.2.2 The theory of magnitudes (Groups 3 and 4)

I include under this heading several principles which are obvious extensions of the Euclidean Common Notions, and which deal with mereology, mereological addition and subtraction, equality, and relations of order among magnitudes. These principles were often looked upon as the main axioms of the *mathesis universalis*, that is to say, of a mathematical theory which aspired to encompass both geometry and arithmetic (or, in other words, a theory of both continuous and discrete quantities). Many of them were later to be applied in the theory of proportions, which has also been conceived by some authors as a form of universal mathematics. A number of them are especially connected with the arithmetical axioms and may in fact be conflated with them; the reason is that numbers themselves were often conceived of as collections of unities (this is the Euclidean definition in *Elements* VII, def. 2), and therefore the rules of composition and division of numbers were stated as mereological operations (cf. *Elements* VII, def. 3–4). The axioms which I have listed under the category of the theory of numbers (Group 7) should therefore be integrated with the principles regarding sum, subtraction and multiplication of magnitudes which are also presented here.

Many of the principles listed here concern the transitivity of some relation or operation (taking as a model CN1), and usually aim at establishing an equivalence relation of some kind. Other principles are directly about the substitutivity of equivalent magnitudes in several contexts (see especially A18, A19). Still other principles are aimed at providing an ordering structure for magnitudes (although these should not be confused with the missing *geometrical* axioms of order), and, to this effect, there were also conceived and stated certain axioms on trichotomy (see A3, A21, A24).

A corruption of the medieval text appears to be the insertion of the Common Notion A1, which seems to be a useless reduplication of CN1.

This first appeared in Adelard's version, which, in turn, lacks axiom CN8*; it is very likely that a trivial translation slip from the Greek into the Arabic, to be found in the Al-Hajjāj tradition, changed the original CN8* into A1 (such a mistake is probably absent from the Isḥāq ibn Ḥunain translation, and therefore also from Gerardo da Cremona's Latin version, which was drawn from this Arabic source).

The mistake, however, spread in the whole tradition following Adelard (Robert of Chester, John of Tynemouth, Campano da Novara). The spelling of CN1 and A1 is in fact slightly different, but it is hard to figure out a proper difference in meaning: in Adelard's prose, CN1 reads "*Si fuerint alique due res alicui rei equales, unaqueque earum erit equalis alteri*," while A1 reads "*Si fuerint due res alicui uni equales, unaqueque earum equalis erit alteri*"; Campano changed the formulation, and his CN1 is "*Que uni & idem sunt equalia, & sibiinvicem sunt equalia*," while his A1 is "*Si fuerint due res uni equales ipse sibiinvicem erunt equales*." Luca Pacioli, in 1509, tried to preserve the received text and explained the difference between the two axioms in terms of their being an abstract and a concrete version of the same principle. In 1543, Niccolò Tartaglia simply eliminated A1 from the list of axioms, reintroducing CN8* (following, in this one instance, Bartolomeo Zamberti's translation from the Greek, which had not suffered from the Arabic textual corruption).

On the other hand, Pierre Herigone's principle A15, stating that things equal to equal things are also equal to one another, may be regarded as a axiom different from CN1, as it expresses a more complex relation than simple transitivity (i.e., if a=ba=b and c=dc=d and b=db=d, then a=ca=c). The same holds for Herigone's M20 and M21 compared with CN8*, and M23 with CN9*.

Axioms M1 and M2, added by Gerardo da Cremona, are taken from An-Nayrīzī's commentary; Proclus (*In Euclidis* 197) attributes them to Pappus. Principle A4 is almost classical and will be repeated very often in the Euclidean tradition, but it is first stated as an axiom in Bradwardine's geometry.

It may be remarked that many of the principles under this heading were introduced during the Renaissance (by Lefèvre, Commandino, and Clavius), and that they were in fact among the first newly conceived axioms for the *Elements*.

Herigone also contributed much to this category, in an attempt to completely formalize the relations of equivalence. In many respects these principles may be considered to constitute the prehistory of the attempts later made toward an axiomatization of algebraic relations.

1.2.3 The theory of proportions and the Axiom of Archimedes (Groups 5 and 6)

The theory of proportions, to be found in Book V of the *Elements*, is probably that part of elementary mathematics which has undergone, in the history of the Euclidean corpus, the greatest amount of modification. It was entirely based on a complex system of definitions, possibly conceived by Eudoxus, which suffered serious textual corruptions during the Middle Ages, and which affected all the medieval Latin translations depending on Adelard (including Campano): see Murdoch (1963) and Molland (1983) for a first survey. On the other hand, an entire new theory was developed in the Arabic world (from a few hints in classical sources), the so-called anthyphairetic theory of ratios, which however (as far as I know) was never axiomatized and reduced to principles (see, for instance, Itard 1961). The original meaning of the theory expounded in Book V of the *Elements* was recovered during the Renaissance, and the restitution of Euclid on this topic was in fact the aim of many philological editions. Once the text was established, several criticism were raised against it, as its meaning remained for a long time quite obscure. Later on, it was widely accepted and began to be regarded as the highest point of the mathematical construction of the *Elements*. In fact, the theory of proportion was adopted as the main mathematical tool to express the laws of Nature at the dawn of the scientific revolution. This, in turn, produced several attempts to reform the Eudoxian theory so as to adapt it to the needs of the new science, and in the first half of the seventeenth century many new principles were envisaged in order to fashion theories of ratios which were grounded on very different foundations. The theory gradually waned in the second half of the century, when algebra and analysis took its place as the tools with which Nature might be mathematized.

Given that the theory was originally founded on a system of definitions, many of the further efforts were also aimed at a reformation of the Euclidean definitions, and the list of axioms presented under this heading do not faithfully represent the actual debates on the matter. The first mathematician to substitute axioms for definitions in the theory of proportions was Francesco Maurolico (in 1567), whose work remained unpublished but probably exerted a certain influence on further studies thanks to its manuscript circulation. We know, for instance, that Clavius studied Maurolico's notes (even though this is not apparent in Clavius's own work on the theory of proportions), while we find traces of Maurolico's axiomatization in Giovan Battista Benedetti (1585), who in turn had a strong influence on Galileo and his school. Borelli himself gained access to Maurolico's manuscripts before writing his *Euclides restitutus* (1658). It will be noted, in any case, that most of the axioms in the theory of proportions come from editions produced in the second half of the seventeenth century (i.e., after Galileo's death), when the theory itself had but very little foundational significance as it had been already superseded by more advanced developments in analysis and algebra. These axiomatizations may nonetheless stand as testimony to the fact that the axiomatic approach (here contrasted with the definitional approach) to the foundations of mathematics was, by the end of the seventeenth century, on the point of triumphing. It should also be noted that many among the first propositions of Book V of the *Elements* were employed as axioms; the general idea was that they were immediate consequences of the difficult, or debatable, Eudoxian definitions and (in this case) could perhaps be substituted for them as more perspicuous principles. Thus, even if we may regard some of the new foundations of the theory of proportions as very remarkable logical accomplishments, it is difficult to find any single axiom of the theory which might be considered as a substantial progress as regards formalization. We should, however, mention the axiom of the existence of the fourth proportional (R1, R17, R25), which was implicitly employed by Euclid in proving *Elements* V, 18, and engendered a wide debate on the very notion of mathematical existence and existential assumptions (for instance, Clavius claimed that a fourth proportional to three given magnitudes has to exist, since its existence does not entail any contradiction—thus equating mathematical existence and logical consistency). The missing principle had been already noticed in the Middle Ages, and Khayyām, for instance, had attempted to prove it (see Rashed and Vahabzadeh 1999, pp. 350–351).

Strictly connected with the theory of magnitudes and the theory of ratios is the so-called Axiom of Archimedes, and the notion of Archimedean quantities, which was first formulated by Euclid (or by Eudoxus) as a foundational definition in the theory of ratios (*Elements* V, def. 4), and later also stated, in a different version, as an assumption (ΛΑΜΒΑΝΌΜΕΝΟΝ) of Archimedes's *De sphaera et cylindro*. This principle was introduced among the axioms of the *Elements* by Commandino (axiom AA1) and was re-stated in several other editions. Its immediate application in a geometrical context was the debate on the angle of contact that had classical sources (Proclus, Simplicius) but especially flamed in the Renaissance and the Early Modern Age, when Clavius, Wallis and many others discussed whether a curvilinear angle (which has no finite ratio with a rectilinear one) might still be considered an angle (and an array of related questions). A few decades after Commandino's axiomatization, in any case, the theory of indivisibles was developed in Italy (by Cavalieri, Torricelli and others), and these non-Archimedean quantities gave rise to a further, wider debate on their admissibility in mathematics; still later, similar concerns were raised about the infinitesimals of the new Calculus. The debate itself, then, on the Archimedean Axiom is mainly to be found outside of the domain of elementary mathematics, and many eighteenth-century treatises on analysis deal with the topic. Nevertheless, it continued to contribute to enlarging the number of the axioms in the *Elements* in various forms. Indeed, among the editions of the *Elements*, the only (very partial) axiomatization of a theory of non-Archimedean quantities is to be found in the edition of the *Elements* produced by Bernard Lamy (1685), and perhaps in the unpublished papers by Roberval, whose axioms I18–I21 (that I have arranged in Group 9 following their most immediate meaning) might also be read as a foundation of a theory of non-Archimedean magnitudes.

1.2.4 The principles of arithmetic (Group 7)

As is well known, Euclid included no principles at all (except definitions) in the arithmetical books of the *Elements*. The reasons for his treating arithmetic differently from geometry in this way are hard to envisage precisely. My guess is that the epistemology of Euclid's age neither required nor allowed the use of subject-specific axioms in any form. The notion of this age was that a perfect science ought to be grounded in definitions and general logical laws alone (that may include laws on quantity, i.e., "common notions") without recurring to any further specific principle. Postulates were allowed in the geometrical books to answer a few sophistical objections and were intended to warrant those auxiliary constructions which were needed in the proofs, but since the arithmetical proofs were considered to have no need of any preparatory ΚΑΤΑΣΚΕΥή, and to be susceptible of being carried out in a purely demonstrative way (that is, through a simple ἀΠόΔΕΙΞΙφ), no postulates were added in Book VII (note, however, that a few propositions of this book require, indeed, that the mathematician is allowed to *take*or *construct* certain numbers; from these few instances, the notion of an arithmetical postulate will later arise). In sum, the very possibility of arithmetic-specific axioms was inconceivable to Euclid (or was, at least, considered by him to be simply wrong), and their first appearance became a possibility only in the late Hellenistic age, if only in the form of assumptions that should (and could) be proven through a further analysis of the matter. In any case, it must be stressed once again that the ancient conception of a system of axioms was very different from our own, and that the need for principles in a given domain was not seen, in Antiquity, as a logical necessity (to rule and define such domain), but rather as a shortcoming characteristic of an imperfect science which needed to make assumptions in order to justify its development; from this point of view, it seems clear that no Greek could ever have had the sense that anything was missing from the arithmetical books of the *Elements* simply by reason of the fact that they contained neither axioms nor postulates.

Be it as it may, however, the conspicuous absence of arithmetical principles in the *Elements*had paramount consequences for the later history of mathematics and epistemology. Whereas the axiomatization of geometry was, in fact, discussed for centuries and hundreds of new principles were invented and tested for it, the foundations of arithmetic remained an issue which was almost entirely neglected until the great formal constructions of the nineteenth century (by Dedekind, Frege and Peano, for instance), with geometry remaining during all this preceding period the grounding discipline for the whole of mathematics. We may recall that Kant himself, in the *Critique of Pure Reason*, still attempted (at the very end of the eighteenth century) a philosophical justification of the fact that geometry has axioms, while arithmetic has no proper principles of its own (*KrV*, A163–165/B204–206). These common epistemological views, which in many respects were just the outcomes of a mathematical tradition (possibly consolidated by the loss of the arithmetical axiomatizations of Late Antiquity), were a further cause of the long delay in the discovery and outline of a transparent structure for the theory of numbers.

It should not be thought, however, that attempts at moving toward an axiomatization of arithmetic were completely absent from the Early Modern Age or in the tradition of Euclid's *Elements*. In fact, the first printed volume of the *Elements* (1482) was based on the medieval edition of Campano da Novara (thirteenth century), who had, in his time, depended on several other Arabic sources. Campano's crucially important edition prefaced Book VII of the *Elements*with four postulates and ten common notions bearing on arithmetic, thus constituting a first attempt at an axiomatization of the theory of numbers. When, a few years later, a sounder text of Euclid's *Elements* was translated into Latin by Bartolomeo Zamberti—which translation, of course, being based this time on a Greek original, presented no principles in Book VII—the new humanistic environment of the Renaissance was quick to reckon the interpolated arithmetical axioms as examples of the barbarisms of a dark age; the greater part, therefore, of the subsequent editions of Euclid (up to the age of Kant) simply stuck to philology and ignored these principles. This notwithstanding, Campano's edition, which was so important in the mathematical studies of sixteenth century, proved able to assert itself in foundational research. The important editions of Euclid by Commandino and Clavius followed the Greek text (rather than Campano's Arabized Latin) but found it useful, for foundational purposes, to interpolate Campano's arithmetical axioms into Book VII. Even if their choice was not followed by several other important editions (which eschewed any textual addition), a minority tradition regarding the *Elements* continued to include arithmetical axioms, and these were sometimes discussed, modified, extended or reduced depending upon the particular foundational opinions of the mathematical editor. In this way, a persistent practice of axiomatic discussion on the theory of numbers was established and, although this discussion remained far less developed than that regarding the foundations of geometry, an unbroken line may be observed to run from Campano down to those important contributions of the nineteenth century which, in many respects, grounded contemporary foundational research.

We may well ask, however, what were Campano's own sources. It is easy enough to establish that Campano, who was more a compiler than a translator, took his fourteen principles of Book VII from two different sources: the *Arithmetica* of Jordanus de Nemore (a treatise which is not immediately connected with the Euclidean tradition) and the translation of the *Elements*, from the Arabic, of John of Tynemouth (both thirteenth century).

Jordanus's life, education and culture (even his nationality) are unknown, and so are his sources (on him, see Høyrup 1988). He was, however, undoubtedly a very good mathematician and may have invented his system solely with a view to providing principles for arithmetic. Two of his principles are already to be found in Proclus (whose work Jordanus probably could not have known, since Proclus's commentary to Euclid is believed to have been rediscovered in the West only at the end the fifteenth century, and to have not been know in its entirety to the Arabic world); these, however, are just interpolations from the text of Euclid (see below), and Jordanus may have re-interpolated them as Proclus (or someone else) did before him. A few others of Jordanus's principles have undoubtedly a Greek flavor, but some others are decidedly formulated in medieval terms (one of them, for instance, refers to the medieval notion of *denominatio*). In any case, it is very unlikely that Jordanus could have had access to any Greek text, and we are obliged to hold that, if he had sources at all, they were probably Arabic. But whatever Jordanus's sources were (if any), it seems likely that he had modified and refashioned them so as to better fit his own use of them and his own conceptual instruments; we cannot say for certain whether he himself conceived the idea of introducing axioms into the domain of arithmetic, or whether found it in an older text. It should lastly be noted that Jordanus fashioned a few arithmetical definitions as well that Campano added in his *Elements*.

Tynemouth's sources, on the other hand, are relatively clear: The arithmetical axioms given by him all come from the Arabic commentary on the *Elements* (which was not itself an edition of Euclid) authored by the Persian mathematician and astronomer An-Nayrīzī (tenth century), who had had, in turn, some access to Heron of Alexandria's lost commentary on Euclid. We do not know whether Heron had provided a proper set of axioms for arithmetic: From An-Nayrīzī's wording, it seems rather that he simply took note of a few assumptions which seemed necessary in order to complete this or that proof of Euclid's, and he may even have proven these further assumptions (see Vitrac 1990–2001, vol. 2, pp. 292–296); from Proclus (*In Euclidis* 196), we know that Heron had in fact *reduced* the number of the Euclidean common notions to three, and it would be strange if Proclus did not mention in this connection Heron's addition of several arithmetical axioms, if indeed this had really occurred. It is possible, but unlikely, that an Arabic edition of the *Elements* had already inserted these assumption into the Euclidean text labeling them as axioms of Book VII. In any case, An-Nayrīzī's commentary had already been translated into Latin by Gerardo da Cremona in the twelfth century, and Tynemouth may have been able to interpolate these Heronian assumptions, as proper arithmetical *axioms,* into his edition of the *Elements*—something that had not been done by either An-Nayrīzī, or Gerardo. In this way, Tynemouth inaugurated the modern practice of regarding the Euclidean theory of numbers as an axiomatic domain.

The question of a possible Greek axiomatization of arithmetic is, in any case, relevant indeed. No Greek arithmetical work (such as Diophantus's, for instance) has come down to us which explicitly mentions any principle besides the definitions. Even if, of course, the greater part of the ancient mathematical texts are lost, this may at first appear to be a confirmation that Euclid's practice in Book VII was followed by the entire Greek tradition. There has come down to us, however, a (highly problematic) contrary instance in Proclus. The Neoplatonic philosopher, in his commentary on Book One of Euclid's *Elements*, discussed several opinions regarding the basis of the distinction between axioms and postulates. One of them consisted simply in the contention that axioms (or "common notions") are principles shared by all mathematical sciences, and thus by both geometry and arithmetic in particular (that is to say, they form the principles of what will be later called the *mathesis universalis*). Another way of spelling out the difference between axioms and postulates, however, is that axioms express states of affairs (e.g., "all right angles are equal"), while postulates license constructions ("to draw a straight line between two points"). If the latter distinction is accepted, Proclus said, we may still conceive that there are general principles of a *mathesis universalis*, and these will be, in their turn, distinguished into general axioms and general postulates, and we will also have specific geometrical postulates and axioms, and specific arithmetical postulates and axioms. Proclus (*In Euclidis* 184) gave as examples of these arithmetical principles such propositions as "Every number is measured by the number one" (axiom), and "A number can be divided into least parts" (postulate). This is possibly the only mention of arithmetical principles that is to be found in the whole Greek mathematical corpus.

It is not easy to understand the actual historical relevance of Proclus's remark, or whether it should be considered as a piece of evidence for the existence of a well-developed (but now lost) system of axioms for arithmetic already in ancient times. On a first reading, this seems not to be the case: Proclus's argument here is clearly dictated by a concern for symmetry (if geometry has postulates and axioms, then arithmetic must have its own as well) and may just represent the philosophical hobby-horse of a metaphysician trying to introduce some order into certain disordered mathematical writings which did not fit well with his elegant epistemological design. If this is the case, we should not expect to find that any Greek mathematician ever developed an axiomatic approach to arithmetic. And yet the examples that Proclus gives are grounded in real mathematical needs: His axiom is an assumption (explicitly) taken for granted by Euclid in *Elements* VIII, 6, while his postulate is assumed in the proof of *Elements* VII, 31. They thus appear to enjoy the same status as those other principles which, already in ancient times, were extrapolated from Euclid's demonstrations, where they appeared as unwarranted, and transformed into axioms (for example, the axiom that "two straight lines do not enclose a space", which Proclus himself considered spurious, seems to have been taken from a similar statement appearing in the proof of *Elements* I, 4). It would be odd if Proclus, when inventing two principles exemplifying the existence of arithmetical axioms and postulates, had happened to hit upon exactly these propositions; it seems more likely that he was taking them from an existing edition of the *Elements*, or a commentary thereon, where they had already been inserted as principles of the arithmetical books. It is hard to guess at Proclus's own source (if any) in this respect: It might have been Heron himself, but several other sources, Hellenistic, Roman, or belonging to Late Antiquity, are also possible. Odd as it may seem that we find no further trace of arithmetical principles in any of the extant mathematical works of Antiquity, we do at least find, for the reasons given, a hint in Proclus that an axiomatization of arithmetic (be it in a systematic or unsystematic form) had been attempted already in ancient Greece. We know that Proclus had planned, and perhaps realized, a commentary on the other books of the *Elements*, and it is possible that much more information about these arithmetical principles was offered in these projected or lost commentaries, but we cannot say more (we have no extant commentary dating from antiquity on *Elements* VII, 31 and VIII, 6). But it should be clear, in any case, that in Proclus's times (and possibly already in the age of Heron) mathematical epistemology had already changed, with respect to Euclid, to such an extent as at least to allow the possibility of an arithmetical axiomatization.

As for those principles themselves which were envisaged in the centuries after Euclid, their relevance is very obvious, and even if they do not attain to (nor aim to attain to) the simplicity and effectiveness of Peano's axioms, they nonetheless formalize a vast array of those properties involved in basic operations on numbers, often with a very high degree of logical abstraction. It has already been noted that they should be read together with most of the general principles on the mereology of magnitudes (Groups 1–4).

1.2.5 Principles on space and situation (Group 8)

I have placed under this heading a few very general principles which are more philosophical than mathematical but which nonetheless enjoy an important geometrical significance. Many of them come from Francesco Patrizi's *Della nuova geometria* (1587) and Pascal's *Introduction à la géometrie* (1655). The importance of these two works consists in their claim that geometry should be regarded as the science of space (S1, S7), a contention which represented a rather new epistemological position in the sixteenth and seventeenth Centuries. Since the days of Eudoxus and Aristotle, throughout antiquity and for a long time afterward, the science of geometry had tended to identify its proper and specific object rather as the notion of *magnitude,* or continuous quantity. As a consequence, spatial notions tended not to be represented in classical mathematical works and are not employed in the *Elements*. This situation persisted on into the Middle Ages, and indeed even on into the Renaissance and the Early Modern Age; the axioms of Groups 2–6 bear witness in fact, to the preeminence of the notion of magnitude over any other general concept. The new view that it was the proper and essential task of geometry to investigate space and spatial relations is a view which is first to be found in Patrizi's work (a book on elementary geometry which is otherwise unremarkable from the mathematical point of view); several decades later it resurfaces in Pascal's unpublished introduction to geometry, which had been, in its turn, inspired by the introduction of a few spatial terms into the mathematical treatises on perspective. Pascal's work influenced Leibniz's attempt to establish a new geometry of space, which the latter named *analysis situs*; finally, this Leibnizian project was among the motivating factors behind a proper development of spatial notions in late-eighteenth- and nineteenth-century geometry. For a survey of this important transformation of geometry, see the collection of essays published in De Risi (2015a); on Patrizi's new geometry, see in particular (De Risi 2014).

It should be noted that projective geometry and non-Euclidean geometry are simply inconceivable without the notion of a geometrical space, and several principles grounding modern geometry in the nineteenth century cannot but refer to spatial or positional concepts (including the above-mentioned axioms of order). The editions of the *Elements* which we consider below do not show evidence of these later outcomes, which fell outside of the domain of commentaries on Euclid, and were largely devised at periods situated some years beyond the chronological boundary we have set. Nevertheless, it is important to observe how spatial notions were indeed employed in a number of works on the foundations of geometry well before the nineteenth century. In particular, we may note that the notion of *situation* (Latin: *situs*), meaning the basic spatial relation obtaining between two objects, is the main tool used by Pascal to build up his own axiomatics (Patrizi also employed it in his definitions of geometrical entities). This notion was derived from the Euclidean treatise on *Data*, that is, from one of the very few essays belonging to ancient mathematics which (very briefly) discusses a spatial concept (i.e., *position*, ΘῐΣΙῳ). The latter work had been translated into Arabic during the Middle Ages, and into Latin (from the Arabic sources) by Gerardo da Cremona; another Latin translation, made directly from the Greek, circulated in the Middle Ages, and was later superseded by Zamberti's translation; the first important editions, though, came only with Hardy (1625) and Gregory (in his Euclidean edition from 1703).

We may also note that, in this Group of axioms, a number of principles also state the tridimensionality of space, an axiom which had been never explicitly assumed before the late Renaissance.

1.2.6 Principles of incidence and continuity (Groups 9 and 10)

Dimensional considerations are nonetheless central in the axioms of incidence, since these latter deal with the intersections of various geometrical figures. The axioms in this category are among the most interesting of our survey, and have survived in more intact forms into modern axiomatizations (such as Hilbert's). Their importance lies in their characterizing very basic geometrical entities, such as a straight line and a plane, the geometrical behavior of which had hardly been adequately captured by ancient or early modern definitions; in fact, in the absence of exact definitions of these notions, the axioms here in question formed the true principles employed in the proofs. While the origin of a number of these axioms is to be found in the Greek commentaries, many more were added by Clavius (who achieved important advances in this connection), and by Borelli, who introduced several axioms which were later to be employed by many other editions.

We may note that Clavius's axioms I5 ("*Two straight lines do not have a common segment*") is from time to time also spelled as "*Two straight lines which coincide in a segment, will coincide in their entirety (if extended far enough),*" which is, however, an equivalent formulation of the former.

It should also be noted that, contrary to Hilbert's neater axiomatization, the present category has no clear boundaries, and a few principles classed here might well also have been classed among the principles of congruence: see, for instance I10, I24, or the series of axioms by Lamy regarding perpendicular lines; postulate I4, which is a simple extension of P2 to a higher dimension, is not, properly speaking, a principle of incidence either; nevertheless, it may legitimately be placed here since it expresses the indefinite (or infinite) extension of planes (it may also be regarded as a specification of EX1). The lack of sharp boundaries is especially evident with respect to the group formed by the principles of continuity, some of which can be distinguished only with difficulty from those of incidence; my basic criterion of distinction here was to include all principles which dealt with intersections with bounded figures among those of continuity (the only exception being C11).

Rather than assuming a single all-encompassing continuity principle (as did, for example, Hilbert) early modern axiomatizations attempted to solve the problem of spatial connection by means of a vast array of properly geometrical principles. Several of them were developed in the course of discussing the proof of *Elements* I, 1, in which a number of mathematicians believed they could identify a demonstrative gap, since Euclid did not prove that two drawn circumferences actually meet in two points. Borelli is, once again, a good example of this kind of discussion, which fills some pages of his *Euclides restitutus*; the first mathematician to conceive an axiom to fill this gap, however, had been Richard in 1645. Other principles were conceived for broader purposes, and in fact they may be considered topological axioms in the proper sense. A French line in the evolution of the continuity axioms started with Fine and was further developed by Richard, Pascal and Roberval. Many axioms deal with the intersections between the boundary of a closed figure and a line which has its extremities inside and outside of the latter figure; some of them may be regarded as early modern forms of the so-called Line–Circle continuity principle, which states that an (infinite) line with a point in a circle cuts the circle in two points. This principle is, in fact, equivalent to the Circle–Circle continuity axiom which is tacitly employed in *Elements* I, 1 in order to establish the assumption that the two circles actually meet (cf. C13). The latter principles are both much weaker than Hilbert's full continuity axiom, but are still demonstrably strong enough to ground all theorems of elementary geometry, as far as continuity claims are concerned. A perspicuous and historically sensible proof that Hilbert's axioms of incidence, order and congruence, together with the Parallel Postulate and Line–Circle continuity, are a viable basis for the performance of all the demonstrations of Euclid's *Elements* is given in Hartshorne (2000). I have placed among the principles of continuity also Roberval's axiom C18, which is a fairly good statement of that famous Axiom of Pasch (1882) which was employed by Hilbert to complete his axioms of order. Even if it is stated by Roberval himself to be a proposition about continuity and intersections (and already hinted at by Richard in his C9, a version of the axiom more similar to the so-called "Crossbar Principle"), this may well be the only instance of a deep principle of order making its appearance before the nineteenth century. It should also be noted that Pasch's Axiom is itself equivalent to the plane separation axiom, stating that an (infinite) straight line cuts a plane into two disjoint parts; the axiom was in fact stated in this form by Kästner: see his I38 (which I have listed among the principles of incidence). Note that Kästner also provides a very neat statement of the Line–Circle continuity axiom C20, thus forming a quite remarkable system of principles in this respect.

1.2.7 Principles of congruence and equality (Groups 11 and 12)

These axioms are concerned with the equality of measure of different figures and thus with the notion of congruence.

The axioms of Group 11 are especially concerned with the rigid motion of figures and provide a (very partial) view on those innumerable debates regarding the use and abuse of movement in geometry which had already begun in the Arabic Middle Ages but which found their official inception in the West with Pelletier's criticism (1557) of Euclid's proof of *Elements* I, 4 by superposition. The group also includes a few postulates which directly allow the reproducibility of a given figure in another place, and may therefore be closely related to the axioms on situation (Group 8), as the latter may also concern the different spatial positions of congruent figures (cf. T5). I have listed principle T15 in this group as well, which states that one may perform constructions keeping the opening of the compass—something that was not allowed by Euclid, but does not increase the deductive power of the *Elements*, as it amounts to assume *Elements* I, 2–3 as an axiom (cf. T13, T22, T26). It may be noted in this connection that I have omitted from the list of Euclidean editions a few geometrical works dealing with issues on constructability and geometrical instruments, such as Giovan Battista Benedetti's *Resolutio omnium Euclidis problematum* (Venezia, 1553), Georg Mohr's *Euclides danicus* (Amsterdam, 1672, in both Danish and Dutch; with the addition of the essay *Euclides curiosus*, Amsterdam, 1673), or Lorenzo Mascheroni's *La geometria del compasso* (Pavia, 1797). All these treatises show how to perform classical Euclidean constructions by means of a compass alone (without straightedge), or with a compass with fixed opening. Despite their clear foundational significance, however, none of them lists any axiom, and their approach to the subject is more practical than axiomatic.

The axioms in Group 12 do not employ constructions or displacements but rather establish abstract criteria for the equality of measure. Clearly, the divide between the two groups is rather artificial, and, looked at from an abstract point of view, they be considered to state very similar propositions; nonetheless, their foundational strategies were clearly different in the Early Modern Age, and I have therefore distinguished between them.

1.2.8 The parallel postulate (Group 13)

The last group of principles lists a few alternatives to the Parallel Postulate which were conceived to replace the infamous Euclidean principle (P5). It should be noted that the list is far from complete, as many attempts to demonstrate the Euclidean Postulate were undertaken by making use of different definitions of parallel lines, the most famous of which is that two straight lines are parallel if they are equidistant (a statement itself equivalent to the Postulate, if one assumes the existence of these equidistant lines). Moreover, several other attempts to prove P5 are contained in monographic essays, such as Wallis's *De postulato quinto* (1663), Saccheri's *Euclides vindicatus* (1733) or Lambert's *Theorie der Parallellinien* (1766). Such works are not included in our list of editions of the *Elements*, and often offered still more and different statements equivalent to P5. The present list is thus quite insufficient even to sketch a short history of the attempts to prove the Parallel Postulate. A few considerations on the axioms are nonetheless in order. Axiom PP1 may be regarded as a principle stating the existence of equidistant straight lines, and it shows a good understanding (by Fine) of the fact that the definition alone would be insufficient to this effect unless one could prove (or assume) that equidistant straight lines are possible (something that is equivalent to the acceptance of P5). Axiom PP2 is a restatement of *Elements* I, 30 on the transitivity of parallelism, a property of parallels also depending on P5 and only true in Euclidean geometry. It was first stated by Proclus (*In Euclidis* 371) and later employed many times in the history of the attempts to prove the Postulate. Axiom PP3 excludes asymptotic parallel lines in hyperbolic geometry (which do not share a common perpendicular), thus establishing Euclidean geometry. Axiom PP4, which originates in the Arabic treatises on this topic, establishes the Euclidean character of a (so-called) Saccheri quadrilateral (these quadrilaterals in hyperbolic or elliptical geometry have the base shorter or longer than the opposite side). Axiom PP5 is taken from Clavius's proof of the Parallel Postulate, which assumed this statement to show that equidistant straight lines may be constructed, and thus Euclidean geometry be true, arguing that the present proposition about the flow of a point was a self-evident proposition. This argument was employed already in the Middle Ages by Arabic-writing authors (and Ibn al-Haytham in particular) and exerted a lasting influence on the attempts to prove P5; explicitly assuming it as an equivalent statement represented a significant step forward toward a full understanding of the subject. Axiom PP6 is a simple restatement of the Parallel Postulate itself, based on the fact that it is possible to prove that, if two straight lines cut by a third line form angles lesser than two rights, the lines will *locally* approach one another; it is thus possible to state P5 in the form of the proposition that, in the latter situation, the lines will also meet. The confusion between the local approach of the two lines (in a neighborhood of the intersection with the transversal) with their global approach was a serious hindrance to the proper understanding of the behavior of hyperbolic lines. Axiom PP7 is a restatement of PP4 on Saccheri's quadrilaterals in a different form. Axioms PP8, PP10 and PP11 are further variations on the theme of equidistance. Mercator's Axiom PP9 had some historical relevance, as it may be regarded as the first instance of the so-called direction theory of parallelism stating that parallel lines share the same direction (here: they do not "incline" toward one another) and therefore the property of parallelism has to be transitive (a theory often employed at the end of the eighteenth century and the beginning of the nineteenth to establish Euclidean geometry). Axiom PP12 is a property which had been implicitly employed by the German mathematician Andreas Segner to prove the Parallel Postulate (as it is false in hyperbolic geometry), and which had been explicitly stated as a principle by his disciple Johann Friedrich Lorenz. It may easily be regarded as an alternative statement of PP2 or PP9 on transitivity. "Playfair's Axiom" PP13 was also well known many centuries before Playfair's edition of Euclid, but it had been never stated as such in substitution of P5 in a version of the *Elements*. Given the immediate perspicuousness of the statement, it enjoyed an enormous success in the following axiomatizations.

I have also included in this category two ancillary principles, PP14 and PP15, which are *not*equivalent to the Parallel Postulate. The latter is a restatement of Ptolemy's proof of *Elements* I, 28 (a converse statement of the Parallel Postulate, which is provable without it) based on symmetry, which Richard found to be so general in its application as to deserve to be included among the axioms. The important axiom PP14 is the so-called Principle of Aristotle (since it was vaguely stated in *De Caelo* A 5, 271bb30–32), which Proclus employed in his commentary on the *Elements* (*In Euclidis* 371) to prove the Parallel Postulate, and which was later included by Commandino (1572) in the axiomatic system of the *Elements* with the purpose of replacing the latter postulate by a simpler statement. Proclus's and Commandino's proofs were, however, flawed, and in fact it is impossible to prove P5 from PP14 alone. Clavius later (1589) attempted to prove PP14 but gave an insufficient demonstration of it, and only Saccheri (1733) provided a complete and direct proof of the validity of PP14 in absolute geometry. This notwithstanding, Aristotle's Principle may play an important role in the foundations of geometry, as it may be proven that, if one accepts some very basic statements about geometrical continuity (the Circle–Circle continuity axiom, for instance), the acceptance of Aristotle's Principle is in fact enough to grant the full Axiom of Archimedes (AA1) for geometrical magnitudes; see Greenberg (1988, 2010), and Pambuccian (1994). Lastly, it may be remarked that the Axiom of Archimedes is also directly involved in the equivalence of the above-mentioned principles, which aim to substitute for P5; for instance, PP4 and PP7 on Saccheri's quadrilaterals are only equivalent to P5 if AA1 (or PP14 and Circle–Circle continuity) are also assumed; the latter fact was, however, only discovered by Dehn at the beginning of the twentieth century and was never referred to in early modern treatments of this topic.

In translating the principles into English, I generally followed the particular formulation which was given to them in their first historical occurrence. This may result in some slight deviation in terminology in related principles introduced by different authors (for instance, someone may say that a magnitude *measures* another magnitude, while someone else may say, in a related principle, that a magnitude *divides* another magnitude, or *is a part of* this latter; the meaning is, however, the same). I have not differentiated between principles which only differ from one another as regards the terminology employed but which agree with one another as regards their mathematical meaning.

Thus, for instance, I have sometimes conflated "magnitude" and "quantity" when these two terms were used by different authors to express the same mathematical notion.

These terminological differences may have, in some cases, important epistemological consequences and are very useful for understanding the general philosophical outlook of a mathematician, or the geometrical tradition to which he belongs. For these further and deeper studies, therefore, the original works should always be consulted.

The principles are listed in the various editions in the same order in which they occur, separated by a comma. A dash means that a single principle in the work is in fact composed of two different principles in my list (e.g., Adelard of Bath has a First Postulate stating that a straight line can be drawn between two points (P1) and can be further extended (P2); this postulate is thus mentioned as P1–P2). Sometimes, a number of principles are added in later sections of an edition of Euclid (for instance, since the Parallel Postulate is only used to prove *Elements* I, 29, many editions do not list it at the beginning, but rather in the middle of the First Book, i.e., after *Elements* I, 28), and these are separated from the others by a semicolon. The symbol (...) after the list of principles means that the author explicitly states that the principles are infinite in number, and that he is only giving a few instances of them.

The intended meaning of a few more obscure principles is given in the short commentary to the edition in which they occur for the first time. I have signaled in square brackets those principles which are somehow problematic within the context of the work in which they appear; a fuller explanation is given in the commentary.

2. THE PRINCIPLES

2.1 The principles of Euclid

Let the following be postulated:

P1:To draw a straight line from any point to any point.

P2:To produce a straight line continuously in a straight line.

P3:To describe a circle with any center and distance.

P4:That all right angles are equal to one another.

P5:That, if a straight line falling on two straight lines make the interior angles on the same side less than two right angles, the two straight lines, if produced indefinitely, meet on that side on which are the angles less than the two right angles.

CN1:Things which are equal to the same thing are also equal to one another.

CN2:If equals be added to equals, the wholes are equal.

CN3:If equals be subtracted from equals, the remainders are equal.

CN4:Things which coincide with one another are equal to one another.

CN5:The whole is greater than the part.

2.2 Further principles in the Greek tradition

P6:*Two straight lines do not enclose a space.

CN6:*If equals be added to unequal things, the wholes are unequal.

CN7:*If equals be subtracted from unequal things, the remainders are unequal.

CN8:*The doubles of the same thing are equal.

CN9:*The halves of the same thing are equal.

*EX1:*Magnitudes are susceptible of the unlimited both by way of addition and by way of successive diminution, but in both cases potentially only.

*EX2:*All parts of a plane, and all parts of a straight line, coincide with one another.

*EX3:*A point divides a line, a line a surface, and a surface a solid.

2.3 Additional common notions and the foundations of the theory of magnitudes

*A1:*If two things are equal to another thing, they are equal to one another. *(Adelard of Bath, twelfth century) [cf. CN1]*

*A2:*Any part is lesser than its whole. *(Jordanus de Nemore, thirteenth century) [cf. CN5]*

*A3:*If two equal quantities are compared to a third of the same genus, they will both be equally greater than the third, or equally lesser, or both equal to the third.*(Campano da Novara, thirteenth century)* \

*A4:*The whole is equal to all the parts taken together. *(Bradwardine, fourteenth century)*

*A5:*Things which are different from the same thing are also different from one another. *(Bradwardine, fourteenth century)*

*A6:*The container is greater than the contained. *(Maurolico, 1567) [cf. CN5]*

*A7:*Given any magnitude, it is possible to take a magnitude greater or lesser than it.*(Clavius, 1574)*

*A8:*First is what precedes everything else in its order. *(Patrizi, 1587)*

*A9:*Whole is what has parts. *(Patrizi, 1587)*

*A10:*Every whole is divided into parts. *(Patrizi, 1587)*

*A11:*Every quantum can be divided, i.e., is divisible. *(Patrizi, 1587)*

*A12:*Every divisible thing can be divided into lesser things. *(Patrizi, 1587)*

*A13:*If a thing is greater, or lesser, than another thing, it is also greater, or lesser, than a thing equal to the latter. *(Clavius, 1589)*

*A14:*If a thing is greater, or lesser, than another thing, something equal to the former is also greater, or lesser, than the latter. *(Clavius, 1589)*

*A15:*Things which are equal to equal things are also equal to one another. *(Herigone, 1634) [cf. CN1]*

*A16:*If a magnitude is greater than another magnitude, and the latter is greater than a third magnitude, the first magnitude is also greater than the third. *(Herigone, 1634)*

*A17:*If a magnitude is lesser than another magnitude, and the latter is lesser than a third magnitude, the first magnitude is also lesser than the third. *(Herigone, 1634)*

*A18:*The substitution of equals does not change the equality. *(Herigone, 1634)*

*A19:*The interpretation does not change the equality. *(Herigone, 1634)*

*A20:*The measure is not greater than the measured thing. *(Herigone, 1634)*

*A21:*Every magnitude is such as it is said to be, if it cannot be said otherwise.*(Herigone, 1634)*

*A22:*Parts which are identical with other parts are also identical among themselves.*(Herigone, 1634) [cf. CN1]*

*A23:*If the same quantity, or equal quantities, are divided into a smaller or a larger number of equal parts, any one of the parts taken in larger number will be lesser than any one of the parts taken in smaller number. *(Richard, 1645)*

*A24:*A magnitude which is not greater or lesser than another magnitude is equal to it.*(Schott, 1661) [cf. the intended meaning of A21].*

*A25:*A part of a greater magnitude is equal to a given lesser magnitude. *(Roberval, 1675)*

*A26:*Every magnitude is equal to itself. *(Prestet, 1675) [cf. also A21]*

*A27:*A proposition, taken in one of its meanings, cannot be both true and false. *(Prestet, 1689)*

*A28:*If the legitimate and natural consequence of a proposition is false, the proposition itself is false, and its contradictory proposition is true. *(Prestet, 1689)*

*A29:*A quantity is equal to another quantity when it is neither greater nor less than that quantity. *(Hill, 1726)*

*A30:*A quantity is greater than another quantity when it is neither equal nor less than that quantity. *(Hill, 1726)*

*A31:*A quantity is less than another quantity when it is neither equal nor greater than that quantity. *(Hill, 1726)*

*A32:*If a magnitude is smaller than another magnitude, it is possible to take in the latter one or more parts equal to the former. *(König, 1758) [cf. A25, unpublished at the time]*

2.4 Principles on mereological composition and multiples of magnitudes

M1:If unequal things are added to equals, the wholes will exceed one another as the added unequal things. (An-Nayrīzī, transl. by Gerardo da Cremona, twelfth century) [cf. Pappus in PROCLUS, In Euclidis 197]

M2:If equals are added to unequal things, the wholes will exceed one another as the initial unequal things. (An-Nayrīzī, transl. by Gerardo da Cremona, twelfth century) [cf. Pappus in PROCLUS, In Euclidis 197]

M3:The equimultiples of the same thing, or of equal things, are equal to one another.(Jordanus de Nemore, thirteenth century) [cf. CN8*]

M4:Things the equimultiples of which are equal are also equal to one another.(Jordanus de Nemore, thirteenth century) [cf. CN9*]

M5:If two equals are the halves of two things, the two things are equal. (Lefèvre d'Étaples, 1496)

M6:If unequal things be added to equals, the wholes are unequal. (Lefèvre d'Étaples, 1496)

M7:If unequal things be subtracted from equals, the remainders are unequal. (Lefèvre d'Étaples, 1496)

M8:If two greater things are added to one another, and two lesser things are added to one another, the sum of the greater things will be greater than the sum of the lesser things. (Lefèvre d'Étaples, 1496)

M9:Any magnitude that measures another magnitude also measures any magnitude that is measured by it. (Commandino, 1572) [cf. N3]

M10:If a magnitude measures both another magnitude and a magnitude that is subtracted from it, it will also measure the remainder. (Commandino, 1572) [cf. N1]

M11:Any magnitude that measures two or more magnitudes also measures the magnitude composed by them. (Commandino, 1572) [cf. N2]

M12:If unequal things are subtracted from equals, the differences will exceed one another as the subtracted unequal things. (Clavius, 1574)

M13:If equals are subtracted from unequal things, the differences will exceed one znother as the initial unequal things. (Clavius, 1574)

M14:If a whole is the double of another whole, and the subtracted part is the double of the corresponding subtracted part, the remainder will also be the double of the other remainder. (Clavius, 1574)

M15:If unequal things are added to unequal things, the greater to the greater and the lesser to the lesser, the wholes will be unequal, and in fact the former greater than the latter. (Clavius, 1589)

M16:If unequal things are subtracted from unequal things, the lesser from the greater and the greater from the lesser, the remainders will be unequal, and in fact the former greater than the latter. (Clavius, 1589)

M17:If two equals are the doubles of two things, the two things are equal. (Clavius, 1589) [cf. M4, M5]

M18:If a half is subtracted from a whole, the remainder will be a half; if more than a half is subtracted, the remainder will be less than a half; if one third is subtracted, the remainder will be the two thirds, etc. (Herigone, 1634)

M19:The double of the greater is greater than the double of the lesser. (Herigone, 1634)

M20:If a thing is the double of a thing equal to another thing, it is also the double of the latter thing. (Herigone, 1634)

M21:If a thing equal to another thing is the double of a magnitude, the latter thing will also be the double of that magnitude. (Herigone, 1634)

M22:The half of the greater is greater than the half of the lesser. (Herigone, 1634)

M23:If a thing is the half of a thing equal to another thing, it is also the half of the latter thing. (Herigone, 1634)

M24:If a thing equal to another thing is the half of a magnitude, the latter thing will also be the half of that magnitude. (Herigone, 1634)

M25:If the parts of a whole are equal to one another, the whole will be a multiple of each part as many times as there are parts. (Herigone, 1634)

M26:If each part of a whole is the double of each part on another whole, the former whole will be the double of the latter. (Herigone, 1634) [cf. M14]

M27:Equimultiples of equimultiples are equimultiples to one another. (Herigone, 1634) [Elements V, 3]

M28:A multiple of a multiple of a magnitude is also a multiple of that magnitude.(Richard, 1645)

M29:If a magnitude measures two other magnitudes, it also measures their sum and their difference. (Borelli, 1658) [cf. N2 and N1]

M30:If a magnitude is contained a certain number of times in another magnitude, a magnitude equal to the former will be contained the same number of times in a magnitude equal to the latter. (Borelli, 1658) [cf. N30 and N31]

M31:Equimultiples of two magnitudes to which other equimultiples of the same magnitudes are added or subtracted, are still equimultiples of the two magnitudes.(Arnauld, 1667) [Elements V, 5 and 6]

M32:If a magnitude is subtracted from itself, the result is nothing. (Prestet, 1675)

M33:Many nothings are equal to nothing. (Prestet, 1675)

M34:A multiple of a greater magnitude is greater than the same multiple of a lesser.(Simson, 1754)

M35:That magnitude whose multiple is greater than the same multiple of another is greater than that other magnitude. (Simson, 1754)

M36:To take any multiple of a given magnitude. (König, 1758)

M37:The sum of the equimultiples of the parts of a magnitude are equal to the equimultiple of that magnitude. (Simpson, 1760) [cf. Elements V, 1]

2.5 The theory of ratios and proportions

R1:As a certain quantity is to any other quantity of the same genus, so will any third quantity be to a certain fourth quantity of the same genus. (Campano da Novara, thirteenth century) [existence of the fourth proportional]

R2:Ratios resulting from equal ratios are equal. (Lefèvre d'Étaples, 1496)

R3:Equal magnitudes have the same ratio to the same magnitude. (Maurolico, 1567) [cf. Elements V, 7]

R4:If two magnitudes have the same ratio to a third magnitude, they are equal to one another. (Maurolico, 1567) [cf. Elements V, 9]

R5:A magnitude has the same ratio to two equal magnitudes. (Maurolico, 1567) [cf. Elements V, 7]

R6:If a magnitude has the same ratio to two magnitudes, the latter are equal to one another. (Maurolico, 1567) [cf. Elements V, 9]

R7:If two magnitudes are equal to two other magnitudes, their ratios are equal.(Maurolico, 1567)

R8:A magnitude greater than another magnitude has a greater ratio to a third magnitude; and vice versa. (Maurolico, 1567) [Elements V, 8]

R9:Of magnitudes which have a ratio to the same, that which has a greater ratio is greater; and that to which the same has a greater ratio is less. (Maurolico, 1567) [Elements V, 10]

R10:Two ratios which are equal to a third ratio are equal to one another. (Maurolico, 1567) [Elements V, 11]

R11:If equal ratios are composed with equal ratios, the composed ratios will also be equal. (Maurolico, 1567) [cf. Elements V, 22

R12:If equal ratios are taken away from equal ratios, the remaining ratios will be equal. (Maurolico, 1567)

R13:Equimultiples of two magnitudes are to one another as the magnitudes themselves are to one another. (Maurolico, 1567) [Elements V, 15]

R14:If unequal ratios are added to equal ratios, the composed ratios will be unequal.(Maurolico, 1567)

R15:If unequal ratios are subtracted from equal ratios, the remaining ratios will be unequal. (Maurolico, 1567)

R16:Equal ratios are both equal, lesser, or greater than another ratio. (Maurolico, 1567)

R17:The ratio that holds between two magnitudes will also hold between a given magnitude and some other magnitude. Moreover, the same ratio will also hold between some magnitude and the given magnitude. (Clavius, 1574) [a different version of R1]

R18:The ratio of a magnitude to another magnitude is composed by the ratios of the parts of the first magnitude to the second magnitude. (Benedetti, 1585) [cf. Elements V, 12]

R19:If a first magnitude have to a second the same ratio as a third to a fourth, and the third have to the fourth a greater ratio than a fifth to a sixth, the first will also have to the second a greater ratio than the fifth to the sixth. (Benedetti, 1585) [Elements V, 13]

R20:The multiple of a magnitude is to this magnitude as the equal multiple of another magnitude is to the latter magnitude. (Benedetti, 1585)

R21:The aggregate of all magnitudes in a continued proportion is a multiple of each of the ratios. (Benedetti, 1585)

R22:Any ratio, divided in any way, is composed by the parts in which it is divided.(Benedetti, 1585)

R23:If two magnitudes have a different ratio to a third, the greater among them will have a greater ratio. (Torricelli, 1647)

R24:A ratio is greater than another ratio if the first magnitude of the first ratio is greater than it had to be to have the same ratio to the second as the third to the fourth. (Torricelli, 1647) [from Galileo]

R25:Given three magnitudes A, B and C, a fourth one, Z, can be given, that has with C the same ratio as A to B. (Tacquet, 1654) [a different version of R1 and R17]

R26:Two ratios in proportion may have (at least) three terms: if one of them is taken twice as an antecedent, or as a consequent, or one time as an antecedent and the other time as a consequent. (Borelli, 1658)

R27:If four quantities are such that the first is greater than the second, but the third is not greater than the fourth, the ratio of the first to the second will be greater than the ratio of the third to the fourth. (Borelli, 1658)

R28:Ratios which have a common consequent, are to each other as are the antecedents.(Arnauld, 1667)

R29:Different multiples of a magnitude are to one another as the equal multiples of any other magnitude. (Arnauld, 1667)

R30:Equimultiples of two magnitudes are to one another as different equimultiples of the same magnitudes. (Arnauld, 1667)

R31:If two ratios are equal, the inverse ratios are also equal. (Arnauld, 1667)

R32:Two continuous magnitudes F and G have the same ratio as the ratio F to H to the ratio G to H, with a common consequent H. (Gottignies, 1669)

R33:The double is to the double as the half is to the half, and so for the other multiples and submultiples; and vice versa. (Fabri, 1669) [cf. R13]

R34:If a magnitude has some ratio to another magnitude, the latter also has some ratio to the former. (Fabri, 1669)

R35:If a magnitude is substituted in a ratio with an equal magnitude, the ratio does not change; and vice versa. (Fabri, 1669) [Elements V, 7]

R36:If a magnitude is substituted in a ratio with an equal magnitude, but the other magnitude is changed to a different magnitude, the ratio also changes. (Fabri, 1669)

R37:If a third ratio is greater or lesser than one of two equal ratios, it is also greater or lesser than the other equal ratio. (Viviani, 1674)

R38:If four magnitudes are in proportion, an equimultiple of the first magnitude is to the second as the same equimultiple of the third magnitude is to the fourth.(Viviani, 1674) [cf. Elements V, 4]

R39:If four magnitudes are in proportion, the first magnitude is to an equimultiple of the second as the third magnitude is to the same equimultiple of the fourth.(Viviani, 1674) [cf. Elements V, 4]

R40:If a ratio is lesser than another ratio, which is itself lesser than a third, the first will also be much lesser than the third. (Viviani, 1674)

R41:A magnitude has a greater ratio to a certain magnitude than to a magnitude greater than it. (Mercator, 1678) [Elements V, 8]

R42:If magnitudes be proportional separando, they will also be proportional componendo. (Mercator, 1678) [Elements V, 18]

R43:If four magnitudes are in proportion, any equimultiples of the first and the third are both greater, or both equal, or both lesser than any equimultiples of the second and the fourth. (Rohault, 1682) [cf. Elements V, def. 5]

R44:If four magnitudes are such that any equimultiples of the first and the third are both greater, or both equal, or both lesser than any equimultiples of the second and the fourth, they are proportional. (Rohault, 1682) [Elements V, def. 5]

R45:If four magnitudes are such that for some equimultiples of the first and the third and some equimultiples of the second and the fourth it may happen that the equimultiple of the first is greater than the equimultiple of the second but the equimultiple of the third is equal or lesser than the equimultiple of the fourth, the four magnitudes are not proportional, and the first ratio is greater than the second. (Rohault, 1682) [Elements V, def. 7]

R46:All ratios of a magnitude to a magnitude equal to it are equal to one another.(Arnauld, 1683)

R47:The ratio of a magnitude to another magnitude is equal to the sum of the ratio of a magnitude smaller than the first to the second and the ratio of their difference to the second. (Arnauld, 1683)

R48:The ratio of a magnitude to the part of another magnitude is greater than the ratio of the first magnitude to the whole second magnitude. (Arnauld, 1683)

R49:Ratios which have a common antecedent, are to one another as is the inverse ratio of the consequents. (Arnauld, 1683)

R50:Given four magnitudes in two ratios, three equalities are possible: the equality of the antecedents, the equality of the consequents, and the equality of the ratios. If any two of these equalities are given, the third is also given. (Arnauld, 1683)

R51:Given four ratios in two proportions, the first ratio of the first proportion is to the second ratio of the first proportion as the second ratio of the first proportion is to the second ratio of the second proportion. (Arnauld, 1683)

R52:A proportion does not change if the ratios are exchanged among them. (Arnauld, 1683)

R53:Lines and surfaces the ratio of which is equal to a ratio between numbers are commensurable; otherwise they are incommensurable. (Lamy, 1685)

R54:If a whole is to another whole as a part of the former is to a part of the latter, the remaining part of the former has the same ratio to the remaining part of the latter. (Newton, 1706)

R55:Given four magnitudes, if the first is the same equimultiple, or the same part, or the same parts of the second, as the third is to the fourth, the ratio between the first and the second will be equal to the ratio between the third and the fourth.(Marchetti, 1709) [Elements VII, def. 20]

R56:If equal ratios are added to unequal ratios, the composed ratios will be unequal.(Marchetti, 1709)

R57:If equal ratios are subtracted from unequal ratios, the remaining ratios will be unequal. (Marchetti, 1709)

R58:Ratios that are double, triple,&c., a half, a third,&c. of equal ratios (or of the same ratio) are equal. (Marchetti, 1709)

R59:Given four magnitudes in proportion, the first and the third magnitudes will both be greater, or will both be lesser, or will both be equal, to the second and the fourth. (Marchetti, 1709)

R60:Quantities having the same ratio to another quantity are equal to one another.(Simpson, 1747) [together with R62, a reformulation of R4]

R61:Equal quantities have the same ratio with the same quantity. (Simpson, 1747) [together with R63, a reformulation of R35]

R62:Quantities to which another quantity has the same ratio are equal to one another.(Simpson, 1747) [together with R60, a reformulation of R4]

R63:The same quantity has the same ratio with equal quantities. (Simpson, 1760) [together with R61, a reformulation of R35]

R64:If two magnitudes are divided into equal parts, the first is to the second as the number of equal parts of the first is to the number of equal parts of the second.(Simpson, 1760).

2.6 The Axiom of Archimedes and the theory of indivisibles

*AA1:*Any magnitude may be multiplied as many times as to exceed any given homogeneous magnitude. *(Commandino, 1572)*

*AA2:*Every solid is formed by an infinite number of parallel planes which have an insensible thickness. *(Lamy, 1685)*

*AA3:*If two solids have the same height and the planes that compose them have the same thickness, the two solids are formed by the same number of planes. *(Lamy, 1685)*

*AA4:*If a straight line arbitrarily small, and a straight or curved line arbitrarily long (but finite), are given, it is possible to take the first line a number of times such that it will exceed the second line. *(Roberval, 1675)*

*AA5:*If two or more quantities differ by a quantity less than any assignable quantity, these quantities will be perfectly equal to one another. *(Deidier, 1739)*

*AA6:*It is possible to take a straight line so small that its square is less than any assigned surface. *(Simpson, 1760)*

2.7 Principles of arithmetic and number theory

N1:A number that divides a whole and the part which is subtracted from it also divides the difference. (John of Tynemouth, twelfth century)

N2:Any number that divides two numbers also divides their sum. (John of Tynemouth, twelfth century)

N3:Any number that divides something also divides everything that is divided by it.(John of Tynemouth, twelfth century)

N4:To take any quantity of numbers equal to any given number. (Jordanus de Nemore, thirteenth century)

N5:To take a number however greater than any given number. (Jordanus de Nemore, thirteenth century)

N6:The series of numbers can be extended to infinity. (Jordanus de Nemore, thirteenth century)

N7 That part is lesser which has a greater denomination; and that part is greater which has a lesser denomination. (Jordanus de Nemore, thirteenth century)

N8:The unity is a part of any number, and denominated by it. (Jordanus de Nemore, thirteenth century)

N9:Every number is a multiple of the unity as many times as the unity is part of it.(Jordanus de Nemore, thirteenth century)

N10:Every number multiplied by the unity results in itself. The unity multiplied by any number results in this number. (Jordanus de Nemore, thirteenth century)

N11:The difference of the extremes is composed of the differences of the same extremes to the mean term. (Jordanus de Nemore, thirteenth century)

N12:To take any quantity of numbers equal to, or multiples of, any given number.(Campano da Novara, thirteenth century) [cf. N4]

N13:No number can be diminished to infinity. (Campano da Novara, thirteenth century)

N14:If a number is greater than another, and their difference is added to the lesser, or subtracted from the greater, the resulting numbers will be equal. (Lefèvre d'Étaples, 1496)

N15:Two numbers that have the same ratio to the same number are equal. (Lefèvre d'Étaples, 1496) [cf. Elements V, 9]

N16:A number may be subtracted from another number as many times as the former divides the latter. (Lefèvre d'Étaples, 1496)

N17:The greater number does not divide the lesser. (Lefèvre d'Étaples, 1496)

N18:If a given number is multiplied by another number and then is divided by it, the result is again the given number. If a given number is divided by another number and then is multiplied by it, the result is again the given number. (Lefèvre d'Étaples, 1496)

N19:If a number measures a part [i.e., a submultiple], it also measures the whole.(Candale, 1566)

N20:The sum of a given number with a number that is greater than another one is greater than its sum with the lesser number. (Pedro de Monzón, 1569)

N21:Numbers which are made by the same quantity of unities are equal. (Pedro de Monzón, 1569)

N22:Two number are equal if their parts with equal denomination are in equal quantity. (Pedro de Monzón, 1569)

N23:If two number are multiplied, the product is to the first as the second is to the unity. (Pedro de Monzón, 1569)

N24:If a number divides another number, the divided number is to the divisor as the quotient is to the unity. (Pedro de Monzón, 1569)

N25:If two numbers are added together which are greater than two other numbers, the sum of the greater ones is greater than the sum of the lesser ones. (Pedro de Monzón, 1569)

N26:The ratio of a number with another number is the same as the ratio of the parts of the two numbers with the same denomination. (Pedro de Monzón, 1569)

N27:If a unity is added to an odd number, the result is an even number. (Pedro de Monzón, 1569)

N28:If a unity is added to an even number, the result is an odd number. (Pedro de Monzón, 1569)

N29:If several ratios among numbers are all equal to the same ratio, they will also be equal to one another. (Billingsley, 1570) [cf. Elements V, 11; and R10]

N30:Those numbers which are the same part (or the same parts) of the same number (or of equal numbers) are equal to one another. (Commandino, 1572)

N31:Those numbers, which have the same (or equal) number of the same part (or parts) are equal to one another. (Commandino, 1572)

N32:The unity divides any number. (Commandino, 1572)

N33:Every number divides itself. (Commandino, 1572)

N34:If two numbers are multiplied, the one measures the product by the other and the latter by the former. (Commandino, 1572)

N35:If a number measures another number by a third one, the third also measures the latter by the former. (Commandino, 1572)

N36:If a number measures another number by a third one, the product of the first by the third, or of the third by the first, results in the second. (Commandino, 1572) [cf. N18]

N37:It is possible to make additions, subtractions, multiplications, divisions, extractions of roots, and squares and cubes of numbers. (Herigone, 1634)

N38:Every property of a number is also a property of a number equal to it. (Herigone, 1634)

N39:It is the same to multiply a whole with a whole, or a whole with each of the parts, or each of the parts with a whole. (Arnauld, 1667) [cf. Elements V, 1]

N40:No magnitude has a divisor greater than itself. (Prestet, 1675)

N41:The greatest common divisor of two parts composing a whole is also the greatest common divisor of one part and the whole. (Prestet, 1675)

N42:The greatest common divisor of a whole and one of its two parts is also the greatest common divisor of the whole and the other part. (Prestet, 1675)

N43:If a magnitude with the sign + is joined to the same magnitude with the sign –, the result is equal to nothing. (Lamy, 1685)

N44:Two numbers are always commensurable the one another. (Lamy, 1685)

N45:If a number is subtracted from another number, the result is a number. (Lamy, 1685)

N46:It is the same thing to multiply 12 times 8, or 8 times 12. (Malézieu, 1705) [cf. Elements VII, 16]

N47:It is the same thing to multiply 12 times 8, or 12 times many parts which, taken together, are equal to 8. (Malézieu, 1705) [cf. N39]

N48:If equal quantities are multiplied by equal quantities, their products will be equal.(Malézieu, 1705)

N49:If equal quantities are divided by equal quantities, their quotients will be equal.(Malézieu, 1705)

N50:Given a number, to add a unity to it. (Newton, 1706)

N51:Given a number, to subtract a unity from it. (Newton, 1706)

N52:If the roots of any two quantities are equal, their squares are also equal. (Hill, 1726)

N53:If the squares of any two quantities are equal, their roots are also equal. (Hill, 1726)

N54:If unequal quantities are multiplied by equal quantities, their products will be unequal. (Deidier, 1739)

N55:If unequal quantities are divided by equal quantities, their quotients will be unequal. (Deidier, 1739)

2.8 General principles of space, situation, and givenness

*S1:*Mathematical sciences are not abstracted from natural things; nor are they in the imagination, nor are they in the intellect; but rather, space is their subject.*(Patrizi, 1587)*

*S2:*Geometry takes into consideration the point, the line, the surface, and the body, in this natural order. *(Patrizi, 1587)*

*S3:*Space is extension, and extension is space. *(Patrizi, 1587)*

*S4:*Every space is a minimum, a maximum, or in the middle of these. *(Patrizi, 1587)*

*S5:*Every space is long; or long and wide; or long and wide and deep. *(Patrizi, 1587)*

*S6:*Given a thing, to take in it a point or a straight line. *(Ricci, 1651)*

*S7:*The object of pure geometry is space, which is considered as extended in three dimensions. *(Pascal, 1655) [cf. S1 and S5]*

*S8:*Space is infinite in each dimension. *(Pascal, 1655)*

*S9:*Space is immovable as a whole and its parts are also immovable. *(Pascal, 1655)*

*S10:*Points may only differ in situation. *(Pascal, 1655)*

*S11:*Lines may differ in situation, magnitude and direction. *(Pascal, 1655)*

*S12:*Surfaces may differ in situation, length, breadth, content, and direction. *(Pascal, 1655)*

*S13:*Equal straight lines only differ in situation. *(Pascal, 1655)*

*S14:*Equal circles only differ in situation. *(Pascal, 1655)*

*S15:*Equal arcs of the same circle only differ in situation. *(Pascal, 1655)*

*S16:*Cords of equal arcs of equal circles only differ in situation. *(Pascal, 1655)*

*S17:*If an end of a straight line is given in position, and the straight line is given in magnitude, the other end of the straight line will fall on the circumference of the circle with that given center and radius. *(Arnauld, 1667)*

*S18:*To assume a figure, whose properties are looked for in a theorem, even though it is not constructed. *(Fabri, 1669)*

*S19:*There exists a solid body. *(Roberval, 1675*

*S20:*A straight line is given in position and magnitude when its endpoints are given.*(Kästner, 1758) [Euclid,* Data *26; cf. also I30].*

2.9 Principles of intersection, incidence and connection

I1:A surface cuts a surface in a line. (An-Nayrīzī, transl. by Gerardo da Cremona, twelfth century) [cf. EX3]

I2:If two surfaces which cut one another are planes, they cut one another in a straight line. (An-Nayrīzī, transl. by Gerardo da Cremona, twelfth century)

I3:A line cuts a line in a point. (An-Nayrīzī, transl. by Gerardo da Cremona, twelfth century) [cf. EX3]

I4:To indefinitely extend any plane. (An-Nayrīzī, transl. by Gerardo da Cremona, twelfth century)

I5:Two straight lines do not have a common segment. (Clavius, 1574) [from PROCLUS, In Euclidis 214; cf. I3]

I6:Two straight lines converging into a point will necessarily intersect one another at that point when produced. (Clavius, 1589)

I7:If a point lies on two straight lines, it will be at their intersection or meeting.(Herigone, 1634)

I8:If two points are on the same plane, the straight line joining them will be on the same plane. (Herigone, 1634)

I9:Two concurrent and intersecting straight lines have no common part.(Grienberger, 1636) [another version of I6]

I10:From any point of a circular arc that is not one of its two ends, two straight lines may be drawn to the ends of the arc, forming a triangle with the chord between the ends. (Richard, 1645)

I11:Given two points, to draw a circle having one of the points as its center and passing through the other point. (Ricci, 1651

I12:Through any line, to draw a plane. (Borelli, 1658)

I13:Two planes do not have a common [plane]segment. (Borelli, 1658)

I14:Two planes do not enclose a solid figure. (Borelli, 1658)

I15:The common section of two planes is a line, and if a line lies on two planes it is their common section. (Borelli, 1658) [cf. I2]

I16:Two intersecting straight lines lie on the same plane. (Borelli, 1658) [Elements XI, 2]

I17:Two intersecting straight lines may have at most one point in common. (Arnauld, 1667) [cf. P6*]

I18:A bounded solid body can be divided into two parts in any direction. Taking together all the divisions, it will be divided into an innumerable number of parts.(Roberval, 1675)

I19:The division of a body may be physical, thus disconnecting the parts of the body, or mathematical, thus preserving its continuity. (Roberval, 1675)

I20:Besides the surfaces that bound a solid body, innumerable other surfaces may be conceived of which bound its parts. (Roberval, 1675)

I21:A bounded surface can be divided into an innumerable number of lines, and a bounded line into an innumerable number of points. (Roberval, 1675)

I22:The existence of a bounded solid body implies the existence of surfaces, the existence of bounded surfaces that of lines, the existence of bounded lines that of points. (Roberval, 1675)

I23:However big a bounded solid body is posited, a bigger one may be conceived of that contains the former as a proper part. The included body may be conceived of as separated from the bigger one by some space. (Roberval, 1675)

I24:A solid body revolved around two fixed points has an entire line fixed. (Roberval, 1675) [cf. T2]

I25:If a line is perpendicular to another line, its extension is also perpendicular to it.(Lamy, 1685)

I26:If a line is perpendicular to another line, the latter is also perpendicular to the former. (Lamy, 1685)

I27:If a straight line perpendicular to a plane moves uniformly in a straight line on the plane, its flow will produce a plane perpendicular to the first plane. (Lamy, 1685)

I28:If a straight line is perpendicular to a plane, a plane passing through the straight line is also perpendicular to the first plane. (Lamy, 1685)

I29:The extremities of a line are points. (Lamy, 1685) [Elements I, def. 3]

I30:A straight line is determined by two points only. (Lamy, 1685)

I31:Only one straight line may be drawn between two points. (Lamy, 1685) [cf. I17]

I32:Two straight lines that have two points in common coincide. (Lamy, 1685) [cf. I17]

I33:A straight line cannot be partly on one plane and partly on another plane. (Lamy, 1685) [Elements XI, 1]

I34:Every triangle lies on a plane. (Lamy, 1685) [Elements XI, 2]

I35:If a part of a plane is congruent to a part of another plane, the two planes are congruent to one another. (Marchetti, 1709) [cf. I13]

I36:A straight line [i.e., a segment] may only be bisected in one point. (Rondelli, 1719)

I37:To draw a straight line perpendicular to any given straight line in any given point. (Simpson, 1747) [Elements I, 11 and 12].

I38:An infinite straight line divides the infinite plane in two parts. (Kästner, 1758).

I39:It is possible to extend a plane through two intersecting lines. (Simpson, 1760)

2.10 Principles of continuity

*C1:*If a straight or curved line is drawn from a point which is within a figure to another point in the same plane which is outside the figure, it will intersect the sides or boundary of the figure. *(Fine, 1532)*

*C2:*If a straight line drawn from any angle of a rectilinear figure meets the opposite side or angle of the figure, it also intersects this latter side or angle. *(Fine, 1532)*

*C3:*If a circumference with a given center and a radius of a given length is produced, and the radius is extended beyond the center to infinity, it will intersect the circumference. *(Richard, 1645)*

*C4:*Only two points on a circumference are on one straight line. That is: if a straight line intersects a circle entering in its area and leaving it, it will intersect the circle at two points only, the one when it enters it and the other when it leaves it.*(Richard, 1645)*

*C5:*If a circumference is described which has an end of a given segment as its center and passes through the other end of the segment, it will also pass through the ends of all segments equal to the given one that are produced from the given center.*(Richard, 1645)*

*C6:*If a circumference is described which has its center on the side or boundary of a given figure, and whose radius falls inside the figure (not on its boundary), it will intersect the figure at least once. *(Richard, 1645)*

*C7:*If a circumference is described which has its center outside a given figure, and whose radius falls inside the figure (not on its boundary), it will intersect the figure at least once. *(Richard, 1645)*

*C8:*If a straight line drawn inside a figure is produced to infinity (if needed) in both directions, it will intersect the boundary of the figure. *(Richard, 1645)*

*C9:*If a straight line divides a rectilinear angle and is produced to infinity, it will intersect a given straight line which is applied *[with its ends]* to the straight lines forming the angle. And a straight line drawn from a rectilinear angle to a point inside its base will divide the angle. And a straight line drawn from a point of the base of a triangle to the opposite angle will divide the angle. *(Richard, 1645)*

*C10:*If a point is taken inside a space which is everywhere bounded by lines, and an infinite straight line passes through the point, it will cut the boundary of the space at at least two points. *(Pascal, 1655)*

*C11:*If two points are taken on the different sides of a straight line, a straight line going from one point to the other will intersect the first line at exactly one point. *(Pascal, 1655)*

*C12:*An infinite straight line passing through a point inside a circle intersects the circumference at exactly two points. *(Pascal, 1655)*

C13:A circumference passing through a point inside a circle and a point outside it intersects the circumference of the circle at exactly two points. *(Pascal, 1655)*

*C14:*If two circumferences both have some points inside one another, they will intersect one another at exactly two points. *(Pascal, 1655)*

*C15:*If a circumference has a point on one side of an infinite straight line and the center on the other side, it intersects the straight line at exactly two points. *(Pascal, 1655)*

*C16:*If the same straight line is inside two figures, the two figures will have a common part and will intersect one another. *(Borelli, 1658)*

*C17:*If twoequalcircles have their centers on the ends of a segment taken as their radius, they meet. *(Schott, 1661)*

*C18:*If a point F is taken in the base BD of a triangle BCD and a straight line FG is drawn inside the triangle and extended enough, it will intersect one of the other sides BC or CD at a point G, or both of them at their common point C (and C is then the same point as G). *(Roberval, 1675)*

*C19:*If an infinite straight line passes inside a plane or a solid figure, it will intersect the boundaries of this figure. *(Roberval, 1675)*

*C20:*If a segment has one end inside a circle and one end outside it, it intersects the circumference at one point. *(Kästner, 1758)*

2.11 Principles of congruence, motion, and transportation

T1: Quantities and their properties may be superposed by an intellectual act. *(Richard, 1645)*

*T2:*Quantities may be conceived to move completely or around something fixed.*(Richard, 1645)*

*T3:*Quantities the divisible boundaries of which are congruent are themselves congruent and equal to one another. Equal straight lines are congruent with one another. Equal rectilinear angles are congruent with one another. Equal circles are congruent with one another. [Equal] arcs of equal circles are congruent with one another. Equilateral and equiangular rectilinear figures are congruent with one another, and equal. *(Richard, 1645)*

*T4:*If the ends of two straight lines are congruent with one another, the straight lines are themselves congruent and equal. *(Richard, 1645)*

*T5:*The mere change of position of a triangle or any other plane and rectilinear figure does not change anything in its sides or angles. *(Richard, 1645)*

T6:Given a thing, to reproduce it. (Ricci, 1651)

T7:Given two things, to superpose them. (Ricci, 1651)

T8:If the boundaries of two plane figures are congruent, the two figures are themselves congruent. (Ricci, 1651)

T9:If two plane figures are congruent, their boundaries are congruent as well. (Ricci, 1651)

T10:Equal straight lines are congruent with one another. (Tacquet, 1654) [cf. T3]

T11:Equal circles are congruent with one another. (Tacquet, 1654) [cf. T3]

T12:Equal angles are congruent with one another. (Barrow, 1655) [cf. T3, and E4]

T13:To draw a straight line equal to a given straight line. (Arnauld, 1667) [Elements I, 2]

T14:Things with equal motions traverse equal spaces in equal times. (Fabri, 1669)

T15:To move the compass keeping its opening. (Fabri, 1669)

T16:To bend a straight line through the movement of the sides. (Fabri, 1669)

T17:If a rectilinear angle is congruent with another rectilinear angle and it is reflected without changing its magnitude or width, it will be still congruent with the second angle. (Roberval, 1675)

T18:To put two straight lines directly one to the other. (Scarburgh, 1705)

T19:Any extended entity you please may move with any given law whatever. (Newton, 1693–1706)

T20:A moving body may halt in any given position whatever. (Newton, 1693–1706)

T21:Extended entities may describe the paths marked out through the courses of moving things or by the position of stationary ones. (Newton, 1693–1706)

T22:To draw through two given points a line congruent to a given line at a distance no less than that of the points, an ordained point of which shall coincide with one or other given point. (Newton, 1693–1706)

T23:To draw any line on which there shall always fall a point which is given according to a precise rule by drawing from points through points lines congruent to given ones. (Newton, 1693–1706)

T24:To extend from any given point a straight line equal to a given straight line.(Marchetti, 1709) [Elements I, 2; cf. T13]

T25:To cut from any given straight line a portion equal to a shorter given straight line.(Marchetti, 1709) [Elements I, 3]

T26:To reproduce any straight line or rectilinear angle. (Simpson, 1747) [Elements I, 2–3 and Elements I, 23].

T27:The magnitudes that are congruent with one another are similar and equal, and the magnitudes that are similar and equal are congruent with one another.(König, 1758)

2.12 Principles of the equality and the comparison of measure

E1:The shortest line between two points is the straight line. (Bradwardine, fourteenth century) [ARCHIM. De sphaera et cylindro I, Ass. 1]

E2:If two more sides are drawn on the base of a triangle, they will be greater or lesser than the previous sides. (Vögelin, 1528) [Elements I, 7]

E3:If two triangles have the two sides equal to two sides respectively, and have the angles contained by the equal sides equal, they will also have the base equal to the base, the triangle will be equal to the triangle, and the remaining angles will be equal to the remaining angles respectively. (Pelletier, 1557) [Elements I, 4]

E4:If one of several equal angles is a right angle, all angles will be right angles.(Herigone, 1634) [cf. PROCLUS, In Euclidis 189]

E5:The distance between two points is the straight line. (Pascal, 1655) [cf. ARCHIM. De sphaera et cylindro I, Ass. 1; cf. E1]

E6:The diameter bisects the circle. (Pascal, 1655) [Elements I, def. 18].

E7:All lines drawn from the center to the circumference of a circle are equal to one another. (Borelli, 1658)

E8:Two circles that have equal radii are equal. (Borelli, 1658)

E9:Two equal circles have equal radii. (Borelli, 1658)

E10:If, from the two ends of a straight line, there is drawn an arc of a circle, as well as two straight lines meeting at a point outside of the circle and not intersecting the circle, and a curve entirely outside the circle, the latter curve and the two straight lines will both be longer than the circle, and the circle will be longer that the initial straight line. (Borelli, 1658)

E11:If two lines on the same plane have their ends in common and are both convex toward the same part, the one which is contained by the other is shorter than it.(Arnauld, 1667) [ARCHIM. De sphaera et cylindro, Ass. 2]

E12:The degrees of equal circumferences are equal. (Arnauld, 1667)

E13:Among all straight lines drawn from the center of a circle, those smaller than the radius have their other end inside the circle, those equal to the radius have their other end on the circumference, and those longer than the radius have their other end outside of the circle. (Arnauld, 1667)

E14:If two straight lines intersect, and two points of the cutting line are singularly equidistant from two points of the other line, then all other points in the cutting line will also be equidistant from those two points of the other line. (Arnauld, 1667)

E15:Squares with equal sides are equal to one another. (Arnauld, 1667)

E16:Rectangles with equal bases and heights are equal to one another. (Arnauld, 1667)

E17:Similar segments of circles on equal straight lines are equal to one another.(Arnauld, 1667) [Elements III, 24].

E18:A line between two points that deviates toward one part of the other from the straight line between the same points is longer than the straight line. (Lamy, 1685)

E19:A polygon is larger than the circle inscribed in it. (Lamy, 1685)

E20:A polygon is smaller than the circle in which it is inscribed. (Lamy, 1685)

E21:If one end of a straight line lies on a plane and is conceived as the center of a circle in the plane, the straight line is perpendicular to the plane if and only if the other end is equidistant from the circumference of the circle. (Lamy, 1685)

E22:A solid figure is bigger than the one to which it is circumscribed and smaller than the one to which it is inscribed. (Lamy, 1685)

E23:Among prisms with equal height, the one with the smaller base is smaller. (Lamy, 1685)

E24:In equal circles, arcs of equal chords are equal and chords of equal arcs are equal.(Lamy, 1685) [Elements III, 23 and 24; cf. E17]

E25:The space ABC formed by the straight lines running from a point A to all the points of a line BC is the smallest space bounded by AB, BC, and AC. (Prestet, 1689)

E26: Equal circles have equal circumferences. (Marchetti, 1709)

E27:If two circles have equal circumferences, they are equal. (Marchetti, 1709)

E28:A greater circle has a greater circumference and greater radii. (Marchetti, 1709)

E29:If a circle has a greater circumference or radius than another circle, it is greater than the latter. (Marchetti, 1709)

E30:In a triangle two sides taken together in any manner are greater than the remaining one. (Marchetti, 1709) [Elements I, 20]

E31:If two straight lines forming an angle have the same length as two other straight lines forming an equal angle, the segment joining the ends of the former lines will be equal to the segment joining the ends of the latter lines. (Simpson, 1747) [Elements I, 4; cf. E3].

E32:If two solids of the same height are cut by planes parallel to their bases and the sections at equal distance from the bases are either equal or in constant ratio, the solids themselves are equal or in the same constant ratio. (Simpson, 1747)

E33:Arcs of a circle that contain equal angles are similar to one another. (König, 1758)

E34:Upright prisms with equal bases and heights are equal to one another. (Simpson, 1760) [cf. Elements XI, 29]

E35:The circumference of a circle is greater than the perimeter of any inscribed polygon, and lesser than the perimeter of any circumscribed polygon. (Simpson, 1760) [cf. E19 and E20]

2.13 The parallel postulate

PP1:If two straight lines lying on the same plane and extended to infinity in both directions do not meet, they are equidistant. (Fine, 1532)

PP2:If a straight line meets one parallel, it will also meet the other parallels. (Ricci, 1651)

PP3:Parallel lines have a common perpendicular. (Tacquet, 1654)

PP4:Two perpendiculars cut off equal segments from each of two parallel lines.(Tacquet, 1654)

PP5:If a straight line perpendicular to a given straight line is translated into a plane remaining with one of its ends on the given straight line, the other end of it will draw with its flow another straight line. (Borelli, 1658)

PP6:If two straight lines extended toward one side come closer and closer, they will eventually meet. (Arnauld, 1667)

PP7:If two lines are parallel, all the perpendiculars from one to the other are equal to one another. (Dechales, 1672)

PP8:If two straight lines diverge more and more when they are extended in one direction, they will converge and meet in the other direction. (Roberval, 1675)

PP9:Two straight lines do not incline toward one another if one of them does not incline toward a third, unless the other also inclines toward it. (Mercator, 1678)

PP10:Two non-equidistant straight lines lying on the same plane, if produced far enough, will meet on one side. (Rondelli, 1719) [cf. PP6]

PP11:If two points on a straight line have different distances to another straight line, the two straight lines, infinitely produced, will meet. (Simpson, 1747) [cf. PP10]

PP12:A straight line passing through a point that is inside an angle cuts at least one of the sides of the angle. (Lorenz, 1773)

PP13:Two straight lines cannot be drawn through the same point, parallel to the same straight line, without coinciding with one another. (Playfair, 1795)

PP14:If two straight lines forming an angle are infinitely produced from a point, their distance will exceed any finite magnitude. (Commandino, 1572)

PP15:If one assumes that two straight lines which form internal angles equal to two right angles when intersected by a third straight line meet in that direction, one should also grant that the two lines always meet when they form the same angles with a third line. (Richard, 1645)

2.14 Hilbert's Axioms of geometry

2.14.1 Axioms of incidence

I, 1:For every two points A, B there existsa line a that contains each of the points A, B.

I, 2:For every two points A, B there exists no more than one line that contains each of the points A, B.

I, 3:There exist at least two points on a line. There exist at least three points that do not lie on a line.

I, 4:For any three points A, B, C that do not lie on the same line there exists a planeααthat contains each of the points A, B, C. For every plane there exists a point which it contains.

I, 5:For any three points A, B, C that do not lie on one and the same line there exists no more than one plane that contains each of the three points A, B, C.

I, 6:If two points A, B of a line a lie in a planeααthen every point of a lies in the planeαα.

I, 7:If two planesα,βα,βhave a point A in common then they have at least one more point B in common.

I, 8: There exist at least four points which do not lie in a plane.

2.14.2 Axioms of order (or Betweenness)

II, 1:If a point B lies between a point A and a point C then the points A, B, C are three distinct points of a line, and B then also lies between C and A.

II, 2: For two points A and C, there always exists at least one point B on the line AC such that C lies between A and B.

II, 3:Of any three points on a line there exists no more than one that lies between the other two.

II, 4: Let A, B, C be three points that do not lie on a line and let a be a line in the plane ABC which does not meet any of the points A, B, C. If the line a passes through a point of the segment AB, it also passes through a point of the segment AC, or through a point of the segment BC. (Pasch's Axiom)

2.14.3 Axioms of congruence

III, 1:If A, B are two points on a line a, and A″is a point on the same or another line a″, then it is always possible to find a point B″on a given side of the line a″through A″such that the segment AB is congruent or equal to the segmentA′B′A′B′. In symbolsAB≅A′B′AB≅A′B′.

III, 2:If a segment A′B″B′and a segmentA″B″A″B″, are congruent to the same segment AB, then the segment A′B″B′ is also congruent to the segmentA″B″A″B″, or briefly, if two segments are congruent to a third one they are congruent to each other.

II, 3:On the line a let AB and BC be two segments which except for B have no point in common. Furthermore, on the same or on another line a″let A′B″B′and B′C″C′be two segments which except for B″also have no point in common. In that case, ifAB≅A′B′AB≅A′B′andBC≅B′C′BC≅B′C′, thenAC≅A′C′AC≅A′C′.

III, 4:Let∠∠(h,k) be an angle in a planeααand a″a line in a planeα′α′and let a definite side of a″inα′α′be given. Let h″be a ray on the line a″that emanates from the point O″. Then there exists in the planeα′α′one and only one ray k″such that the angle∠∠(h,k) is congruent or equal to the angle∠∠(h″, k′k′) and at the same time all interior points of the angle∠∠(h″, k′k′) lie on the given side of a″. Symbolically∠∠(h,k)≅∠≅∠(h″, k'). Every angle is congruent to itself, i.e.,∠∠(h,k)≅∠≅∠(h,k) is always true.

2.14.4 Axiom of parallels

IV.1:Let a be any line and A a point not on it. Then there is at most one line in the plane, determined by a and A, that passes through A and does not intersect a. (Playfair's Axiom)

2.14.5 Axioms of continuity

V, 1:If AB and CD are any segments then there exists a number n such that n segments CD constructed contiguously from A, along the ray from A through B, will pass beyond the point B. (Axiom of Archimedes)

V, 2:An extension of a set of points on a line with its order and congruence relations that would preserve the relations existing among the original elements as well as the fundamental properties of line order and congruence that follow from Axioms I–III and from V-1 is impossible. (Axiom of Linear Completeness)

3. THE EDITIONS OF THE *ELEMENTS*

3.1 Greek manuscripts of the *Elements* used by Heiberg

P (Vatican, Early ninth century; partially belonging to a non-Theonine tradition)

αἰτήματα: P1, P2, P3, P4, P5, P6*.

κοιναὶ ἔννοιαι: CN1, CN2, CN3, CN6*, CN7*, CN8*, CN9*, CN4, CN5.

B (Oxford, 888 AD)

αἰτήματά ἐστι πέντε: P1, P2, P3, P4, P5.

κοιναὶ ἔννοιαι: CN1, CN2, CN3, CN6*, CN8*, CN9*, CN4, CN5, P6*.

F (Firenze, tenth century)

αἰτήματά ἐστι πέντε: P1, P2, P3, P4, P5, P6*.

κοιναὶ ἔννοιαι: CN1, CN2, CN3, CN6*, CN8*, CN9*, CN4, CN5.

B (Bologna, eleventh century)

αἰτήματα: P1, P2, P3, P4, P5.

κοιναὶ ἔννοιαι: CN1, CN2, CN3, CN6*, CN8*, CN9*, CN4, CN5, P6*.

V (Wien, twelfth century)

αἰτήματα πέντε: P1, P2, P3, P4, P5, [P6*].

κοιναὶ ἔννοιαι: CN1, CN2, CN3, CN6*, CN7*, CN8*, CN9*, CN4, CN5, P6*.

P (Paris, twelfth century)

αἰτήματα: P1, P2, P3, P4, P5.

κοιναὶ ἔννοιαι: CN1, CN2, CN3, CN6*, CN8*, CN9*, CN4, CN5, P6*.

Q (Paris, twelfth century)

αἰτήματα: P1, P2, P3, P4, P5, P6*.

κοιναὶ ἔννοιαι: CN1, CN2, CN3, CN6*, CN7*, CN8*, CN4, CN5

3.2 Medieval translations in Latin manuscripts

[BOETHIUS], *Geometria*.

Book I

Petitiones, sive Postulata: P1, P2, P3, P4, P5.

Communes animi conceptiones: CN1, CN3, CN2, CN4.

·· Cassiodorus (*Institutiones* II, VI 3) states that Severinus Boethius (early sixth century) translated Euclid into Latin. A few fragments of this translation were preserved in the Middle Ages and merged with other geometrical material (mostly for practical use). However defective, "Boethius's" geometry remained for several centuries the only access to Euclid's *Elements* in the West. Although a complete critical edition of the Boethian fragments is still missing, all the present evidence agrees upon the above system of principles. Boethius's *Geometria altera* (the most important item attributed to him on this topic) was printed in Venice in 1492, and later appended to the edition of the Sacrobosco's *Sphaera* published in Paris in 1500 (with several reprints, ed. Lefèvre d'Étaples), but we will not follow this trend any further. A modern critical edition of the *Geometria altera* is to be found in Folkerts (1970).

ADELARDOF BATH, first half of twelfth century.

Book I

Petitiones quinque: P1–P2, P3, P4, P5, P6*.

Scientia universaliter communis: CN1, CN2, CN3, CN6*, A1, CN9*, CN4, CN5.

·· Translated from a modified version of the Arabic translation of Al-Hajjāj (first half of the ninth century). It is the earliest still-extant translation of Euclid into Latin. Axiom A1 is missing in some manuscripts. Edition in Busard (1983).

HERMANNVON CARINTHIA, first half of twelfth century.

Book I

Quinque modi investigandi: P1–P2, P3, P4, P5, P6*.

Sapientia: CN1, CN2, CN3, CN6*, CN9*, A1, CN4, CN5.

·· Translated from a modified version of the Arabic translation of Al-Hajjāj, but probably independent from Adelard's tradition. Book I only survives in one manuscript and it may have been modified since Hermann's translation. A discussion on Hermann's sources is to be found in Brentjes (2001b). Edition in Busard (1967, 1977).

ROBERTOF CHESTER, first half of twelfth century.

Book I

Peticiones sunt quinque: P1–P2, P3, P4, P5, P6*.

Communes animi concepciones: CN1, CN2, CN3, CN6*, A1, CN9*, CN4, CN5.

·· Formerly attributed to Adelard, the authorship of Robert of Chester is still doubtful, but the text seems to be posterior to Adelard's version. It also depends on Hermann von Carinthia's version. Some manuscripts only have the statements of the propositions without the demonstrations, while others have hints on how carry on the proofs, and still others show the complete text. It was one of the most widespread editions of Euclid in the Middle Ages. The system of principles follows Adelard's version. A few manuscripts add the axiom CN8* before CN9*. Edition in Busard and Folkerts (1992).

GERARDODA CREMONA, *Liber Euclidis philosophi*, mid. twelfth century.

Book I

Ea in quibus necesse est convenire: P1, P2, P3, P4, P5.

Communes animi conceptiones: CN1, CN2, CN3, CN6*, CN7*, CN8*, CN9*, CN4, CN5, P6*.

·· An important edition of the *Elements* translated from the Arabic, which follows a modified version of the Arabic translation made by Isḥāq ibn Ḥunain and revised by Thābit ibn Qurra in the late ninth century (generally considered mathematically more reliable than the version by Al-Hajjāj). Edition in Busard (1984). A comprehensive study of Gerardo's version is to be found in de Young (2004).

GERARDODA CREMONA, [*Anaritii in decem libros priores elementorum Euclidis commentarii*], mid. twelfth century.

Book I

Petitiones: P1, P2, P3, P4, P5, [P6*].

Propositiones per se note: CN1, CN2, CN3, CN6*, CN7*, CN8*, CN4, CN5, P6*, M1, M2, I1, I2, I3, P2–I4.

·· A translation from the Arabic of the commentary to the *Elements* written by An-Nayrīzī (tenth century), who was in turn reading Simplicius's and Heron's lost Greek commentaries to Euclid. An-Nayrīzī reports that Simplicius considered P6* to be an interpolation; An-Nayrīzī also gives a proof of it. The original Arabic version also has CN9* after CN8*. Edition in Curtze (1899), improved in Tummers (1994) for Books I–IV; English translation and commentary in Lo Bello (2003a, b). On the relation between the Arabic original by An-Nayrīzī and Gerardo's translation, see Brentjes (2001a).

ANONYMOUS, second half of twelfth century.

Book I

Unlabeled principles: P1, P2, P3, P4, P5; CN1, CN2, CN3, CN6*, CN8*, CN9*, CN4, CN5, P6*.

·· Possibly the first medieval translation made from the Greek. The final editor is not known, but probably the translation is the work of several scholars working at the court first of Frederick II and then of his son Manfredi in Sicily. Most of the text seems to follow a Greek manuscript close to Heiberg's manuscript **B**, with several discrepancies. The system of principles seems to follow manuscripts **V** or **P**. Critical edition in Busard (1987).

JOHNOF TYNEMOUTH, late twelfth century or beginning of the thirteenth century.

Book I

Petitiones sunt quinque: P1–P2, P3, P4, P5, P6*.

Communes animi conceptiones: CN1, CN2, CN3, CN6*, A1, CN9*, CN4, CN5, A3, R1. (...)

Book VII

Communes animi conceptiones: N1, N2, N3. (...)

·· Translation from the Arabic probably made by John of Tynemouth (Johannes de Tinemue), formerly attributed to Adelard and following the same tradition of the Arabic text. Tynemouth made use of Adelard's system of principles. However, he claimed that the common notions are infinite in number, and gave axioms A3 and R1 in Book I as examples of them. Tynemouth also stated a few arithmetical common notions in Book VII, which form the first basis for an axiomatization of arithmetic in Euclid's *Elements*. They were to be found in An-Nayrīzī's commentary to the *Elements* (see Curtze 1899, p. 191), who in turn attributed them to the commentary of Heron after *Elements* VII, 2 (without explicitly saying whether these assumptions, needed to prove *Elements* VII, 3, were to be considered as axioms). Gerardo da Cremona's translation of An-Nayrīzī mentioned this point, but did not include the latter propositions as axioms of Book VII. The statement about the infinite number of common notions is also to be found in other anonymous manuscripts in the tradition of Robert of Chester. These thirteenth-century manuscripts (that may predate, or not, Tynemouth's version) also added axioms A3 and R1 to Book I, but had no principles in Book VII; they have been edited in Busard (1996). The dating of Tynemouth's version is controversial; for an educated guess that it is to be dated to the thirteenth century, see Knorr (1990). The edition of Tynemouth's text is in Busard (2001).

JORDANUSDE NEMORE, *De elementis arithmetice artis*, first half of the thirteenth century.

Petitiones sunt tres: N4, N5, N6.

Communes animi conceptiones sunt octo: A2, N7, M3, M4, N8, N9, N10, N11.

·· Jordanus's *Arithmetica* was the most important medieval treatise on arithmetic, but it is not an edition of the *Elements*. The material treated in it comes from the arithmetical Books of the *Elements* (in the version of Gerardo), as well as from Boethius's *De institutione arithmetica*(itself a translation of a Neopythagorean tract by Nichomacus of Gerasa) and several Arabic sources. Jordanus's axiomatics, however, seems to be his own creation (as it is not attested to in his known sources), and it influenced the further tradition of the arithmetical Books of the *Elements* (starting with Campano's edition). While Regiomontanus and Maurolico tried in vain to publish Jordanus's *Arithmetica*, it eventually went to print in 1496, edited (and modified) by Lefèvre d'Étaples (see below). Edition in Busard (1991).

CAMPANODA NOVARA, around 1255–1259.

Book I

Petitiones sunt quinque: P1–P2, P3, P4, P5, P6*.

Communes animi conceptiones: CN1, CN2, CN3, CN6*, A1, CN9*, CN4, CN5; A3, R1. (...)

Book VII

Petitiones: N12, N5, N6, N13.

Communes animi conceptiones: A2, M3, M4, N8, N7, N9, N10, N2, N3, N1.

·· This edition of Euclid is not, properly speaking, a new translation but rather a recension of various earlier Latin editions. The main text used in this version seems to be one in the tradition of Robert of Chester, but Campano emended and integrated it, drawing on An-Nayrīzī (via Gerardo da Cremona), John of Tynemouth, and the above-mentioned Latin translation made from the Greek text. Most of the arithmetical principles in Book VII come from Jordanus de Nemore's *Arithmetica*, while the others are drawn from Tynemouth's edition. Campano adds that R1 (the existence of the fourth proportional) is universally valid in continuous quantities, but it may be applied to numbers only when the second quantity is a multiple of the first, since numbers cannot be diminished to infinity (cf. N13). This N13 in Book VII is probably an original addition by Campano, and its main aim seems to be to allow a stop in the decomposition of a number into its factors (cf. *Elements* VII, 31); on

Campano's system of arithmetical principles, see Rommevaux (1999). Printed in 1482 (see below). Edition in Busard (2005).

THOMAS BRADWARDINE, *Geometria speculativa*, around 1320–1335.

Book I

Petitiones ab Euclyde ponuntur quinque: P1–E1, P3, P4, P5, P6*.

Communes scientie multe sunt, sed sufficiant novem: A4, CN5, CN1, A5, CN2, CN3, CN6*, CN7*, CN4. (...)

·· Bradwardine's treatise is a commentary on a few propositions of Euclid's *Elements* and several other mathematical results taken from various sources (including Archimedes). Bradwardine's Latin text generally follows Campano as far as the *Elements* are concerned, but also modified it under several respects. In particular, Bradwardine employed Archimedes's famous axiom (from *De sphaera et cylindro*), stating that the straight line is the shortest line, to ground the geometry of the *Elements*; and possibly rejected postulate P2 on the extendibility of a straight line on metaphysical grounds, i.e., since the universe is finite (he offered a discussion on the topic in his major theological treatise, *De causa dei*). Bradwardine also introduced principle A4 that will be widely employed in further axiomatizations and represented the grounding principle of many early modern (extensional) mereological theories. He also added principle A7, which seems to be false as it is spelled and received no further attention in subsequent treatments of the matter. It should be noted that, in the fourth part of the book, Bradwardine attempted to prove a few assumptions among those mentioned by An-Nayrīzī (in Gerardo da Cremona's translation), such as I1, I2 and I3. Bradwardine had probably written an *Arithmetica speculativa*, even though scholars are still in doubt about the exact version of it that should be ascribed to him (several works of this kind have been attributed to Bradwardine in the past); none of these works, however, contains axioms for the theory of numbers. The *geometria speculativa* was printed in 1495 (see below), and can now be read in Molland (1989). See also the essays by Molland (1978) (on Bradwardine's *geometria*) and Busard (1971, 1998) (on Bradwardine's and Johannes de Mures's *arithmetica*).

3.3 Printed editions

CAMPANODA NOVARA, *Preclarissimus liber elementorum Euclidis perspicacissimi, in artem geometrie incipit quamfoelicissime*, Venezia, Ratdolt, 1482.

Book I

Petitiones sunt quinque: P1–P2, P3, P4, P5, P6*.

Communes animi conceptiones: CN1, CN2, CN3, CN7*, CN6*, A1, CN9*, CN4, CN5, A3, R1. (...)

Book VII

Petitiones: N12, N5, N6, N13.

Communes animi conceptiones: A2, M3, M4, N8, N7, N9, N10, N2, N3, N1.

·· A printed edition of the medieval translation, and the first of all scientific treatises to go into print. The editor added CN7* among the common notions, which is absent in the known manuscript tradition.

THOMAS BRADWARDINE, *Breve compendium artis geometrie*, ed. Pedro Sánchez Ciruelo, Paris, Guy Marchant, 1495.

Book I

Petitiones ab Euclyde ponuntur quinque: P1–E1, P3, P4, P5, P6*.

Communes scientie multe sunt, sed sufficiant novem: A4, CN5, CN1, A5, CN2, CN3, CN6*, CN7*, CN4. (...)

·· A printed edition of Bradwardine's *geometria speculativa*, edited by the mathematician and theologian Ciruelo. The volume was reprinted several times in the following decades and exerted a certain influence on many authors, such as Pacioli, Bovelles, Vögelin and others.

JACQUES LEFÈVRED'ÉTAPLES, *Arithmetica decem libris demonstrata*, Paris, Higman, 1496.

Dignitates: A2, N7, M3, M4, N8, N9, N10, N11, N14, CN1, CN3, M7, M8, CN2, M6, A4, N15, N16, R2, M5.

Petitiones: N4, N5, N6, N13, N17, N18.

·· A printed edition of Jordanus's *Arithmetica*. Lefèvre d'Étaples (Faber Stapulensis), however, added several arithmetical axioms and postulates, which are partly new and partly taken from the medieval Euclidean tradition. The treatise had a big influence on the cossist tradition, and for instance the celebrated *Arithmetica integra* (1544) by Michael Stifel (1487–1567) depended heavily on Lefèvre's version of Jordanus. We will not follow further this line of development, which falls outside of the tradition proper of the *Elements*.

GIORGIO VALLA, *De expetendis et fugiendis rerum opus*, Venezia, Manuzio, 1501.

Book I

Petitiones: P1, P2, P3, [P4], P5.

Axiomata: CN1, CN2, CN3, CN6*, CN7*, CN8*, CN9*, CN4, CN5, P6*.

·· Several books of Valla's posthumous encyclopedic work deal with the content of the *Elements*, which Valla could read in the Greek and of which he translated about a hundred of propositions. He mentioned no principles for arithmetic (since he found none in the Greek manuscripts), but provided a list of principles for geometry (Book X, Chapter 110). Integrating the Greek manuscript of the *Elements* with Proclus's commentary, he stated that P4 can in fact be proven. Note that even if their sources were different, Valla's system of principles is the same as Gerardo da Cremona's (but different from Campano's).

BARTOLOMEO ZAMBERTI, *Euclidis megarensis philosophi platonici Mathematicarum disciplinarum Janitoris: Habent in hoc volumine quicumque ad mathematicam substantiam aspirant: elementorum libros XIII. cum expositione Theonis*, Venezia, Tacuino, 1505.

Book I

Postulata: P1, P2, P3, P4, P5.

Communes sententiae: CN1, CN2, CN3, CN6*, CN7*, CN8*, CN9*, CN4, CN5, P6*.

·· The first printed translation made from the Greek, and thus providing the system of principles of some Greek manuscripts. Zamberti's edition is the heir of Valla's endeavor (see above), and he often reproduced the latter translation. Zamberti's system of principles is in fact the same as Valla's. Zamberti was highly critical regarding the "barbarisms" of Campano's translation from the Arabic and initiated a long-lasting quarrel on the superiority of the Greek manuscript tradition over the Arabic one.

LUCA PACIOLI, *Euclidis megarensis philosophi acutissimi mathematicorumque omnium sine controversia principis opera a Campano interprete fidissimo translata*, Venezia, Paganino, 1509.

Book I

Petitiones: P1–P2, P3, P4, P5, P6*.

Communes animi conceptiones: CN1, CN2, CN3, CN7*, CN6*, A1, CN9*, CN4, CN5, A3, R1. (...)

Book VII

Petitiones: N12, N5, N6, N13.

Communes animi conceptiones: A2, M3, M4, N8, N7, N9, N10, N2, N3, N1.

·· An emended version of Campano (1482), with Pacioli's remarks. It was written in 1485 and possibly printed already in 1486. Pacioli may also have translated the *Elements* into Italian, but this work is now lost.

JACQUES LEFÈVRED'ÉTAPLES, *Euclidis Megarensis Geometricorum elementorum libri XV. Campani Galli transalpini in eosdem commentariorum libri XV. Theonis Alexandrini Bartholomaeo Zamberto Veneto interprete, in tredecim priores, commentariorum libri XIII*, Paris, Henri Estienne, 1516.

·· A celebrated synoptic edition of Campano and Zamberti, with a comparison of their respective systems of principles. It was reprinted several times (see for instance the edition published in Basel, Herwagen 1537).

JOHANNES VÖGELIN, *Elementale geometricum ex Euclidis geometria*, Wien, Singren, 1528.

Book I

Postulata: P4, P5, P6*, E2.

Communes sententiae: CN1, CN2, CN3, CN6*, CN7*, CN8*, CN9*, CN4, CN5, A4.

·· Vögelin's version is not a proper edition of the *Elements* but rather a collection of a few important theorems taken from Books I, III, V and VI. It is remarkable that Vögelin did not include P1, P2 and P3 among the principles, whereas he added *Elements* I, 7 (=E2) as a postulate, since (he claimed) it is difficult to prove and his aim was to write a book for students. Reprinted in 1536 with a preface by Melanchton.

ORONCE FINE, *Protomathesis*, Paris, 1532.

Book I

Postulata, seu Petitiones: P1, P2, P3, P4, PP1, C1, C2. (...)

Axiomata, vel Effata, seu Communes sententiae: CN4, CN1, M3–M4, CN2–CN3, CN6*–CN7*, P6*, CN5–A4.

·· A treatise on several pure and applied mathematical sciences that does not follow the deductive order of the *Elements* (it begins with an essay on arithmetic). Fine allowed no axioms for the theory of numbers, but offered one of the very first reworkings of the geometrical axiomatics. Fine explicitly claimed that PP1 and C1 are needed in the proofs of the *Elements*. The Parallel Postulate (P5) is not included among the postulates (but implicitly employed many times), even though Fine stated that the latter principles are infinite in number, and the ones that he mentions at the beginning of the treatise are just a few examples. See also Fine (1536).

SIMON GRYNAEUS, ΕΥΚΛΕΙΔΟΥ ΣΤΟΙΧΕΙΩΝ, Basel, Herwagen, 1533.

Book I

Αἰτήματα: P1, P2, P3.

Κοιναὶ ἔννοιαι: CN1, CN2, CN3, CN6*, CN7*, CN8*, CN9*, CN4, CN5, P4, P5, P6*.

·· The *editio princeps* of the Greek text, and one of the most influential works in the history of mathematics. It was printed along with Proclus's commentary to Book I. The modified division between postulates and common notions will have a great importance in the following tradition, and it is probably owed to Proclus's opinion on the matter, who attributed it to Geminus (*In Euclidis* 181–184). Such division is already to be found in a Greek manuscripts used by Grynaeus for his edition (the *Marc. gr.* 301), dating from the fifteenth century, while the alleged source of the latter manuscript (following Heiberg's reconstruction of this line of transmission), dating from the fourteenth century, followed the standard arrangement; it may be possible, therefore, that the modified list of axioms and postulates had its origins in the work of the fifteenth-century Byzantine scribe that produced the *Marc. gr.* 301. On the making of Grynaeus's edition, see Oosterhoff (2014).

Oronce Fine, *In sex priores libros geometricorum elementorum Euclidis Megarensis demonstrationes*, Paris, Simon de Colines, 1536.

Book I

Postulata: P1, P2, P3, P4, P5.

Communes sententiae: CN1, CN2, CN3, CN6*, CN7*, CN8*, CN9*, CN4, CN5, P6*.

·· Zamberti's system of principles.

Niccolò Tartaglia, *Euclide Megarense philosopho, solo introduttore delle scientie mathematice*, Venezia, Rossinelli, 1543.

Book I

Petitioni: P1, P2, P3, P4, P5, P6*.

Commune sententie: CN1, CN2, CN3, CN6*, CN7*, CN8*, CN9*, CN4, CN5.

Book VII

Petitioni: N12, N5, N6, N13.

Commune conceptioni dell'animo: A2, M3, M4, N8, N7, N9, N10, N2, N3, N1.

·· The first printed translation of the *Elements* into Italian, although several manuscripts with previous translations are known, mostly dating from the first decades of the sixteenth century; see Pagli (2003). Tartaglia translated from Campano. He acknowledged that Campano also included A3 and R1 among the "*commune sententie*," but claimed that this was a mistake (esp. for R1). He also removed A1 as a useless reduplication of CN1.

PIERREDELA RAMÉE, *Euclides*, Paris, Grandin, 1545.

Book I

Postulata: P1, P2, P3, P4, P5.

Communes sententiae: CN1, CN2, CN3, CN6*, CN7*, CN8*, CN9*, CN4, CN5, P6*.

·· Zamberti's system of principles. It is possible that a previous edition had been published already in 1541. The volume contains no proofs of the theorems, just the statements of the propositions. Ramus was subsequently to publish, besides his celebrated *Scholae mathematicae*, also a treatise on *Arithmetica* (1557) and a *Geometria* (1669), neither of which contains any axiom. Ramus's mathematical works enjoyed many translations in modern languages and several abridgements in the sixteenth and seventeenth Centuries. On Ramus's edition of Euclid, see Loget (2004).

ANGELO CAIANI, *I quindici libri degli Elementi di Euclide, di greco tradotti in lingua thoscana*, Roma, Blado, 1545.

Book I

Domande: P1, P2, P3.

Comuni pareri: CN1, CN2, CN3, CN6*, CN7*, CN8*, CN9*, CN4, CN5, P4, P5, P6*.

·· Grynaeus's system of principles. Like the contemporary edition by La Ramée, Caiani only stated the theorems without proving them.

JOACHIM CAMERARIUS, *Euclidis elementorum geometricorum libri sex, conversi in Latinum sermonem*, Leipzig, Valentin, 1549.

Book I

Petita: P1, P2, P3.

Communes notitiae: CN1, CN2, CN3, CN6*, CN7*, CN8*, CN9*, CN4, CN5, P4, P5, P6*.

·· Grynaeus's system of principles. The volume was edited by Rheticus.

JOHANN SCHEYBL, *Euclidis Megarensis philosophi et mathematici excellentissimi, Sex libri priores de geometricis principijs*, Basel, Herwagen, 1550.

Book I

Postulata: P1, P2, P3.

Communes notitiae: CN1, CN2, CN3, CN6*, CN7*, CN8*, CN9*, CN4, CN5, P4, P5, P6*.

·· Grynaeus's system of principles. The geometrical section of the book is introduced by a treatise on algebra.

ROBERT RECORDE, *The Pathway to Knowledg, containing the First Principles of Geometrie*, London, Wolfe, 1551.

[Book II]

Grauntable requestes: P1, P2, P3, P4, P5, P6*.

Common sentences: CN1, CN2, CN3, CN7*, CN6*, CN8*, CN9*, CN4, CN5, A4.

·· Recorde's book is not an edition of Euclid, but rather a rearrangement of some propositions of the first four books of the *Elements* for practical use, and the first printed translation into English of this material. After giving the definitions and solving a few classical geometrical problems in Book I (written already in 1546), Recorde began Book II with a list of principles taken from Campano and other sources (as in the case of A4); he made no real use of these principles, though, as the *Pathway* does not contain any proof of the geometrical statements (see above the 1545 Euclidean edition by La Ramée, which may have influenced Recorde's work). While this translation was soon superseded by the one by Billingsley and Dee (1570), it exerted a certain influence on further English editions of Euclid (especially in practical mathematics). Recorde had published his English treatise on practical arithmetic, *The Grounde of Artes*, in 1542; the latter, however, contains no principles. On Recorde's life and works, see Roberts and Smith (2012) and Barany (2010); on the British tradition of practical mathematics, see Taylor (1954, 1966).

VALENTIN NABOTH, *Euclidis Megarensis mathematici clarissimi elementorum geometricorum liber primus*, Köln, Birkmann, 1556.

Book I

Petita: P1, P2, P3.

Communes notitiae: CN1, CN2, CN3, CN6*, CN7*, CN8*, CN9*, CN4, CN5, P4, P5, P6*.

·· Grynaeus's system of principles. The volume contains a selection of theorems and problems from Book I of the *Elements*, without proofs, and a few other propositions chosen from Books II, III and VI. Naboth (aka Nabod, or Nabodus) was a famous astrologer.

JACQUES PELETIERDU MANS, *In Euclidis Elementa Geometrica Demonstrationum Libri sex*, Lyon, Tornes, 1557.

Book I

Postulata: P1, P2, P3, P4, [P5].

Animi notiones: CN1, CN2, CN3, CN6*, CN7*, CN8*, CN9*, CN4, CN5, [P6*]; E3. (...)

·· Zamberti's system of principles as a starting point (but the main text is based on Campano). Peletier claimed that P5 is in fact a kind of definition of non-parallel (converging) lines and should not be placed among the postulates. He also stated that P6* should be ranged among the *postulata* rather than the *animi notiones*. Finally, he famously claimed that Euclid's proof of *Elements* I, 4 (=E3) should be rejected, as it employs motion and superposition, and proposed to count the statement among the innumerable implicit principles of geometry.

[JEAN MAGNIEN, PIERREDE MONTDORÉ, ST. GRACILIS], *Euclidis Elementorum libri XV, Graecè et Latinè*, Paris, Cavellat, 1557.

Book I

Postulata: P1, P2, P3.

Communes notiones: CN1, CN2, CN3, CN6*, CN7*, CN8*, CN9*, CN4, CN5, P4, P5, P6*.

·· Grynaeus's system of principles. A preface by a certain St. Gracilis explains that this edition was first conceived and prepared by Jean Magnien, and later completed, after the latter's untimely death, by Gracilis himself. Book X of the *Elements* follows the edition of it given by Pierre de Montdoré (*Euclidis elementorum liber decimus, Petro Montaureo interprete*, Paris, Vascosan 1551). The edition enjoyed several reprints and was especially used in Jesuit colleges. Like La Ramée (1545) and Caiani (1545), it gave no proofs of the propositions.

XYLANDER (WILHELM HOLTZMANN), *Die Sechs ersten Bücher Euclidis, vom Anfang oder Grund der Geometrj*, Basel, Kündig, 1562.

Book I

Petitiones, seu Postulata: P1, P2, P3.

Communes notitiae: CN1, CN2, CN3, CN6*, CN7*, CN8*, CN9*, CN4, CN5, P4, P5, P6*.

·· Grynaeus's system of principles. The first printed translation of the *Elements* into German. Before Xylander, at least one other important mathematical volume had been published in German, *Das erst Buch der Geometria*, written by Wolfgang Schmidt and printed in Nürnberg in 1539.

The latter book, however, did not follow the Euclidean order of the propositions, and only dealt with a few geometrical problems (rather than theorems); it included no axioms whatsoever.

CONRAD DASYPODIUS, *Euclidis quindecim Elementorum Geometriae primum*, Strasbourg, Mylius, 1564.

Book I

Postulata: P1, P2, P3.

Communes notiones: CN1, CN2, CN3, CN6*, CN7*, CN8*, CN9*, CN4, CN5, P4, P5, P6*.

·· Grynaeus's system of principles. Dasypodius (Konrad Rauchfuss) edited several books of the *Elements*, in both Greek and Latin, between 1564 and 1579, without altering the list of principles. See also the important Herlin & Dasypodius (1566).

PIERRE FORCADEL, *Les six premier livres des Elements d'Euclide traduicts et commentez par Pierre Forcadel de Bezies*, Paris, Marnef, 1564.

Book I

Demandes: P1, P2, P3.

Communes sentences: CN1, CN2, CN3, CN6*, CN7*, CN8*, CN9*, CN4, CN5, P4, P5, P6*.

·· Grynaeus's system of principles. The first printed translation of the *Elements* into French. The following year, Forcadel also published his edition of the arithmetical books of the *Elements*. The latter edition has no axioms, but in discussing the proof of *Elements* VII, 1, Forcadel claimed the necessity to assume N17, N2, N1.

CHRISTIAN HERLIN, CONRAD DASYPODIUS, *Analyseis geometricae sex librorum Euclidis*, Strasbourg, Rihel, 1566.

Book I

Postulata: P1, P2, P3.

Communes notiones: CN1, CN2, CN3, CN6*, CN7*, CN8*, CN9*, CN4, CN5, [P4], [P5], P6*.

·· Grynaeus's system of principles. A famous and influential edition that reshaped into syllogisms the proofs of Euclid. Books I and V were written by Herlin, while his pupil Dasypodius, who had already published several partial editions of the *Elements*, completed the volume. In the appended *Commentaria*, Dasypodius stated that P4 and P5 are provable and should not be counted among the *communes notiones*.

FRANÇOISDE FOIX-CANDALE, *Euclidis Megarensis mathematici clarissimi elementa geometrica libri XV ad germanam geometriae intelligentiam è diversis lapsibus temporis iniuria contractis restituta*, Paris, Royer, 1566.

Book I

Postulata: P1, P2, P3.

Communes sententiae: CN1, CN2, CN3, CN6*, CN7*, CN8*, CN9*, CN4, CN5, P4, P5, P6*.

Book VII

Communes sententiae: N19, N1.

·· Grynaeus's system of principles in Book I. Candale, however, stated that CN8* and CN9* should be interpreted as referring to any multiple and submultiple (cf. M3 and M4, in Jordanus and Campano); a similar remark is found in PROCLUS, *In Euclidem* 196–197. Candale also thought it necessary to add a few arithmetical principles, but found Campano's list of them too long and reduced it to two axioms only.

FRANCESCO MAUROLICO, *Euclidis elementorum compendia*, 1567 (unpublished).

Book I

Postulata: P1–P2, P3, P4, P5, P6*.

Communes sententiae: CN1, CN2–CN3–CN6*–CN7*, CN8*–CN9*, CN4, CN5–A6.

Book V

Communes animi conceptiones: R3, R4, R5, R6, R7, R8, R9, R10, R11, R12, R13, R14, R15, R16.

·· Maurolico was highly dissatisfied with the editions of Euclid printed in the Renaissance, and from the 1530s on he raised several criticisms of both the Campano and Zamberti traditions. Between 1532 and 1534, he prepared new versions of Books II, V, VII, VIII, IX and X of the *Elements* which treated the Euclidean material with great freedom, and never went to print. In the same years, he also reworked the stereometric books of the *Elements* which were later published among Maurolico's *Opuscula mathematica* (1575), and he wrote the *Arithmeticorum libri duo* (also published in 1575). None of these works presents any principle besides the definitions. In 1567, however, Maurolico prepared the abridged version of a number of Euclidean books, the *Euclidis elementorum compendia*, with the aim of merging together Zamberti's and Campano's versions (something that Lefèvre had only partially done in 1516) and adding a few new proofs and results. The latter edition, which also remained unpublished but circulated in manuscript form, has a set of axioms in Book I and V. The new axiomatization of the theory of proportions, in particular, represented an important starting point for further studies on the topic. Note that axioms R11 and R12 on the composition of ratios are completely new, and treat multiplications and divisions among ratios. The formal statement of A10 is that if A:B=D:EA:B=D:E and B:C=E:FB:C=E:F, then A:C=D:FA:C=D:F (composition ΔΙ'ΪΣΟΥ, or *ex aequali*). The axiom will be employed again by Benedetti, Viviani, Marchetti, and others. The edition of Maurolico's text is in Garibaldi (2002). On Maurolico's axiomatization of Book V, see Sutto (2000).

PEDRO JUANDE LASTANOSA (DE MONZÓN), *Elementa Arithmeticae ac Geometriae*, Valencia, Huete, 1569.

Dignitates arithmeticae: CN1, N7, N8, N9, A4, N20, N21, N22, CN2, CN3, CN6*, CN7*, N10, N14, N23, N24, N15, N25, N26, N16.

Petitiones seu postulata [*arithmeticae*]: N6, N13, N27, N28, N4, N17.

Axiomata, seu dignitates [*geometriae*]: CN1, CN2, CN3, CN6*, CN7*, CN8*, CN9*, CN4, CN5–A4.

Postulata [*geometriae*]: P1, P2, P3, P4, P5, C1, P6*. (...)

·· The book does not follow the Euclidean order and only covers a few elementary results in arithmetic and geometry. Monzón gave, however, two complete systems of principles for the two sections of his book, mostly drawing on Lefèvre d'Étaples (1496) and Fine (1532), but with a few additions of his own.

HENRY BILLINGSLEY, JOHN DEE, *The elements of geometrie of the most ancient Philosopher Euclide of Megara, faithfully (now first) translated into the English tongue*, London, Daye, 1570.

Book I

Peticions, or requestes: P1, P2, P3, P4, P5, P6*.

Common sentences: CN1, CN2, CN3, CN6*, CN7*, CN8*, CN9*, CN4, CN5.

Book VII

Common sentences: N7, M3, M4, N1, N3, N2, N29.

·· The first complete translation of the *Elements* into English, with a Preface by John Dee. Billingsley adopted Campano's system of principles in Book I, as emended by Tartaglia, and a modification of Campano's axiomatics in Book VII. He added principle N29 of his own, stating the transitivity of the equality of ratios; it is an arithmetical version of *Elements* V, 11. The edition set the standards for further translations of

Euclid into English. Thomas Rudd's edition of the *Elements* (London, Leybourn 1651), for instance, followed quite closely Billingsley's edition, and endorsed the same system of principles. On the Euclidean tradition in Britain, see Barrow-Green (2006).

FEDERICO COMMANDINO, *Euclidis elementorum libri XV una cum Scholiis antiquis*, Pesaro, Franceschini, 1572.

Italian version as *De gli elementi d'Euclide libri quindici con gli scolii antichi*, Urbino, Frisolino, 1575.

Book I

Postulata (*postulati, overo dimande*): P1, P2, P3, P4, [P5].

Axiomata, seu communes notiones (*assiomi, overo communi notitie*): CN1, CN2, CN3, CN6*, CN7*, [CN8*], CN9*, CN4, CN5, P6*; PP14.

Book V

Notiones communes (*communi notitie*): M3, M4.

Book VII

Petitiones (*petitioni, overo dimande*): N12, N5, N6–N13.

Notiones communes (*communi notitie*): M3, M4, N30, N31, N8, N32, N33, N35, N36, N34, N2, N3, N1.

Book X

Notiones communes (*communi notitie*): AA1, M9, M10, M11.

·· A crucially important edition based on the Greek text by Grynaeus (1533) but double-checked against other, more reliable Greek manuscripts. Commandino remarked that P6* is not self-evident enough to be ranged among the axioms, and for this reason Campano had put it among the *petitiones*. He stated that P5 should be proven and removed from the list of postulates, and gave Proclus's proof of it, relying on the new assumption PP14 (Latin ed. p. 19v; Italian ed. p. 21v). Moreover, Commandino noted that CN8* can be deduced by reiterated applications of CN2, and is thus redundant as a principle (cf. PROCLUS, *In Euclidis* 196–197). This notwithstanding, Commandino added axioms M3 (a generalization of CN8*) and M4 in Book V, taking them from Campano's edition of Book VII; he also employed the same two principles in Book VII, just as Campano did. While in the Latin edition these axioms refer generally to "things" that are multiplied, in the Italian edition of the *Elements* Commandino stated a modified version of axioms M3 and M4 in Book V, which only applies to continuous magnitudes, and another version of the same axioms in Book VII, which only applies to numbers. In Book X, Commandino added a version of Archimedes's Axiom (AA1) and three new axioms (M9, M10, M11) which are modifications of the arithmetical axioms N3, N1 and N2.

CHRISTOPH CLAVIUS, *Euclidis elementorum libri XV*, Roma, Vincenzo Accolto, 1574.

Book I

Petitiones, sive Postulata: P1, P2, P3, A7.

Communes notiones, sive Axiomata, sive Pronunciata, sive Dignitates: CN1, CN2, CN3, CN6*, CN7*, CN8*, CN9*, CN4, CN5, P4, [P5], P6*, I5, M1, M2, M12, M13, A4, M14.

Book V

Axioma: R17.

Book VII

Postulata, sive petitiones: N12, N5, [N6–N13].

Axiomata, sive pronuntiata: M3, M4, N30, N31, N32, N33, N34, N35, N36, N2, N3, N1.

Book X

Postulatum, sive petitio: AA1.

Axiomata, sive pronunciata: M11, M9, M10.

·· Another very important edition, which remained, for two centuries, the basis of many further axiomatizations of the *Elements*. Clavius claimed that all the principles can be proven from the definitions. Among them, P5 is complex enough to have to be removed from the list of axioms and to be considered a theorem. Clavius stated that M1 and M2 are taken from Pappus, quoted in PROCLUS, *In Euclidem* 197 (but they were already known to Campano through Gerardo da Cremona's translation of An-Nayrīzī's commentary). Axiom R17 (the existence of the fourth proportional) is added at the beginning of Book V. Clavius mostly followed Commandino's system of principles in Books VII and X. Principle N6–N13 is not listed among the arithmetical postulates, but added in a following remark. Principle AA1 was an axiom in Commandino and became a postulate in Clavius. See below the different list of geometrical principles in the 1589 edition of Clavius's masterwork.

RODRIGO ZAMORANO, *Los seis libros primeros dela geometria de Euclides*, Sevilla, La Barrera, 1576.

Book I

Peticiones: P1, P2, P3, P4, P5.

Comunes sentencias: CN1, CN2, CN3, CN6*, CN7*, CN8*, CN9*, CN4, CN5, P6*.

·· Zamberti's system of principles. The first translation of the *Elements* into Spanish.

GIOVAN BATTISTA BENEDETTI, *In quintum Euclidis librum*, in *Diversarum speculationum mathematicarum, & physicarum Liber*, Torino, Belvilacqua, 1585.

Book V

Postulata: R11–R12, CN3–CN2, R3–R5, R4–R6, R8, R9, R10, R18, R19, R20, R21, R22.

·· A reworking of Book V of the *Elements* with a new system of principles, which exerted a certain influence on seventeenth-century axiomatizations of the theory of proportions. Benedetti's first and second axioms put together the standard Euclidean common notions on the composition of magnitudes (their sum and difference), with Maurolico's new axioms on the composition of ratios (their multiplication and division), with the clear aim of justifying the latter by the reference to the former. Benedetti's following axioms are simply the Euclidean propositions of *Elements* V, 7–13. They are followed by R20 that makes explicit the connection between ratios among magnitudes and ratios among numbers, and by two more axioms that seem to define continuous proportion. On Benedetti's theory, see Bodriga (1926) and Giusti (1993).

FRANCESCO PATRIZI, *Della nuova geometria*, Ferrara, Baldini, 1587.

[Book I]

Supposizioni: S1, S2, S3, S4, S5.

Assiomi: A8, A9, A10, CN5, A11, A12.

·· This book is not, properly speaking, an edition of the *Elements*, but rather an independent logical treatment of the first Euclidean results regarding straight lines, angles and triangles (covering more or less the Euclidean material up to *Elements* I, 29). The book is important inasmuch as it is the first treatise to discuss space as the object of geometry, as well as for developing a quasi-formal treatment of this field (cf. Herlin & Dasypodius, 1566). Patrizi's geometrical "suppositions" are mostly taken from his own essays on the metaphysics of space (*De rerum natura libri II priores. Alter de spacio physico, alter de spacio mathematico*, Ferrara, Baldini 1587), while his "axioms" are modified Euclidean common notions.

CHRISTOPH CLAVIUS, *Euclidis elementorum libri XV*, Roma, Bartolomeo Grassio, 1589215892.

Book I

Petitiones, sive Postulata: P1, P2, P3, A7.

Communes notiones, sive Axiomata, sive Pronunciata, sive Dignitates: CN1–A13–A14, CN2, CN3, CN6*–M15, CN7*–M16, CN8*–M17, CN9*–M5, CN4, CN5, I5, I6, P4, [P5], P6*, M1, M2, M12, M13, A4, M14.

Book V

Axioma: R17.

Book VII

Postulata, sive petitiones: N12, N5, [N6–N13].

Axiomata, sive pronuntiata: M3, M4, N30, N31, N32, N33, N34, N35, N36, N2, N3, N1.

Book X

Postulatum, sive petitio: AA1.

Axiomata, sive pronunciata: M11, M9, M10.

·· With respect to the 1574 edition, Clavius's second edition added, in Book I, a number of common notions on the comparison of magnitudes, as well as the new axiom I6, which Clavius employed in the proof of P5. The main difference between the two editions is given, in fact, by the lengthy discussion of the Parallel Postulate, that was engendered by Clavius's new acquaintance with the work of Nasīr ad-Dīn (see below). The following editions of Clavius's commentary (the last one being in Clavius's *Opera mathematica* from 1612) made no further changes in the system of principles.

NASĪRAD-DĪNAT-TŪSĪ, *Euclidis elementorum geometricorum libri tredecim ex traditione doctissimi Nasiridini Tusini*, Roma, Typographia Medicea, 1594.

Book I

al-usul al-mawdu'a: P6*, P2, P4, P5.

al-'ulum al-muta'arifa: CN1, CN2, CN3, CN6*–CN7*, CN8*, CN9*, CN4, CN5.

·· The Persian mathematician and scientist Nasīr ad-Dīn at-Tūsī (1201–1274) wrote a commentary on the *Elements* (largely inspired by Ibn al-Haytham's work), as well as an important treatise on the proof of the Parallel Postulate. After his death a disciple of his compiled, in 1298, a further commentary on Euclid that drew on Nasīr ad-Dīn's work and was later attributed to Nasīr ad-Dīn himself. Many centuries later, three manuscripts containing the two commentaries (the one by Nasīr ad-Dīn, and the other by his disciple, later called the pseudo-Tūsī) came into possession of the Medici family, along with many other Arabic works. A publishing house, the Oriental Press or Typographia Medicea, was then established in Rome under the direction of Giovan Battista Raimondi (1536–1614) with the aim of publishing several of these works. The commentary of the pseudo-Tūsī (still attributed to Nasīr ad-Dīn) was published in Arabic (with a Latin title) in 1594. This famous Arabic edition could not exert much influence in Europe: yet Nasīr ad-Dīn's proof of the Parallel Postulate was read (still in manuscript form) by Clavius, engendering the 1589 edition of his commentary (see above), and was later translated into Latin by Edward Pocock (1661) and published as an appendix to Wallis's *Algebra* (1693). This was the only fragment of the pseudo-Tūsī's commentary that enjoyed some diffusion in a Western language. His more general system of principles in Book I, apparently, was not able to affect any further study on the topic. While the list of axioms of this edition is quite standard, the geometrical postulates are widely discussed and connected with metaphysical considerations; P1 and P3 on the construction of lines and circles are expunged from the list of proper principles. On Nasīr ad-Dīn's commentary on Euclid, its sources, and its relation with the pseudo-Tūsī's edition, see de Young (2003, 2009). A complete discussion of the Medici edition is to be found in Cassinet (1986).

JAN PIETERSZ DOU, *De ses eerste Boecken Euclidis*, Leiden, Bouwensz, 1606.

Book I

Eenige stucken tot dese konst behoorende: P1, P2, P3.

Ghemeene bekantenissen: CN1, CN2, CN3, CN6*, CN7*, CN8*, CN9*, CN4, CN5, P4, P5, P6*.

·· Grynaeus's system of principles. The first translation of the *Elements* into Dutch. Dou's book follows the German version by Xylander (1562) with a few modifications. The same system of principles was to be used by the following Dutch edition of the *Elements* edited by Frans van Schooten, Sr. (*De propositien vande XV. Boucken der Elementen Euclidis*, Leyden, Basson, 1617).

PIETRO ANTONIO CATALDI, *I primi sei libri de gl'Elementi d'Euclide ridotti alla Prattica*, Bologna, Cochi, 1613.

Book I

Petitioni, overo domande: P1, P2, P3, [P4], [P5], P6*.

Comuni concessioni, o comuni notitie: CN1, CN2, CN3, CN6*, CN7*, [CN8*], CN9*, CN4, CN5–A4.

Book V

Petitione: M14.

Comuni concessioni: CN1, CN2.

·· Cataldi claimed that P4, P5 and CN8* (but probably also CN9*) can be proven and should not be counted among the principles.

DENIS HENRION, *Les quinze livres des elements d'Euclide traduicts de Latin en François*, Paris, Joallin, 1615.

Book I

Petitions, ou demandes: P1, P2, P3.

Communes sentences: CN1, CN2, CN3, CN6*, CN7*, CN8*, CN9*, CN4, CN5, P4, P5, P6*.

Book VII

Communes sentences: M3, M4, N30, N31, N8, N32, N34, N35, N36, N2, N3, N1.

Book X

Demande: AA1.

Communes sentences: M11, M9, M10.

·· Grynaeus's system of principles in Book I, and a simplification of Clavius's (and Commandino's) principles in the other books. The volume enjoyed great success and was reprinted several times.

PIERRE HERIGONE, *Cursus mathematicus. Cours mathematique*, Paris, Le Gras, 1634.

Book I

Petitiones, sive Postulata (*Petitions, ou Demandes*): P1, P2, P3, A7.

Communes notiones, sive Axiomata, sive Pronunciata, sive Dignitates (*Communes notions, ou communes sentences, ou maximes*): CN1, A15, A13, A14, A16–A17, A18, A19, CN2, CN3, M18, CN6*, M6, M15, CN7*, M7, M16, CN8*, M19, M20, M21, CN9*, M22, M23, M24, CN4, CN5, A20, I5, I6, P4, E4, P5, P6*, I7, I8, M1, M2, M12, M13, A4, M25, M14, M26, A21.

Book V

Axioma: M27.

Book VII

Postulata: N12, N5, N37.

Axiomata: N38, A22, N30, N31, N32, N33, N34, N35, N36, N2, N3, N1.

Book X

Postulatum: AA1.

Axiomata: M11, M9, M10.

·· A complete reworking of the system of principles of the *Elements*, Herigone's work was a milestone in the development of Euclidean axiomatics, and several of its principles were employed many times in the following years. The intended interpretation of a few of the added principles is the following. Axiom A15 should guarantee the transitivity of equality, and refers to things which are equal (not just "the same things"). Axiom A18 is a principle of substitutivity *salva aequalitate*: if A+B=C+DA+B=C+D and B=DB=D, then A+D=C+BA+D=C+B. The example that Herigone gives for A19 is the following: if AF is the square of AB, CG is the square of CD, and AF is equal to CG, then by A19 the square of AB is equal to the square of CD. The example of A21 is that if A>BA>B and A<BA<B are both false, then A=BA=B; the axiom was later to be stated in this form (A24, in Schott 1661). Axiom M27 means that if A is an equimultiple of B as E is of F (i.e., A:B=E:FA:B=E:F), and C is an equimultiple of D as E is of F, then A is an equimultiple of B as C is of D. Axiom E4 on the right angles seems to come from PROCLUS, *In Euclidis* 189, where it is said that Pappus had denied this principle, stating that curvilinear right angles may be different to one another.

CHRISTOPH GRIENBERGER, *Euclidis sex primi elementorum geometricorum libri cum parte undecimi ex majoribus Clavij commentarijs in commodiorem forma contracti*, Graz, Widmanstad, 1636.

Book I

Petitiones, seu Postulata: P1, P2, P3.

Axiomata, seu Pronunciata: CN1, CN2, CN3, CN6*, CN7*, CN8*, CN9*, CN4, CN5, I9, P4, P5, P6*, A4, M14.

·· Another Jesuit edition of the Elements, Grienbergers's book presented a simplification of Clavius's axiomatics, which substituted I5 and I6 with axiom I9, and deleted M1, M2, M12, and M13 from Clavius's list. Grienberger had no axioms in Book V.

MARIN MERSENNE, *Universae geometriae, mixtaeque mathematicae synopsis*, Paris, Bertier, 1644.

Book I

Postulata: P1, P2, P3, P4, P5.

Axiomata, seu Communes notiones: CN1, CN2, CN3, CN6*, CN7*, CN8*, CN9*, CN4, CN5, P6*.

Book VII

Petitiones: N12, N5, N6–N13.

Communes notiones: M3, M4, N30, N31, N8, N32, N33, N35, N36, N34, N2, N3, N1.

Book X

Communes notiones: AA1, M9, M10, M11.

·· Mersenne's treatise of mathematics offers a simplification of Commandino's axiomatics, which includes no axioms in Book V and does not further discuss the principles in Book I. Mersenne's book only states the theorems, without giving the proofs of them. The book also contains several other mathematical works given in a similarly abridged form.

CLAUDE RICHARD, *Euclidis elementorum geometricorum libros tredecim*, Antwerp, Verdus, 1645.

Book I

Petitiones, sive Postulata: P1, P2, P3, A7, T1, T2.

Axiomata, seu Pronunciata, seu Dignitates, seu Communes notiones: CN1–A13–A14–A15–A16–A17, CN2, CN3, CN6*–M15, CN7*–M16, CN8*–M17, CN9*–M5, CN5–A6, CN4–T3, I5, I6, P4, P5–PP15, P6*, M1, M2, M12, M13, A4, [M14], T4, C3, C4, I10, C5, C6, C7, C8, C9, T5.

Book V

Axiomata: R17, A23, M28; [A4, CN2, CN6*, CN3, CN1, CN7*, A13, M1].

Book VII

Petitiones: N12, N5.

Axiomata: M3, M4, N30, N31, N32, N33, N34, N35, N36, N2, N3, N1.

Book X

Petitio: AA1.

Axiomata, sive pronunciata: M11, M9, M10.

·· Jesuit textbook. Richard enriched Clavius's axiomatization, which he took as a starting point for his emended edition of Euclid. Axioms T1 and T2 are the first attempt to explicitly axiomatize the use of movement in geometry, and depend on Clavius's discussion with Peletier (in CLAVIUS 1589 and following editions, commenting *Elements* III, 16). The movements described by T2 are translations and rotations around a point or an axis. The rigidity of the figure during movement is intended to be guaranteed by the new axiom T5. Axioms CN8*–M17 and CN9*–M5 are in fact stated in a more general way which applies to any multiple and submultiple (not just the double and the half). Richard gave a complete proof of his T3, relying on superposition (but, following Clavius, he thought that a proven axiom is still an axiom). Richard stated that Proclus's and Clavius's proofs of P5 rely on more obscure principles than P5 itself, and thus he claimed that it should be regarded as an axiom rather than as a theorem (as Clavius had stated). Principle PP15 comes from Ptolemy's proof of *Elements* I, 28 (see PROCLUS, *In Euclidis* 362–363): If two straight lines form internal angles equal to two right angles on the one side, they cannot meet; otherwise there would be no reason why they should not meet on the other side as well (where they also form internal angles equal to two right angles), and this is impossible as they would enclose a space. Axiom PP15 states the symmetry requirement. Richard expunged M14 from the list of principles, since it was introduced by later commentators to make explicit an assumption of the Euclidean proof of *Elements* III, 20, while Richard gave another proof of the same proposition that does not recur to such a principle. Axioms C5, C6, C7, and C8 are about "Line–Circle" continuity. Axiom C9 is a form of the so-called Pasch's Axiom. The system of principles employed by Richard in Book V largely consists in new definitions, but he also had a list of eleven axioms that simply repeat (with small changes) a few of the common notions of Book I, adapting them to continuous magnitudes and ratios; the only principles to which nothing corresponds in Book I, besides R17, are A23 and M28; the other principles are modified versions of A4, CN2, CN6*, CN3, CN1, CN7*, A13, M1.

EVANGELISTA TORRICELLI, *De proportionibus liber*, 1647 (unpublished).

Suppositiones et Axiomata: R10, R3–R4, R5–R6, R23, R24, R20.

·· Torricelli's treatise on the theory of proportions reworked Galileo's own theory, which was dictated to him by Galileo and first published by Viviani in 1674 (as the Fifth Day of the *Discorsi*). Even though Torricelli's book was never published, it enjoyed some manuscript circulation, and influenced several writers. Axioms R23 and R24 give in fact a different characterization of proportionality than the Euclidean definition. Torricelli's book on proportions was first published in Vol. 1 of Torricelli's *Opere* (see p. 306), and can now be read with a critical edition and a commentary in Giusti (1993).

JOHN WALLIS, *In Elementa Euclidis Praelectiones*, 1651 (unpublished).

Book I

Postulata: P1, P2, P3.

Axiomata: CN1, CN2, CN3, CN6*, CN7*, CN9*, CN8*, CN4, CN5, P4, [P5], P6*.

·· The Bodleian Library preserves a manuscript with Wallis's Oxford lectures on Euclid's *Elements* from 1651. They are still unpublished, and contain a remarkable amount of notes and philosophical discussions on the principles and theorems of Book I of the *Elements*. Among them, we find Wallis's alleged proof of the Parallel Postulate, that he published in the essay *De postulato quinto*, appended to the 1693 Latin edition of his *Algebra*. In his lectures, Wallis made use of Grynaeus's text and system of principles, even though he discussed Clavius's list as well, stating that the axioms added by the Jesuit are rather theorems (as a matter of fact, Wallis provided a few proofs of other axioms as well).

GIOVANNI RICCI, *De gli Elementi di Euclide li primi sei Libri*, Bologna, Longhi, 1651.

Book I

Postulati, overo Dimande: P1, P2, I11, S6, T6, T7.

Assiomi, overo communi sentenze: CN1, A13, A14, CN2, CN3, CN6*, M15, CN7*, M16, CN8*, CN9*, CN4, CN5, P6*, A4, P4, I6, T8, T9, PP2, A21.

Book II

Assioma: A1.

·· The aim of the author was to produce a text for students. Postulate I11 is a modification of P3, while T6 is intended to guarantee the possibility of reproducing any figure (Ricci employed it in *Elements* I, 5 in order to construct an isosceles triangle congruent with the given one), and T7 is a version of Richard's T1. In Ricci, each axiom A13 and A14 is split in two different principles (referring to the greater or the lesser relation respectively). Axiom T8 is a simplified version of Richard's T3. Axiom PP2 is a substitute for the Parallel Postulate (P5).

ANDREAS TACQUET, *Elementa geometriae planae et solidae*, Antwerp, Meurs, 1654.

Book I

Postulata: P1, P2, P3.

Axiomata: CN1–A13, CN2, CN3, CN6*, CN7*, CN8*–CN9*, CN4, T10–T11, CN5, P4, PP3, PP4, P6*, [I5].

Book V

Axioma: R25.

·· Jesuit textbook. It only contains Books I–VI and XI–XII of the *Elements*, while Books VII–IX were published by Tacquet in 1656 (see below). Tacquet's edition of the *Elements* modified several proofs and offered a new theory of proportions in Book V, as well as a proof of the Parallel Postulate. It enjoyed a broad circulation and was regarded as a model for subsequent reformulations of elementary geometry. Tacquet added axioms T10 and T11 as partial converses of CN4 (cf. Richard's T3). He explicitly recused P5 (the Parallel Postulate) from the list of axioms. Tacquet's new theory of parallels is grounded on the definition of parallel lines as equidistant straight lines, and on his new axioms PP3 and PP4. In fact, each of these three principles is enough to guarantee the provability of the Parallel Postulate (P5). Assuming the existence of equidistant straight lines (as Tacquet did) is in fact equivalent to assuming P5 itself. Axioms PP3 and PP4 are false in hyperbolic geometry as well, and from each of them P5 can be deduced, while both of them can in turn be proven, assuming the existence of equidistant straight lines. Tacquet's intended interpretation of PP4 is that *every* perpendicular to a straight line is also perpendicular to a parallel to it (in hyperbolic geometry, ultraparallel lines only have *one* common perpendicular). After the statement of I5, Tacquet gave a proof of it. The axiom R25 of Book V is simply the existence of the fourth proportional already employed by Clavius and others, in a clearer wording. In dealing with irrational magnitudes and ratios in Book V, Tacquet added three more axioms, which he took to be provable and which had in fact been proven by Grégoire de Saint Vincent (Tacquet's teacher) in his *Opus geometricum quadraturae circuli et sectionum coni* (1647). Further editions of Tacquet's book (among which an emended version from 1665) did not modify the system of principles. The book enjoyed a broad diffusion in English translation (see for instance William Whiston's edition from 1714).

Isaac Barrow, *Euclidis Elementorum Libri XV breviter demonstrati*, Cambridge, Nealand, 1655.

Book I

Postulata: P1, P2, P3.

Axiomata: CN1, CN2, CN3, CN6*, CN7*, CN8*, CN9*, CN4 [T10, T12], CN5, I5, I6, P4, P5, P6*, M1, M2, M12, M13, A4, M14.

Book V

Axioma: M27.

Book VII

Postulata: N12, N5, N37.

Axiomata: N38, A22, N30, N31, N32, N33, N34, N35, N36, N2, N3, N1.

Book X

Postulatum: AA1.

Axiomata: M11, M9, M10.

·· A famous abridged edition of the *Elements*, with new proofs by Barrow. Barrow's system of principles in Book I is a slight simplification of Clavius's 1589 edition. Barrow stated that CN4 can be converted in the case of straight lines and angles (cf. Tacquet's axiom T10, to which he added T12). This notwithstanding, in his *Lectiones mathematicae* from 1665 he explicitly criticized Tacquet, claiming that Euclid's CN4 may be converted in all cases. The meaning of the converse of CN4, according to Barrow, should be that "Things that are equal to one another *may* coincide with one another", where the modal verb should express the possibility to dissect the first thing into pieces (and even in infinitesimal or indivisible pieces) and recompose them again in another thing which is congruent to the second. He wanted, in sum, to ground the equality of measure on the congruence of parts (see in particular his *Lect.* XIII). Barrow's lectures were posthumously published in 1685 and reprinted in a celebrated edition by Whewell (Cambridge University Press, 1860). Barrow's system of principles in Books V, VII, and X follows Herigone (1634).

BLAISE PASCAL, *Introduction à la géométrie*, around 1655 (unpublished).

[Book I]

Principes: S8, S9, S10, S11, E5, S12.

Théorèmes connues naturellement: S13, E8–S14, S15, S16, E6, I3, C10, C11, C12, C13, C14, C15.

·· The sketch of an introduction to geometry written by Pascal around 1655. Pascal went no further in the project, nor did he publish his introduction, since Arnauld was then working on his *Nouveaux éléments* (published in 1667) and Pascal decided that a further edition would have been useless. Pascal's draft was later copied by Leibniz and was found among Leibniz's papers in Hannover (the original manuscript is now lost); it is published in PASCAL, *Oeuvres*, vol. 9, pp. 292–294. Axiom S16 is the converse of *Elements* III, 24. Axiom E6 is taken from a Euclidean assumption in *Elements* I, def. 18: there had been several attempts to prove this proposition, from Antiquity to the Early Modern Age.

ANDREAS TACQUET, *Arithmeticae theoria et praxis*, Leuven, Coenestenius, 1656.

Book VII

Axiomata: M3, M4, N30, N31, N8, N9, N34, N36, N5, N2, N3, N1.

·· Jesuit textbook. Tacquet's edition of Books VII, VIII, and IX of the *Elements*, complementing his edition of the geometrical books, already appeared in 1654 (see above). A treatise on practical arithmetic follows the Euclidean books.

GIOVANNI ALFONSO BORELLI, *Euclides restitutus, sive prisca geometriae elementa*, Pisa, F. Onofrio, 1658; Roma, Mascardi, 1679216792. Italian ed. *Euclide rinnovato ovvero gli antichi elementi della geometria,* Bologna, Ferroni, 1663.

[Book I]

Petitiones, seu Postulata: P1, P2, P3, A7.

Axiomata, seu Pronunciata: CN1–A13–A14, CN2, CN3, CN6*–CN7*, CN8*–CN9*, M29, CN5, A4, CN4, I9, P6*, E7, C16; PP5.

[Book II]

Axioma: E8–E9.

[Book III]

Axiomata: M9, R17, M30, R26, R27.

[Book V]

Axioma: E10.

[Book VI]

Postulata: I12, I4.

Axiomata: I13, I14, I15, I8, I16.

[Book IX]

Axiomata: M2–M13, M12.

·· Borelli's *Euclides restitutus* attempted a new foundation of the *Elements* through the introduction of several new axioms, new proofs and a new ordering of the classical theorems. Book I deals with elementary geometrical theorems similar to *Elements* I, while Book II is the theory of the circle (similar to *Elements* III); Book III discusses the theory of proportions (*Elements* V); Book V a few results in inscribing and circumscribing polygons (cf. *Elements* IV) and their similarity (*Elements* VI); Book VI has several points in common with *Elements* XI; Book VIII deals with arithmetic (*Elements* VII, VIII and IX). Book IX deals with irrational quantities (with a few connections with *Elements* X). In Book I, axiom M29 is a simple extension of N2. Axioms E7 and C16 are used by Borelli to prove *Elements* I, 1, which, in his opinion, was in need of a further continuity assumption. Axiom E7 is used in the following Books as well. Axiom PP5 is added at p. 32, just before proving P5 and *Elements* I, 29; it is in fact equivalent to the Parallel Postulate. Borelli's definition of parallel lines is that of straight lines such that any perpendicular to one of them is also a perpendicular of the others and such that all these perpendiculars are equal to one another (a property that in fact follows from the first condition); axiom PP5 is intended to guarantee their existence. In Book III, axiom R17 on the existence of the fourth proportional is split into two different axioms. The system of principles in Book III, completed with a set of new definitions for equiproportionality, is intended to radically reform the Eudoxian theory of proportions (see Palladino 1991; Giusti 1993). Axioms I13, I14 and I15 are extensions of the usual axioms on the straight line, while I8 is explicitly referred to *Elements* XI, 1, where it is implicitly assumed by Euclid; I16 is referred to *Elements* XI, 2. Borelli had no axioms on arithmetic in his Book VIII. Borelli's work was translated into Italian by Domenico Magni in 1663, in an abridged edition that did not include Books VIII and IX. This edition had (in the remaining Books) the same system of principles than the first edition, with the only exception that Magni, contrary to Borelli's opinion, claims that CN4 can be converted (i.e.: everything which is equal can be superposed; cf. the edition by Barrow, 1655, above). A second Latin edition of the work appeared in 1679, again without Books VIII and IX. In it, Borelli eliminated the continuity principle C16 from the axioms of Book I, supplying a new (less rigorous) proof of *Elements* I, 1 which only employs E7.

Claude François Milliet Dechales, *Huict Livres des Elements d'Euclide rendus plus faciles*, Lyon, Coral, 1672.

Book I

Demandes: P1, P2, P3.

Axiomes, ou Principes: CN1–A13, CN2, CN3, CN6*, CN7*, CN8*–CN9*, CN4, T10–T12, CN5, P4, PP7, P6*, I9.

Book V

Axioma: R25.

·· Jesuit textbook. This edition only contains Books I–VI and XI–XII of the *Elements* (cf. Tacquet, 1654), thus eliminating the arithmetical treatises (VII–IX), the tenth book on irrational magnitudes, and the thirteenth book on the Platonic solids; this new arrangement of the matter was later followed in many eighteenth-century editions. Dechales's system of principles derived from Tacquet (1654), but substituted a new axiom PP7 to prove the Parallel Postulate (it is in fact equivalent to it). Dechales's book enjoyed several reprints and translations. A Latin version of it was inserted into Dechales's monumental *Cursus seu Mundus mathematicus*, the first edition of which was printed in 1674, followed by an important posthumous enlarged edition in 1690. The system of geometrical principles remained, however, unchanged in these later editions. The *Cursus* also added a few books on arithmetic, which are introduced by five arithmetical axioms. These books, however, do not follow the Euclidean development and their (very practical) principles only concern the decimal number system.

CASPAR SCHOTT, *Cursus mathematicus, sive Absoluta omnium mathematicarum disciplinarum encyclopaedia*, Würzburg, Hertz, 1661.

Book I

Petitiones, sive Postulata: P1, P2, P3, A7.

Axiomata, sive Pronunciata: CN1, CN2, CN3, CN6*, CN7*, CN8*, CN9*, CN4, CN5, P4, [P5], P6*, I5, I6, I7, I8, M1, M2, M12, M13, A4, M14, M26, A24, E8, C17, A16.

·· Jesuit textbook. Book III of Schott's *Cursus* includes the first six Books of the *Elements*. Schott's axioms for Euclid's Book I, mostly follow Clavius, with a few additions from Herigone and a couple of new principles. Schott claimed to have proven P5 with a simple (and wrong) conversion of *Elements* I, 27 (p. 73). He had no axioms in *Elements* V.

ANTOINE ARNAULD, *Nouveaux Élémens de géométrie*, Paris, Savreux, 1667.

[Book I]

Axiomes de l'égalité et inégalité: CN5, A4, CN1, CN2, CN3, CN7*, CN6*, M3, M4.

[Book II]

Axiomes: R20, R13, R29, R30, M31, R31.

[Book V]

Axiomes: E11, P1, P2, I5, I17, PP6; P3, T13, E8, E12, S16–E17, E13, S17; E14.

[Book XIII]

Axiomes: E15, E16, N39.

·· Arnauld's text on elementary geometry was one of the most important books on the subject in the seventeenth century. It did not follow the Euclidean order or axiomatics, but wanted to improve upon it (cf. Tacquet's and Borelli's works). It was written around 1655–1656 in connection with Pascal's *Introduction* (see above), but printed only in 1667; a second, modified edition was published in 1683 (see below). It was clearly influenced by Clavius (1589), Herigone (1634), and Tacquet

(1654), among others. Book I deals with arithmetic and adds several unproven rules to the Euclidean common notions. Books II–IV deal with the theory of proportions and incommensurable magnitudes, and begin with an axiomatization of the classical Eudoxian theory (which was to be modified in the second edition of the volume). Book III ends with two simpler axioms on equal ratios which are, however, not employed in the book, and are removed in the second edition. Books V–XV deal with plane geometry, all axioms being at the beginning of Book V with the exception of the three used in Book XIII for the calculation of areas. The interpretation of axiom R29 is that nA:mA=nB:mBnA:mA=nB:mB (for any A, B), while R30 states that nA:nB=mA:mBnA:nB=mA:mB; they are generalizations of R20 (nA:A=nB:BnA:A=nB:B) and R13 (nA:nB=A:BnA:nB=A:B), respectively. Axiom I17 is a formulation of the classical P6*, but has in Arnauld a simpler statement. Axiom PP6 is a substitute for the Parallel Postulate (it rules out asymptotic parallel lines); Arnauld, however, also attempted a proof of P5 independent from this axiom in Book VI. The intended meaning of E12 is that the "degrees" of a smaller circumference are smaller than the degrees of a larger circumference, since both lines are divided in the same number of degrees (i.e., 360). Axiom E14 is connected with Arnauld's definition of a perpendicular, which is a line that satisfies the condition of the axiom (in two points).

GILLES-FRANÇOISDE GOTTIGNIES, *Elementa geometriae planae*, Roma, Angelo Bernabò, 1669.

Postulata: P1, P2, P3.

Axiomata: CN1–A13–A14, CN2, CN3, CN6*, CN7*, CN8*–CN9*, CN4, T10–T11, CN5, P4, P6*, I5, R32, R28.

·· Gottignies's book is not, properly speaking, an edition of the *Elements*, but rather an introduction to geometry following Euclid as closely as possible. Gottignies assumed that P5 had been proven by Clavius.

HONORÉ FABRI, *Synopsis geometrica*, Lyon, Molin, 1669.

Axiomata: CN1–CN8*–CN9*, CN2–CN3–CN6*–CN7*, CN5–A4–M14–A24, R33, I17–I5, T14, CN4, R34, R35, R36.

Postulata: P1, P3, P2–T15, A7, T2, S18, T16.

·· Jesuit textbook. Once again, this is not, properly speaking, an edition of the *Elements*, but rather an introduction to geometry following Euclid as closely as possible. I have simplified somewhat Fabri's verbose axioms R33, R35, R36. Axiom T14 is introduced in order to employ motion in geometry. Fabri claimed that CN4 can be unrestrictedly converted. Fabri's constructive postulate T15 is stated among a list of operations that can be accomplished with geometrical tools and it does not seem to have real foundational relevance. Axiom T2 is again employed in order to make use of rigid motion in geometry, while the new T16 was intended to guarantee non-rigid motion in a few occurrences. The Parallel Postulate is proven from the definition of equidistant lines (assumed as a real definition).

VINCENZO VIVIANI, *Quinto Libro degli Elementi di Euclide, ovvero Scienza Universale delle Proporzioni spiegata colla dottrina del Galileo*, Firenze, Condotta, 1674.

Assiomi, ovvero Comuni notizie: R38, R39, R35, R36, R16–R37, R10, R4, R40, R11.

Domanda: R17.

·· Since Torricelli's book on the theory of proportion (written in 1647) had not been published, this edition by Viviani of Galileo's Fifth Day of the *Discorsi* is the first printed essay containing Galileo's reform of Book V of the *Elements*. It is prefaced by Viviani's own attempt to render more rigorous the theory of proportions, which includes ten new principles. Viviani was to publish a fuller edition of the *Elements* in 1690, in which, however, he was to adhere to a more standard axiomatization, following Zamberti for the system of principles in Book I and including no axioms in Book V.

GILLES PERSONNEDE ROBERVAL, *Éléments de géométrie*, around 1673–1675 (unpublished).

[Book I]

Axiomes: A4, CN5, CN1, CN2.

Postulats: S19, I18, I19, I20, I21, I22, I23, T2, I24, P1, A7, A25.

[Book II]

Axiomes: CN4, T17, E1.

Postulat: C18.

[Book III]

Postulats: P3, AA4, C19.

[Book IV]

Axiome: PP8.

[Book VIII]

Axiome: E19

·· Roberval died before publishing the text, which is a new systematization of elementary geometry (plane and solid) in eight books. Several postulates in Book I are more of a philosophical than of a geometrical nature, while others have a more recognizably mathematical appearance. Among them, S19 and I22 together should guarantee the existence of all mathematical objects. Principles I18–I21 may also be employed in a theory of indivisibles. Axiom I23, stating the possibility to include a "space" between any two bounded regions, seems to state a modern separability property (e.g., the so-called topological separability axiom T4T4, stating that space is "normal"). Axiom I24 is the condition for the definition of a straight line as a rotational axis. Book II has, besides the axioms, two postulates: The first of them is, however, incompletely formulated; the second, C18, may be regarded as a form of Pasch's Axiom. Postulate AA4 in Book III is a form of the Archimedean Axiom. Axiom PP8 in Book IV is a substitute for P5. The edition of Roberval's manuscript is in Jullien (1996).

JEAN PRESTET, *Elemens des mathematiques, ou principes generaux de toutes les sciences qui ont les grandeurs pour objet*, Paris, Pralard, 1675.

Axiomes: A26, M32, A4, CN5, M33, CN1, CN2, CN3, CN6*, CN7*; M3, M4, N40, M11, M10, N41, N42.

·· Jesuit textbook. Prestet's book is a treatise on arithmetic and algebra with a few geometrical examples. It does not follow the demonstrative order of the *Elements*, but contains a few axioms in the first two Books which may be considered an extension of the Euclidean Common Notions. Prestet also added several other unproven practical principles (which he called "*demandes*" or simply "*regles*") that teach how to manipulate the algebraic symbolism; they are not included here, as they are too numerous and depart widely from the Euclidean style. A second edition appeared in 1689 (see below).

MERCATOR (NIKOLAUS KAUFFMANN), *Euclidis Elementa Geometrica Novo Ordine ac Methodo fere demonstrata*, London, Martyn, 1678.

Axiomata: I5–P6*, P4, PP9, CN4–T10–T12, CN5–A4, CN1, CN8*–CN9*, CN2–CN3, CN6*–CN7*; R35, R8, R41, R10, R42.

Principia practica, sive Postulata: P1, P2, P3.

·· Mercator's book is not, properly speaking, an edition of the *Elements*, but rather a new arrangement of Euclid's mathematics. Mercator stated the first nine axioms as principles of plane geometry, and the further five axioms as principles of the theory of proportions. He tried to separate theorems and problems, and claimed that all theorems can be proven relying on the axioms alone, while the problems require the three postulates (or practical principles). Mercator's definition of parallel lines is that of straight lines which do not incline toward one another; the notion of inclination is left undefined. Axiom PP9 states the transitivity of parallelism, and this may be the first instance of the use of this principle to ground the Parallel Postulate. The entire theory may be considered a first step toward a direction theory of parallelism (developed in the nineteenth century). Mercator took *Elements* V, 18 (=R42) as a principle, as Euclid's proof of it requires the assumption of the existence of the fourth proportional; he preferred to directly assume the Euclidean proposition rather than accepting axiom R1 or R17 (employed by Clavius and others to prove the proposition).

VITALE GIORDANODA BITONTO, *Euclide Restituto, ovvero gli antichi elementi geometrici ristaurati e facilitati*, Roma, Angelo Bernabò, 1680.

Book I

Postulati: P1, P2, P3, A7.

Assiomi: CN1–A13, CN2, CN3, CN6*, CN7*, CN8*, CN9*, CN4, CN5, P4, [P5], [P6*], [I5].

Book VII

Postulati: N12, N5.

Assiomi: M3, M4, N30, N31, N32, N33, N34, N35, N36, N2, N3, N1.

Book X

Postulato: AA1.

Assiomi: M11, M9, M10.

·· Giordano's system of principles is strongly dependent on Clavius. Giordano attempted to prove the geometrical axioms P5, P6* and I5 in the course of Book I, and his defective proof of the Parallel Postulate (pp. 45–66) is especially worthy of notice. A second edition in 1686 did not change the system of principles.

JACQUES ROHAULT, *Les six premiers Livres des Elemens d'Euclide*, in *Ouvres posthumes*, ed. C. Clerselier, Paris, Desprez, 1682.

Book I

Demandes: P1, P2, P3, A7.

Axiomes: CN1, CN2, CN3, CN6*, CN7*, CN8*, CN9*, CN4, A4, CN5, P4, P6*, I5, I6, M1, M2, M12, M13, M14.

Book V

Axiomes: R43, R44, R45.

·· One of the most influential treatises on Cartesianism, it also included a shortened version of the *Elements*. In Book I, Rohault mostly followed

Clavius's axiomatization. He did not include P5 among the axioms, nor did he give a proof of it, but rather stated it as an evident truth before proving *Elements* I, 29 (vol. 1, p. 60). Rohault added three axioms in Book V which are a simple variation of the Euclidean definitions of the equality and inequality of ratios, Rohault having himself defined these in a different way.

ANTOINE ARNAULD, *Nouveaux Élémens de géométrie*, Paris, Desprez, 1683216832.

[Book I]

Axiomes de l'égalité et inégalité: CN5, A4, CN1, CN2, CN3, CN7*, CN6*, M3, M4.

[Book II]

Raisons égales naturellement connues: R46, R20, R13, R29, R30.

Axiomes: R18, R47, R48, R28, R49, R10, R4, R50, R51, R52.

[Book V]

Axiomes: E11, P1, P2, I5, I17, PP6; P3, T13, E8, E12, S16–E17, E13, S17; E14.

[Book XIII]

Axiomes: E15, E16, N39.

·· The second edition of Arnauld's 1667 work. Several changes were made in this second edition in Books II–IV on the theory of magnitudes and proportions, mostly following Nonancourt's *Euclides logisticus*, which had been published in 1652 and which Arnauld became familiar with in 1679–1680. The new system of principles in Book II is a modification of the 1667 edition with Nonancourt's own axioms. Several further editions were printed, with no changes in the axiomatics. A modern critical edition, which underlines the differences between the 1667 and 1683 editions, is to be found in Descotes (2009).

BERNARD LAMY, *Les elemens de geometrie*, Paris, Pralard, 1685.

Principes generaux, ou Axiomes: CN5, A4, CN1, CN2, CN3, CN7*, CN6*, CN4, CN8*, CN9*, N43.

[Book I]

Propositions, ou Demandes: I29, I3, E18, P1, P2, I31, I32, I30, I17; P3, E24, E6, E13, E8; E14, I25, I26; PP1.

[Book II]

Demandes: E19, E20.

[Book III]

Demandes: N44, N45, R53.

[Book IV]

Axiomes: I4, I33, I16, I34, I15, E21, I27, I28; E22, E23, AA2, AA3.

·· Lamy's volume is a textbook mostly based on Arnauld (1683). It added a few geometrical axioms, and in particular several principles on the straight line that are almost identical to one another (P1, I31, I32, I30, I17) in an attempt to formalize the incidence properties of this line. It includes no axiomatization for the theory of proportions (except for a few hints in one appendix, and the three postulates on the commensurability of ratios in Book III). Book IV deals with solid geometry. Axiom E14 is in fact expressed by Lamy as four distinct propositions. Axioms AA2 and AA3 are related to the method of indivisibles.

JEAN PRESTET, *Nouveaux elemens des mathematiques, ou principes generaux de toutes les sciences qui ont les grandeurs pour objet*, Paris, Pralard, 1689[2].

Axiomes, ou notions communes, et principes generaux du corps entier des Mathematiques: CN5, A4, M33, CN2, CN3, CN6*, CN7*, M12, A13, A27, A28.

Axiomes, ou notions communes, et principes generaux du corps entier de la Geometrie: P1–P2–P6*, CN4, E25.

·· The widely popular second edition of Prestet's *Elements* (1675) appeared in two largely augmented and revised volumes in 1689. It changed the system of axioms for the theory of magnitudes, and added a few principles for geometry, which is treated of cursorily in the book.

[JOHN KEILL], *Euclidis elementorum libri priores sex, item undecimus et duodecimus ex versione latina Federici Commandini in usum juventutis academicae*, Oxford, Theatrum Sheldonianum, 1701.

Book I

Postulata: P1, P2, P3.

Axiomata: CN1, CN2, CN3, CN6*, CN7*, CN8*, CN9*, CN4, CN5, P6*, P4, P5.

Book V

Axiomata: M3, M4.

·· A very successful edition in Britain. The first Latin edition appeared anonymously, but a number of the many subsequent Latin and English editions stated that the editor was John Keill. The Latin text mostly follows Commandino, but the system of principles in Book I is the one by

Grynaeus. In the Latin editions, the axioms in Book V are those by Commadino, but in the English translation they were later suppressed.

DAVID GREGORY, ΕΥΚΛΕΙΔΟΥ ΤΑ ΣΩΖΟΜΕΝΑ, *Euclidis quae supersunt omnia*, Oxford, Theatrum Sheldonianum, 1703.

Book I

αἰτήματα, *Postulata*: P1, P2, P3.

κοιναὶ ἔννοιαι, *Communes notiones, sive Axiomata*: CN1, CN2, CN3, CN6*, CN7*, CN8*, CN9*, CN4, CN5, P4, P5, P6*.

·· An new edition of the Greek text of the *Elements* and the other works of Euclid, together with a new Latin translation. The Greek of the *Elements* is largely based on Grynaeus's edition, but it is improved in several respects. The system of principles, however, is left unmodified.

EDMUND SCARBURGH, *The English Euclide, being The First Six Elements of Geometry, translated out of the Greek*, Oxford, Theatrum Sheldonianum, 1705.

Book I

Postulates: P1, P2–[T18], P3.

Axioms: CN1, CN2, CN3, CN6*, CN7*, CN8*, CN9*, CN4, CN5, P4, [P5], P6*, PP1.

·· A further Oxonian edition of Euclid from the beginning of the eighteenth century, this one is provided with an extensive commentary by Scarborough. He claimed that Postulate T18 was not stated by Euclid, since it is somehow included in P2, but that it should, nonetheless, be explicitly given. He thought that the Parallel Postulate is too complicated to be assumed as a principle, and preferred to add the alternative axiom PP1 on equidistance. Scarborough also offered a thoroughgoing discussion of the theory of proportions in *Elements* V, but he did not add any further axiom to this effect.

NICHOLASDE MALE´E´ZIEU, *Elemens de geometrie de Monseigneur le Duc de Bourgogne*, Paris, Boudot 1705.

[Book I]

Axiomes, ou verités connues d'elles mêmes: CN5, A6, A4, CN1, CN2, CN3, N46, N47, N30, N48, N49.

·· Malézieu was the mathematics teacher of the Duke of Burgundy (who, in 1705, was a child of only fourteen but who appears nonetheless as the author of this book). The volume is not an edition of Euclid, but rather a textbook in elementary mathematics which enjoyed a certain diffusion in the eighteenth century, largely inspired by the "revisionist" French attitude toward Euclid which had previously informed Richard's and Arnauld's work, and the latter's dismissive opinion about a strictly formal treatment of mathematics. In particular, Malézieu's work has no geometrical postulates at all, and simply assumes the possibility of constructions as obvious in the proofs of the propositions (see below Legendre, 1794, for a similar treatment). It only has a few common notions that are mainly employed in the arithmetical book which opens the volume, and which clearly have to be interpreted in a much more general way than they are actually stated (i.e., N46 and N47 state the commutative and distributive properties of multiplication in general). Many subsequent French books on elementary geometry in the eighteenth century simply removed any axiomatic system for the demonstrations, assuming in the body of the proofs any principle that was needed; among them, we may mention the celebrated books by Joseph Sauveur, *Geometrie élémentaire et pratique*, ed. Le Blonde, Paris, Rollin 1753 (posthumous, but widely circulating in manuscript form already for decades by this date) and by Alexis Claude Clairaut, *Elemens de geometrie*, Paris, Lambert & Durand 1741. See also below, Deidier (1739), which also has only a few common notions.

ISAAC NEWTON, [*The Science of Axiomatc Deduction*], around 1693–1706 (unpublished).

Geometriae libri duo

Postulata: P1, T22, T23.

[*Arithmetic and Geometry*]

Axiomata: CN1, CN2, CN3, R4, A4, R54.

Postulata: N50, N51, … .

De constructione problematum geometricorum

Postulata: T19, T20, T21.

·· A few drafts dealing with the principles of geometry have been found among Newton's unpublished papers and can now be read in volumes 7 (p. 388) and 8 (pp. 197, 200) of Whiteside et al. 1967–1981. The draft of some *Geometriae libri duo* has been tentatively dated by the editor to the early 1690s, possibly 1693. In it, Newton added to the first Euclidean postulate two more principles which had a broader scope than those in the *Elements* (Newton explicitly claimed that T22 implies P2 as a particular case, while T23 implies P3). Principle T22, in fact, is a modification of *Elements* I, 2-3 on the transportation of segments (cf. T15, T26) and adds nothing to the deductive power of the *Elements* (Newton thought, as did many others, that such propositions are so elementary as to be better assumed as principles). Principle T23, on the other hand, far exceeds the standard Euclidean constructions, making it possible to draw algebraic curves of higher degree and therefore to solve "all plane and solid problems" (i.e. those which require conic sections), and others as well (since reiterating constructions by T23 one can draw curves of degree greater than two). In a later passage, Newton admitted that other ways of solving solid problems may also be postulated at the beginning of geometry, such as the classical NEŶΣIω construction, but these other methods do not allow the drawing of conic sections and other algebraic curves, and therefore T23 should be preferred. Even the basic assumption of cutting a cone with a plane ("*datum conum dato plano secare*", p. 382) is more complicated than his own postulate. The mathematician François Viète had previously postulated the NEŶΣIω construction in his *Supplementum geometriae* (1593); I have not included it among the present list of works, since it is neither an edition of Euclid nor does it state any further mathematical principle. Two further Newtonian drafts on axiomatics may belong to the years 1705–1706. A first one on arithmetic and geometry begins with a list of Common Notions, followed by two arithmetical postulates, and then breaks off. A second one on the construction of geometrical problems seems to envisage a geometrical axiomatics based on mechanical principles, as Newton himself had famously claimed in the Preface of the *Principia*.

ANGELO MARCHETTI, *Euclides reformatus*, Livorno, Celsi, 1709.

Book I

Petitiones, seu Postulata: P1, P2, T24, T25, P3, A7.

Axiomata, seu Pronunciata: CN1, A13, CN8*–CN9*, CN2, CN3, M6, M7, CN5, A4, M11, T4–T8, P4, I5, E6, E7, E8–E26, E9–E27, PP5, E28, E29, CN4, E30; PP1, AA1.

[Book III]

Axiomata: R35, R8, R55, R10, R16, R37, R11, R12, R14, R56, R15, R57, R58, R59, R17.

[Book VI]

Postulata: I12, I4.

Axiomata: I35, I14, I15, I8, I16.

·· Marchetti mainly followed Borelli's axiomatization in Book I, but with several twists. He justified the assumption of *Elements* I, 20 (=E30) among the axioms, stating that it follows from the definition of a straight line as the shortest line between two points, which he considered the only viable definition of a straight line. Axiom PP5 is equivalent to the Parallel Postulate in Borelli's axiomatization, but Marchetti also added PP1 (p. 25) to the same effect. Axiom AA1 is added in p. 42. Marchetti's axiomatization of the theory of proportions is to be found in his Book III. It amounts to a profound revision of the Euclidean theory, but it is not entirely new, as it was prepared by several other works of the Galileian School (Torricelli, Borelli, Viviani, and others). Marchetti himself had already published his axiomatization in his booklet *La natura della proporzione e della proporzionalità* in 1695. Axiom R55 is a kind of restatement of *Elements* V, def. 5, for commensurable magnitudes (cf. also *Elements* V, 3). Marchetti's Book VI is on solid geometry. On Marchetti's theory of proportions see Giusti (1993).

GEMINIANO RONDELLI, *Sex priora Euclidis elementa, quibus accesserunt undecimum, & duodecimum*, Bologna, Longhi, 1719.

Book I

Petitiones, seu Postulata: P1, P2, P3, S6, T7, A7.

Axiomata: CN1–A13–A14, CN2, CN3, CN6*–M15, CN7*–M16, CN8*–M17, CN9*–M5, CN4, CN5, I5, I6, P4, [P5], P6*, M2, M1, M13, M12, A4, M14, A16, A24, PP2, PP10, I36.

Book V

Axioma unicum: R25.

·· A first edition of this book had been published by Rondelli in 1684, using Clavius's system of principles. The 1719 edition, however, added several axioms. Some of them were introduced by Rondelli, while others were taken from Ricci (1651). Rondelli held the Bologna University chair of mathematics which had been held by Ricci some years before. Rondelli proved the Parallel Postulate (P5), grounding his demonstration in the equivalent principle PP2. He also added another principle equivalent to P5, namely PP10, which, however, he did not employ in his proof of P5.

HENRY HILL, *The Six First, Together with the Eleventh and the Twelfth Books of Euclid's Elements, Demonstrated after a New, Plain, and Easie Method*, London, Pearson, 1726.

Book I

Postulates: P1, P2, P3.

Axioms: CN1, A13, A14, A29, A30, A31, CN8*, CN9*, N52, N53, CN2, CN3, N48, N49, M2, M13, CN5, A4, CN4, P6*, P4, P5.

·· A quite typical modern edition of Euclid, that added a few algebraic principles and performed several proofs by symbolic reasoning. The entire *Elements* V on the theory of proportions is reduced to a few calculations.

GUIDO GRANDI, *Elementi geometrici piani e solidi di Euclide posti brevemente in volgare*, Firenze, Tartini, 1731.

Book I

Dimande: P1, P2, P3.

Assiomi: CN1, CN2–CN3, CN6*–CN7*, CN8*–CN9*, CN4, CN5, P4, [P5], P6*, I5.

·· Grandi's edition of the *Elements* is very classical, and draws on Grynaeus's system of principles, adding to it a rather standard axiom by Clavius (I5). At the time of Grandi's edition of Euclid, Gerolamo Saccheri had already written his *Euclides vindicatus* (published in 1733), in which he disproved Clavius's proof of the Parallel Postulate (P5). Even though Grandi was in personal relations with Saccheri, however, he proved P5 following Clavius's demonstration.

MANOELDE CAMPOS, *Elementos de geometria plana, e solida, segundo a ordem de Euclides*, Lisboa, Rita Cassiana, 1735.

Book I

Postulados: P1, P2, P3.

Axiomas: CN1, CN2, CN3, CN6*, CN7*, CN8*, CN9*, CN4, CN5, P4, P5 [PP3], PP4, P6*, I5.

·· A Potuguese edition of the *Elements*, employing a rather common choice of principles (see Grandi above). Campos stated that Tacquet's axiom PP3 might be substituted for the Parallel Postulate (P5); in any case, he also added axiom PP4, which is itself equivalent to P5.

ANTOINE DEIDIER, *La science des géomètres*, Paris, Jombert, 1739.

Book I

Axiomes: CN5, A4, CN2, CN3, N48–N49, CN6*–CN7*–N54–N55, CN8*–CN9*, AA5, CN1.

·· A quite famous French edition in the eighteenth century, which does not follow Euclid's text but rather systematizes it, Deidier's version of the *Elements* has no geometrical principles, and grounds its demonstration on the definitions of geometrical figures. See above Malézieu (1705) for a similar example.

SAMUEL KLINGENSTIERNA, *Euclidis elementorum libri sex priores una cum undecimo et duodecimo*, Upsala, Höjer, 1741.

Book I

Postulata: P1, P2, P3.

Axiomata: CN1, CN2, CN3, CN6*, CN7*, CN8*, CN9*, CN4, CN5, P6*, P4, P5.

Book V

Axiomata: M3, M4.

·· An Euclidean edition by one of the greatest Swedish mathematicians. The system of principles follows Grynaeus in Book I, but the additional axioms in Book V bear trace of Keill's edition from 1701. A Swedish translation of Euclid was made a few years later by Strömer (see below).

MÅRTEN STRO¨O¨MER, *Euclidis Elementa eller Grundeliga inledning til geometrien, til riksens ungdoms tienst på swenska språket utgifwen*, Upsala, Kiesewetter, 1744.

Book I

Postulater: P1, P2, P3.

Axiomer: CN1, CN2, CN3, CN6*, CN7*, CN8*, CN9*, CN4, CN5, P6*, P4, P5.

·· Grynaeus's system of principles. Possibly the most widely read (and republished) edition of the *Elements* in Swedish.

THOMAS SIMPSON, *Elements of Plane Geometry*, London, Farrer & Turner, 1747.

Book I

Postulates, or Petitions: P1, P2, P3, I37–T26.

Axioms, or Self-evident Truths: CN1, CN5, A4, CN2, CN3, CN6*–CN7*, P4, PP10, E31.

[Book IV]

Axioms: R60, R61, R62, R8, R10.

[*Treatise of Regular Solids*]

Postulatum: E32.

·· A textbook rearranging the geometrical results of the *Elements*. Simpson's volume substituted for the Parallel Postulate (P5) a principle based on equidistance (PP10), and assumed a few basic or controversial theorems of the *Elements* (Book One, propositions 2, 3, 4, 11, 12, 23) as axioms. Axiom I37–T26 allows the transportation of segments and angles (cf. Hilbert's Axioms III, 1, 4, and 6, for a similar axiomatic treatment), but Simpson incidentally added that it should also ground the possibility to construct parallel lines (in his definition, equidistant lines) by parallel displacement of a segment, therefore assuming a second time the Parallel Postulate. Simpson's Book IV is a treatment of the Euclidean theory of proportions in *Elements* V. Simpson added in the same volume a treatise on solid elementary geometry (*Elements* XI and XII), with its own new postulate. A second, revised edition of Simpson's *Elements* appeared in 1760 (see below).

ROBERT SIMSON, *Euclidis Elementorum libri priores sex item undecimus et duodecimus ex versione latina Federici Commandini*, Glasgow, Foulis, 1756.

English version as *The Elements of Euclid, viz. the first Six Books together with the Eleventh and Twelfth. In this Edition, the Errors, by which Theon, or others, have long ago Vitiated these Books, are Corrected, and some of Euclid's Demonstrations are Restored*, Glasgow, Foulis, 1756.

Book I

Postulata (*Postulates*): P1, P2, P3.

Axiomata (*Axioms*): CN1, CN2, CN3, CN6*, CN7*, CN8*, CN9*, CN4, CN5, P6*, P4, [P5].

Book V

Axiomata (*Axioms*): M3, M4, M34, M35.

·· Grynaeus's system of principles in Book I. An important edition of the geometrical books of Euclid, Simson's volume also attempted to restore the original text of the *Elements* after the many corruptions and modifications added in the previous early modern editions. To this effect, Simson's commentary is put at the end of the book, so that it would not interfere with the original text, and the system of principles is the most conservative possible. In *Elements* I, Simson placed the Parallel Postulate posterior to the other axioms, since he believed that it should, in fact, be ranged among the theorems, and gave the old proof of it by Proclus (even though it had been refuted several times in the preceding literature). In *Elements* V, he added four axioms taken from, or inspired by, Campano to deal with the theory of proportions. The book enjoyed several reprints and translations into other languages (including German and Portuguese), and was used by several other British authors as the foundational text for their editions; see for instance the editions by John Playfair (1795, below), and Alexander Ingram (Edinburgh, Pillans 1799). The French edition by Castillon (Berlin, Société typographique 1767) also followed Simson's *Elements*, and made use of the same system of axioms (with an even more naïve proof of the Parallel Postulate).

SAMUEL KO¨O¨NIG, *Elemens de geometrie contenant les six premiers livres d'Euclide*, Den Haag, Scheurleer, 1758.

Book I

Demandes: P1, P2, P3.

Axiomes ou notions communes: CN1, CN2, CN3, CN6*, CN7*, CN8*, CN9*, CN5, T27, P4, [P5–PP2], P6*.

Book II

Axiomes: A4, E16.

Book III

Axiomes: E8, E33.

Book V

Demandes: M36, A32, [N37].

·· König's edition of the *Elements* attempted to offer a quasi-formal account of the Euclidean proofs. König employed Grynaeus's system of principles in Book I, but added several axioms in the subsequent books. The only modification in the axiomatization of *Elements* I is the substitution, for CN4, of T27. The latter principle is itself quite classical, as Euclid himself had employed the terms "similar and equal" (ἼΣΑ ΚΑῚ ὍΜΟΙΑ) to mean congruent figures in *Elements*XI, def. 10. In the Early Modern Age this formula was often used to explicitly define congruence, but König may be the first to have employed it as an axiom. The Parallel Postulate (P5) is included among the axioms, but the further discussion shows that König wanted to replace it with axiom PP2, which he found simpler and from which he proved P5. Book V does not have axioms, but rather two postulates; it also has an Appendix on logarithms which departs from the Euclidean text, and in which König added a *demande* stating that it is possible to make multiplications and divisions of any magnitude (something similar to axiom N37). In a second edition printed in 1762, the mathematician J.J. Blassière added a version of Books XI and XII in the style of König—without, however, adding any further principle.

ABRAHAM GOTTHELF KA¨A¨STNER, *Anfangsgründe der Arithmetik, Geometrie, ebenen und sphärischen Trigonometrie und Perspectiv*, Göttingen, Vandenhoeck, 1758.

[Book I]

Forderungen: P1–P2–S20, P3.

Grundsätze: I17–P6*, I5–I6, CN4, T10–T12, P4, E6, C20, I38; P5; I33.

·· Not an edition of Euclid's *Elements*, but the most important German textbook on elementary mathematics in the eighteenth century. Kästner did not axiomatize arithmetic, but endorsed a highly composite system of axioms for geometry, which he took from several previous Euclidean editions (Kästner was a remarkable historian of mathematics). He also added a few practical "postulates". The *Anfangsgründe* went through many editions in Germany and deeply influenced research on the foundations of geometry.

THOMAS SIMPSON, *Elements of Geometry*, London, Nourse, 1760.

Book I

Axioms, or Self-evident Truths: CN1, CN5, A4, CN2, CN3, CN6*–CN7*, P4, I17, PP11, E31.

Postulates, or Petitions: P1, P2, P3, [I37–T26].

[Book IV]

Axioms: R63, R61, R60, R62, R8, R41, R10, R64, M37.

[Book VII]

Axiom: E34.

Postulate: I39.

[Book VIII]

Postulates: AA1, AA6, E35.

·· A deeply revised edition of Simpson's work from 1747, this volume rearranges in eight books the geometrical results of the *Elements*. With respect to the first edition, Simpson slightly modified his version of the Parallel Postulate (now PP11), and while he still assumed rigid motion through postulate I37–T26, he deleted any reference to parallel lines in it, and even claimed that it can be proven with a standard Euclidean procedure. He also introduced the classical axiom I17. Simpson's Book IV extended the axiomatic basis of the theory of proportions; in particular, his axiom M37 is the distributive property of addition and multiplication. Simpson's Book VII is a reworking (with different principles) of his treatise on solid geometry, while his newly written Book VIII deals with the method of exhaustion and proves some results of *Elements* X. In it, axiom AA6 is a formulation of Archimedes's Axiom, while axiom E35 is a modification of Lamy's E19 and E20 (which refer to the areas of a circle and a polygon, while Simpson's E35 refers to circumferences and perimeters). The volume also contains a few sections further developing elementary geometry with new methods, as well as many historical and technical notes on Euclid's theorems.

JOHANN FRIEDRICH LORENZ, *Euklid's Geometrie, oder die sechs ersten Bücher der Elemente nebst dem eilften und zwölften aus dem Griechichen übersetz*, Halle, Weisenhaus, 1773.

Book I

Forderungen: P1, P2, P3.

Grundsätze: CN1, CN2, CN3, CN6*, CN7*, CN8*, CN9*, CN4, CN5, P4, P5, P6*; [PP12].

·· Grynaeus's system of principles. Translated into German directly from Grynaeus's Greek text, Lorenz's edition enjoyed a wide popularity in Germany by the end of the eighteenth and the beginning of the nineteenth century, and it was consistently referred to by several studies on the theory of parallels. In the preface to Lorenz's edition, the mathematician Andreas Segner discussed the provability of P5, proposing a demonstration of it which implicitly relied on the new principle PP12. Lorenz, in his later *Grundriß der reinen und angewandten Mathematik, oder der erste Cursus der gesamten Mathematik* (Helmstädt, Fleckeisen, 1791), reworked the entire system of axioms, and explicitly included PP12 among the *Grundlehren*; the latter book, however, was not an edition of Euclid. Lorenz's 1773 *Geometrie* went through many reprints and inspired the following German editions of Euclid by Michelsen (Berlin, Maßdorff 1791), Hauff (Marburg, Akademie 1797) and Hoffmann (Mainz, Kupferberg 1829), among others, which endorsed the same system of principles.

JOHN BONNYCASTLE, *Elements of Geometry*, London, Johnson, 1789.

Book I

Postulates: P1, P2, P3, PP2.

Axioms: CN1, CN2, CN3, CN6*, CN7*, CN8*, CN9*, A4, CN4.

·· A common English edition of the geometrical books of the *Elements* which alters from time to time the Euclidean deductive structure. Principle PP2 is a substitute for P5 and is, surprisingly, stated as a postulate rather than an axiom.

ADRIEN-MARIE LEGENDRE, *Éléments de géométrie avec des notes*, Paris, Didot, 1794.

Book I

Axiomes: CN1, CN2, CN3, N30, N31, CN5, A4, I31, CN4, [P5]. (...)

·· One of the most famous Euclidean editions ever, Legendre's *Elements* does not follow the Greek text but develops the same results in a different order and with different proofs. Legendre did not include any postulate in his list of principles, and stated that the axioms which he gave at the beginning of the first book are merely illustrative of an infinity of other similar axioms. At later times, people complained that he had erased the distinction between axioms and postulates [see for instance the opinion of Grassmann and Houël in Voelke (2005), p. 63], even though this does not seem to have been Legendre's aim. The book went through many editions, with several subsequent modifications, and in most of them Legendre attempted (in various ways) to prove the Parallel Postulate (P5). The system of axioms also changed since the second edition (see below). Legendre's book was later translated into several languages.

John Playfair, *Elements of Geometry, containing the first six books of Euclid with two books on the geometry of solids*, Edinburgh, Bell & Bradfute, 1795.

Book I

Postulates: P1, P2, P3.

Axioms: CN1, CN2, CN3, CN6*, CN7*, CN8*, CN9*, CN4, CN5, P4, PP13.

Book V

Axioms: M3, M4, M34, M35.

·· The edition of the *Elements* by Playfair is mainly based on Simson (1756), who had tried to restore a text closer to the Greek original. Playfair, however, allowed a few modifications (especially in the proofs of Book V) for the sake of clarity and rigor. Among these, the introduction of axiom PP13 ("Playfair's Axiom") as a substitute for the Parallel Postulate (P5) was especially successful. The latter principle had already been recognized as equivalent to P5 by Ibn al-Haytham and others, even though it had never been used explicitly to this effect in an edition of the *Elements*. On Playfair's edition, see Ackerberg-Hastings (2002).

Adrien-Marie Legendre, *Éléments de géométrie avec des notes*, Paris, Didot, 179922.

Book I

Axiomes: CN1, CN5, A4, I31, CN4, [P5]. (...)

·· The second edition of Legendre's celebrated *Éléments de géométrie* offered a significant simplification of the system of axioms employed in the first one from 1794. Further editions of this work followed the present list.

Vincenzo Flauti, *I primi sei libri degli elementi di Euclide*, Napoli, Stamperia Reale, 1810.

Book I

Postulati: P1, P2, P3, P4, [P5].

Assiomi, o nozioni comuni: CN1, CN2, CN3, CN6*, CN7*, CN8*, CN9*, CN4, CN5, P6*.

Book V

Postulato: R1.

·· An edition of the *Elements* which enjoyed broad diffusion in Italy, and which went through many reprints as one volume of Flauti's *Corso di geometria elementare e sublime*. It is very rich in historical and mathematical notes, and in some editions, Flauti claimed to have proven the Parallel Postulate (P5) using an old argument by Nasīr ad-Dīn (in further editions, however, he acknowledged his mistake here).

FRANÇOIS PEYRARD, ΕΥΚΛΕΙΔΟΥ ΤΑ ΣΩΖΟΜΕΝΑ, *Euclidis quae supersunt, Les oeuvres d'Euclide*, Paris, Patris, 1814.

Book I

ΑἸΤΉΜΑΤΑ, *Postulata, Demandes*: P1, P2, P3, P4, P5, P6*.

ΚΟΙΝΑῚ ἜΝΝΟΙΑΙ, *Notiones communes, Notions communes*: CN1, CN2, CN3, CN6*, CN7*, CN8*, CN9*, CN4, CN5.

·· The crucially important edition of Euclid's works by Peyrard gives the original Greek text and a translation into Latin and French. It was the first edition to contain a thoroughgoing discussion of several extant Greek manuscripts, and, in particular, it mainly followed the text of the Vatican manuscript **P** (see above), found by Peyrard himself, which appears to represent the oldest version of the *Elements* which survived into the modern age. Peyrard emended several mistakes in the previous Greek editions by Grynaeus (1533) and Gregory (1703), as well as in Commandino's translation (still considered as counting among the best editions of Euclid); yet, in several places he preferred Grynaeus's version to the text found in his newly discovered Greek manuscript. It was preceded by an incomplete edition of the *Elements* in 1804 by Peyrard which did not make use of the Vatican manuscript. The 1814 edition was prefaced by an appraisal of it by Lagrange, Legendre and Dalambre.

In the following years, several older editions were reprinted many times, but Peyrard's text and system of principles gained a growing acceptance by the scholarly community. Another Latin edition based on Peyrard's text was prepared by Niede in 1825, and in 1826 it was translated into English by Philips. In 1824, Camerer and Hauber published, in Germany, a very influential and heavily annotated edition of the *Elements* also based on Peyrard's manuscript (and axioms). In 1826–1829, E.F. August published another edition of the Greek text, also following Peyrard's manuscript and correcting the latter's edition in a few places. Many years later (1883–1916), Heiberg and Menge built their edition of Euclid on the same Vatican manuscript found by Peyrard (even though they emended and integrated it in several places), and this edition is still the most authoritative reconstruction of the text of the *Elements* that we have.

ACKNOWLEDGMENTS

Open access funding provided by Max Planck Society (Max Planck Institute for the History of Science).

REFERENCES

1. Acerbi, F., and B. Vitrac. 2014. *Héron d'Alexandrie. Metrica*. Pisa: Serra.
2. Ackerberg-Hastings, A. 2002. Analysis and synthesis in John Playfair's Elements of Geometry. *British Journal for the History of Science* 35: 43–72.CrossRefMATH
3. Barany, M.J. 2010. Translating Euclid's diagrams into English, 1551–1571. In *Philosophical aspects of symbolic reasoning in early modern mathematics*, ed. A. Heeffer, and M. Van Dyck. London: College Publications.
4. Barrow-Green, J. 2006. "Much necessary for all sortes of men": 450 years of Euclid's Elements in English. *BSHM Bulletin* 21: 2–25.CrossRefMATH
5. Bodriga, G. 1925–1926. Giovanni Battista Benedetti filosofo e matematico veneziano del secolo XVI. *Atti del Regio Istituto Veneto di Scienze, Lettere e Arti* 85: 585–754.
6. Brentjes, S. 1997–1998. Additions to Book I in the Arabic traditions of Euclid's Elements. *Studies in History of Medicine and Science* 15: 55–117.
7. Brentjes, S. 2001a. Two comments on Euclid's Elements? On the relation between the Arabic text attributed to al-Nayrīzī and the Latin text ascribed to Anaritius. *Centaurus* 43: 17–55.CrossRefMATH
8. Brentjes, S. 2001b. Observations on Hermann of Carinthia's Version of the Elements and its relation to the Arabic transmission. *Science in Context* 14: 39–84.CrossRefMATH
9. Busard, H.L.L. 1967. The translation of the Elements of Euclid from the Arabic into Latin by Hermann of Carinthia (?). *Janus* 54: 1–140.MATH
10. Busard, H.L.L. 1971. Die Arithmetica speculativa des Johannes de Muris. *Scientiarum Historia*13: 103–32.
11. Busard, H.L.L. 1977. *The translation of the Elements of Euclid from the Arabic into Latin by Hermann of Carinthia (?) Books VII–XII*. Amsterdam: Mathematisch Centrum.MATH
12. Busard, H.L.L. 1983. *The first Latin translation of Euclid's Elements commonly ascribed to Adelard of Bath*. Toronto: Pontifical Institute of Mediaeval Studies.MATH
13. Busard, H.L.L. 1984. *The Latin translation of the Arabic verion of Euclid's Elements commonly ascribed to Gerard of Cremona*. Leiden: Brill.MATH
14. Busard, H.L.L. 1987. *The mediaeval Latin translation of Euclid's Elements*. Stuttgart: Steiner.MATH
15. Busard, H.L.L. 1991. *Jordanus de Nemore. De elementis arithmetice artis. A medieval treatise on number theory*. Stuttgart: Steiner.MATH
16. Busard, H.L.L. 1996. *A thirteenth-century adaptation of Robert of Chester's version of Euclid's Elements*. München: Institut für Geschichte der Naturwissenschaften.MATH
17. Busard, H.L.L. 2001. *Johannes de Tinemue's redaction of Euclid's Elements, the so-called Adelard III version*. Stuttgart: Steiner.MATH
18. Busard, H.L.L. 1998. Zwei mittelalterliche Texte zur theoretischen Mathematik: Die Arithmetica speculativa von Thomas Bradwardine und die Theorica numerorum von Wigandus Durnheimer. *Archive for History of Exact Sciences* 53: 97–124.CrossRef
19. Busard, H.L.L. 2005. *Campanus of Novara and Euclid's Elements*. Stuttgart: Steiner.MATH
20. Busard, H.L.L., and M. Folkerts. 1992. *Robert of Chester's (?) Redaction of Euclid's Elements, the so-called Adelard II Version*. Basel: Birkhäuser.CrossRefMATH
21. Cassinet, J. (ed.). 1986. Cahiers d'histoire des mathématiques de Toulouse, 9.
22. Curtze, M. 1899. *Anaritii in decem libros priores elementorum Euclidis commentarii ex interpretatione Gherardi Cremonensis*. Leipzig: Teubner.

23. De Risi, V. 2007. *Geometry and monadology. Leibniz's analysis situs and philosophy of space*. Basel: Birkhäuser.MATH
24. De Risi, V. 2014. Francesco Patrizi e la nuova geometria dello spazio. In *Locus-spatium*, ed. D. Giovannozzi, and M. Veneziani, 269–327. Olschki: Firenze.
25. De Risi, V. (ed.). 2015a. *Mathematizing space. The objects of geometry from Antiquity to the Early Modern Age*. Basel: Birkhäuser.MATH
26. De Risi, V. 2015b. *Leibniz on the parallel postulate and the foundations of geometry*. Basel: Birkhäuser.MATH
27. Descotes, D. (ed.). 2009. *Géométries de Port-Royal*. Paris: Champion.
28. de Young, G. 2003. The Tahrir of Euclid's Elements by Nasit al-Din al-Tusi: Redressing the balance. *Farhang* 15–16: 117–143.
29. de Young, G. 2004. The Latin translation of Euclid's Elements attributed to Gerard of Cremona in relation to the Arabic transmission. *Suhayl* 4: 311–383.MATH
30. de Young, G. 2009. The Tharir Kitab Usul Uqlidis of Nasit al-Din al-Tusi: Its sources. *Zeitschrift für Geschichte der Arabisch- Islamischen Wissenschaften* 18: 1–71.MATH
31. Folkerts, M. 1970. *Boethius Geometrie II*. Stuttgart: Steiner.MATH
32. Garibaldi, A.C. (ed.). 2002. Francesco Maurolico. Euclides et geometrica quaedam. In *Francisci Maurolyci opera mathematica*. http://www.maurolico.unipi.it/ edizioni/ euclide/intro.htm.
33. Giusti, E. 1993. *Euclides Reformatus. La teoria delle proporzioni nella scuola galileiana*. Torino: Bollati Boringhieri.
34. Greenberg, M.J. 1988. Aristotle's axiom in the foundations of geometry. *Journal of Geometry*33: 53–57.CrossRefMATH
35. Greenberg, M.J. 2010. Old and new results in the foundations of elementary plane Euclidean and non-Euclidean geometries. *The American Mathematical Monthly* 117: 198–219.CrossRefMATH
36. Hallett, M., and U. Majer (eds.). 2004. *David Hilbert's lectures on the foundations of geometry 1891–1902*. Berlin: Springer.MATH
37. Hartshorne, R. 2000. *Geometry: Euclid and beyond*. New York: Springer.CrossRefMATH
38. Høyrup, J. 1988. Jordanus de Nemore, 13th century mathematical innovator: An essay on intellectual context, Achievement, and Failure. *Archive for History of Exact Sciences* 38: 307–363.
39. Itard, J. 1961. *Les livres arithmétiques d'Euclide*. Paris: Hermann.MATH
40. Jullien, V. 1996. *Eléments de géométrie de G.P. de Roberval*. Paris: Vrin.
41. Knorr, W.R. 1990. John of Tynemouth alias John of London: Emerging portrait of a singular medieval mathematician. *The British Journal for the History of Science* 23: 293–330.CrossRefMATH
42. Knorr, W.R. 1996. The wrong text of Euclid: On Heiberg's text and its alternatives. *Centaurus*36: 208–276.CrossRefMATH
43. Lo Bello, A. 2003a. *The commentary of Al-Nayrizi on Book I of Euclid's Elements of Geometry*. Leiden: Brill.MATH
44. Lo Bello, A. 2003b. *Gerard of Cremona's translation of the commentary of Al-Nayrizi on Book I of Euclid's Elements of geometry*. Leiden: Brill.MATH
45. Loget, F. 2004. La Ramée critique d'Euclide. Sur le Prooemium mathematicum (1567). *Archives internationals d'histoire des sciences* 54: 3–28.CrossRefMATH
46. Molland, A.G. 1978. An examination of Bradwardine's geometry. *Archive for History of Exact Sciences* 19: 113–175.CrossRefMATH
47. Molland, A.G. 1983. Campanus and Eudoxus; or, trouble with texts and quantifiers. *Physis* 25: 213–225.MATH
48. Molland, A.G. 1989. *Thomas Bradwardine. Geometria speculativa*. Stuttgart: Steiner.
49. Murdoch, J.E. 1963. The medieval language of proportions: Elements of the interaction with greek foundations and the development of new mathematical techniques. In *Scientific change*, ed. A.C. Crombie. London: Heinemann.
50. Oosterhoff, J. 2014. The Fabrist Origins of Erasmian Science: Mathematical Erudition in Erasmus' Basle. *Journal of Interdisciplinary History of Ideas* 3: 1–37.

51. Pagli, P. 2003. Volgarizzazioni manoscritte degli Elementi di Euclide. In *Il sogno di Galois*, ed. R. Franci, P. Pagli, and A. Simi, 53–90. Siena: Centro Studi della Matematica Medioevale.
52. Palladino, F. 1991. Sulla teoria delle proporzioni nel Seicento. Due macchinazioni notevoli: Le sezioni dei razionali del galileiano G.A. Borelli; Le classi di misura del gesuita A. Tacquet. *Nuncius* 6: 38–81.CrossRef
53. Pambuccian, V. 1994. Zum Stufenaufbau des Parallelenaxioms. *Journal of Geometry* 51: 79–88.CrossRef
54. Pascal, B. 1904–1914. *Oeuvres*, ed. L. Brunschvicg, P. Boutroux, and F. Gazier, Paris: Hachette.
55. Riccardi, P. 1974. *Saggio di una bibliografia Euclidea*. Hildesheim, Olms (1887–189211).
56. Rashed, R., and B. Vahabzadeh. 1999. *Al-Khayyām mathématicien*. Paris: Blanchard.MATH
57. Roberts, G., and F. Smith (eds.). 2012. *Robert Recorde. The Life and Times of a Tudor Mathematician*. Cardiff: University of Wales Press.MATH
58. Rommevaux, S. 1999. La proportionnalité numérique dans le livre VII des Éléments de Campanus. *Revue d'histoire des mathématiques* 5: 83–126.
59. Rommevaux, S., A. Djebbar, and B. Vitrac. 2001. Remarques sur l'Histoire du Texte des Éléments d'Euclide. *Archive for History of Exact Sciences* 55: 221–95.CrossRefMATH
60. Schlimm, D. 2013. Axioms in mathematical practice. *Philosophia Mathematica* 21: 37–92.CrossRefMATH
61. Steck, M. 1981. *Bibliographia Euclideana*. Hildesheim: Gerstenberg.MATH
62. Sutto, J.-P. 2000. Le compendium du 5ee livre des Éléments d'Euclide de Francesco Maurolico. *Revue d'histoire des mathématiques* 6: 59–94.
63. Taylor, E.G.R. 1954. *The mathematical practitioners of Tudor & Stuart England*. Cambridge: CUP.MATH
64. Taylor, E.G.R. 1966. *The mathematical practitioners of Hanoverian England*. Cambridge: CUP.MATH
65. Torricelli, E. 1919–1944. *Opere*, ed. G. Loria, and G. Vassura. Faenza: Montanari.
66. Tummers, M.J.E. 1994. *The Latin translation of Anaritius' commentary on Euclid's Elements of Geometry, Books I–IV*. Nijmegen: Ingenium.
67. Tummers, M.J.E. (ed.). 2014. Albertus Magnus. Super geometriam Euclidis. In *Opera omnia*, vol. 39, Münster, Aschendorff.
68. Vitrac, B. 1990–2001. *Euclide. Les Éléments*. Paris: PUF.
69. Voelke, J.-D. 2005. *Renaissance de la géométrie non euclidienne entre 1860 et 1900*. Bern: Lang.
70. Volkert, K. 2015. *David Hilbert. Grundlagen der Geometrie*. Berlin: Springer.
71. Whiteside, D.T., et al. (ed.). 1967–1981. *The mathematical papers of Isaac Newton*. Cambridge: CUP.

CHAPTER 8

The Resolution of the Great 20th Century Debate in the Foundations of Mathematics

Edgar E. Escultura

GVP-Professor V. Lakshmikantham Institute for Advanced Studies, GVP College of Engineering, JNT University Kakinada Madurawada Visakhapatnam, India

ABSTRACT

The paper resolves the great debate of the 20th century between the three philosophies of mathematics—logicism, intuitionism and formalism—founded by Bertrand Russell and A. N. Whitehead, L. E. J. Brouwer and David Hilbert, respectively. The issue: which one provides firm foundations for mathematics? None of them won the debate. We make a critique of each, consolidate their contributions, rectify their weakness and add our own to resolve the debate. The resolution forms the new foundations of mathematics. Then we apply the new foundations to assess the status of Hilbert's 23 problems most of which in foundations and find out which ones have been solved, which ones have flawed solutions that we rectify and which ones are open problems. Problem 6 of Hilbert's problems—Can physics be axiomatized?—is answered yes in E. E. Escultura, Nonlinear Analysis, A-Series: 69(2008), which provides the solution, namely, the grand unified theory (GUT). We also point to the resolution of the 379-year-old Fermat's conjecture (popularly known as Fermat's last theorem) in E. E. Escultura, Exact Solutions of Fermat's Equations (Definitive Resolution of Fermat's Last Theorem), Nonlinear Studies, 5(2), (1998). Likewise, the proof of the 274-year-old Goldbach's conjecture is in E. E. Escultura, The

New Mathematics and Physics, Applied Mathematics and Computation, 138(1), 2003.

Keywords: Axiom of Choice, Banach-Tarski Paradox, Goldbach's Conjecture, Logicism, Constructivism, Fermat's Conjecture, Field Axioms, Formalism, Qualitative Modelling, Rational Thought, Self-Reference

1. INTRODUCTION

Technical mathematics—the mathematics in the various specialized fields of mathematics—has expanded tremendously, especially, nonlinear analysis but work on the foundations of mathematics has stopped by the latter half of the 20th Century. Although the only living founder of logicism and pioneer in foundations, Bertrand Russell, was still active in the 1960s, he was consumed by his opposition to nuclear weapons and the Vietnam War during which time he founded and convened the Bertrand Russell Peace Foundation which tried the United States for crimes against humanity in Vietnam (he passed away in 1970 at the age of 97). We revisit the foundations of mathematics and resolve the great debate of the 20th Century among logicism [1] , intuitionism [2] and formalism [3] founded, respectively, by Bertrand Russell (1872-1970) and A. N. Whitehead (1861-1947) [4] [5] , L. E. J. Brouwer (1881-1966) [6] and David Hilbert (1862-1946) [7] . The issue: which one provides firm foundations for mathematics [8] ? None of them won the debate. Therefore, we consolidate their contributions and add our own to resolve the debate and call the resolution the new foundations of mathematics. We apply the new foundations to David Hilbert's 23 problems of mathematics [9] most of which problems in foundations and assess which ones are solved, which solutions are flawed in which case we rectify them and which ones are open problems. Problem 6—Can physics be axiomatized?—is answered yes in 2008 and the solution is The Grand Unified Theory [10] . We solve other long standing unsolved problems including the 379-year-old Fermat's conjecture (also known as Fermat's last theorem (FLT) [11] and the 274-year-old Goldbach's conjecture [12] . The paper summarizes the requirements for avoiding uncontained ambiguity and paradox or contradiction.

We introduce qualitative mathematics, the complement of computation and measurement and qualitative model of rational thought [13] , part of the main contribution of [14] , consisting of this activity:

Making conclusions, visualizing, abstracting, thought experimenting, learning, creating mathematical spaces, intuition, imagination, trial and error to sift out what is correct, negating what is known to gain insights into the unknown, altering premises to draw out new conclusions, thinking backwards and devising strategies and techniques that yield results.

2. THE GREAT 20TH CENTURY DEBATE

Why is the resolution of this great debate important? Important mainly because technical mathematics cannot solve problems in foundations and most long standing unsolved problems, e.g., FLT and Goldbach's conjecture, including most of Hilbert's 23 problems, belong to this category. To resolve the issue of contention in the debate we look at the sources of ambiguity and paradox or contradiction.

2.1. Sources of Ambiguity

Ambiguity erodes the foundations of mathematics. Therefore, we identify the sources of ambiguity. Ambiguity and contradiction are intertwined because when there is ambiguity in a mathematical space one can find contradictory statements or a counterexample (which is a contradiction) as shown by the work of Imre Lakatos [15] where he found a counterexample at every step of Cauchy's proof of Euler's formula on the relationship between the vertices, edges and faces of a polyhedron including his own remedy to the contradiction. We find ways to avoid them.

2.1.1. Vacuous Concept

One such source of contradiction is vacuous (ill-defined or non-existent) concept. Consider the concept, the root of the equation $x^2 + 1 = 0$ (among the real numbers) denoted by $i = \sqrt{-1}$. Since the equation has no root the concept i is vacuous, does not exist. Consequently,

(1) $i = \sqrt{-1} = \sqrt{1/-1} = 1/i = -i$, from which follows 1 = 0 and i = 0 which collapse both the real and complex number systems. The concept is vacuous for another reason: the mapping $\sqrt{a}$ is defined only on non-negative real number a where a is a perfect square. In fact, the nth root of a, i.e., $\sqrt[n]{a}$ is defined only if a is a perfect nth power of some real number.

Another contradiction arising from a vacuous concept called Perron Paradox [11] follows:

Let N be the largest integer. Since the integers are linearly ordered by the relation <, one and only one of the following holds: N < 1, N = 1, 1 < N. The first inequality is out, so is the second since N < N^2 and N would not be the greatest integer. Therefore, N = 1 which is false.

The culprit: N is vacuous, does not exist.

The concept vacuousity applies to vacuous propositions and problems. This is the reason a careful mathematician pays particular attention to a suitable theorem on the existence of a solution of a problem before actually solving the problem. The constructivist requires not only existence but also an algorithm for computing the solution. That algorithm makes sense only when there is a solution.

2.1.2. Infinity

In traditional mathematics a set is infinite if it can be put into one-one-correspondence with a proper subset. We modify this definition by identifying the concept infinity with its essential property of inexhaustibility. We take inexhaustibility as the defining quality of infinity and this clearly includes the traditional definition. For example, if we count the digits of a nonterminating decimal and label the digits we have already counted by, say, $x_1, x_2, \cdots, x_n$, then a sequence, $x_1, x_2, \cdots, x_n, \cdots, n = 1, 2, \cdots,$ is generated that has no last element and the counting is never complete. This is an ambiguity. It is not the case with a terminating decimal where there is a last element so that its digits are finite. This is an example of countable infinity denoted by ¥. It is a concept that pervades mathematics and the only type of infinity that exists as we shall see later. It is neither a real number nor a counting number and, naturally, the binary additive and multiplicative operations do not apply to ¥ and if real numbers are added to a given real number one at a time ¥ can never be reached. Thus, there is no boundary between the real numbers and infinity that can be crossed. In other words, none of the binary operations in the real number system applies to it.

Since the binary operations addition and multiplication are defined only on terminating decimals the nonterminating decimals are ill-defined and, therefore, ambiguous in the real number system. Therefore, any concept defined in terms of nonterminating decimal is ambiguous, ill-defined. An example of such ambiguous concept is irrational number (irrational for short) defined as nonperiodic nonterminating decimal. Furthermore, periodicity or non-periodicity of a nonterminating decimal is not verifiable. Thus, an irrational number has at least two layers of ambiguity. This ambiguity is illustrated by the fact that the sum of $\sqrt{3}$ and $\sqrt{2}$ is unknown. Moreover, not every rational numbers (rationals for short), i.e., the quotients of two integers, is a real number; only when the divisor has no prime factor other than 2 and 5 is the quotient a terminating decimal, a rational and a real number. Thus, the rationals coincide with the terminating decimals and they are the only defined real numbers.

In the new real number system R^* of which the real number system R is a subspace, the concept nonterminating decimal is defined for the first time [16] [17] but has contained ambiguity. Ambiguity of a concept is contained if it is defined, i.e., it exists, and is approximated with a desired error by a defined concept. As example, the infinite set of decimal digits of a nonterminating decimal is contained and its ambiguity is likewise contained since if we cut or truncate a nonterminating decimal at the nth decimal digit, the margin or fraction of error is less than 10^{-n}. Here, the truncated number is a terminating decimal which is defined in the real number system. The error is the upper bound of the value or norm of the discarded tail of the nonterminating decimal in the truncation. The maximum error occurs when the discarded digits are all 9 s. In a mathematical space this type of ambiguity can be contained or isolated and does not cause a problem; in fact, containing an ambiguity or reducing an error to a desired number is a standard technique in approximation theory. Therefore, it is admissible. Even the definition of a rational as quotient (rational for short) as quotient of two integers is ambiguous. In fact, division of an integer by a prime other than 2 and 5 is ambiguous since the quotient is a nonterminating decimal. What is correct is that the rationals coincide with the terminating decimals. Even the alternative definition of a rational as periodic decimal is incorrect. The periodic decimal 0.99... is a counter example to it since neither addition nor multiplication is defined on it.

Ambiguity of a concept means that not all of its properties can be verified. It follows that any proposition including axiomor theorem involving an ambiguous concept is ambiguous, i.e., its conclusion is uncertain. In particular, a proposition involving the universal quantifier every, or the existential quantifier there exists applied to an infinite set is ambiguous due to the inexhaustibility of the latter. For example, suppose we want to prove that "every element of an infinite set S has property P". Start with an element x_1 and suppose it has this property (otherwise, the statement is outright false), then take another element x_2 and check if it has this property, etc. Then since S is inexhaustible verification of the truth of this statement is never complete, i.e., the statement is ambiguous. Similarly, by the same algorithm but starting with an element that does not have this property, it may not be possible to prove that there exists an element of S that has this property which is an ambiguity. Every infinite mathematical space is presently tainted with this type of ambiguity from the definition of limit of real analysis through the field axioms of the real number system [16] . Thus, the real number system is presently ambiguous. Consequently, FLT being formulated in it is also ambiguous and has no proof. It was this ambiguity of the real number system that catalyzed its critique-rectification as the new real number system [15] [16] , the reformulation of FLT in it and its resolution in 1998 that proved it false [18] . In some cases, it may not be possible to choose a particular element x_1 of an infinite set, e.g., in the construction of a non-measurable set the axiom of choice was invoked [19] [20] . This axiom is flawed when applied to infinite set because of the ambiguity of the latter.

Consider the axiom of choice, one of the axioms of the Zermelo-Fraenkel set theory or ZFC that says [20] .

Given any set of mutually disjoint nonempty family of sets there exists at least one set that contains exactly one element in common with each of the nonempty sets.

The elements of the set are mapped one-one into the family of mutually disjoint sets and the mapping is called the choice function.

This seems obvious but it is tainted with ambiguity by the existential quantifier when applied to infinite set. The axiom of choice or its variants—the completeness axiom and Dedekind cut—is involved everywhere in analysis, e.g., as one of the field axioms of the real number system and in proving limits (which involves the existential and universal quantifiers) and existence of a nonmeasurable set [11] . Moreover, there is no flawless proof of existence of non-measurable set. In other words, measurability is not an issue at this time. Therefore, real analysis can be simplified by dropping the requirement of measurability in analysis.

Consider Cantor's diagonal method [21] [22] in proving that the real numbers form a nondenumerable set. This method is flawed by the application of the existential quantifier "there exists" on infinite set. Moreover, the set of off diagonal elements generated by Cantor's diagonal method is a countable union of countable sets which is countable. Furthermore, it uses the indirect proof which is self-referent and ambiguous. Therefore, Cantor's diagonal method is flawed. With the failure of Cantor's diagonal method, the continuum hypothesis [23] [24] also collapses. This resolves problem 1 of Hilbert's 23 problems that seeks a transfinite number between a denumerable set and the continuum. The problem is vacuous and has no solution since the continuum in the sense of Cantor does not exist aside from the use of the indirect proof and the axiom of choice in the proof.

A recent surprise is the Banach-Tarski paradox [25] which has been used as a counterexample to the axiom of choice. First of all, the axiom of choice is ambiguous on its own terms when applied to infinite set and does not need a counterexample to confirm it.

The Banach-Tarski paradox is one of the most celebrated paradoxes in mathematics. One version of it states that given any two subsets A and B of R^3, which are bounded and have non-empty interior, it is possible to 'cut' A into a finite number of pieces which can be moved by rigid motions (translations and rotations) to form exactly B. This has many amusing consequences; it is most commonly stated as the assertion that a baseball can be cut into a finite number of pieces which can then be reassembled without distortion to form a ball as big as the Earth. The proof is in [25] ; we note the following important points about it:

(1) In the proof of Proposition 3.1, Section 3, p. 4, which is part of the proof of the paradox, is the expression,

$$\left(1/3^n\right)\left(a\sqrt{2},b,c\sqrt{2}\right), \tag{1}$$

involving division of an integer by a prime other than 2 and 5; both the quotient and the product are ill-defined because they involve decimals which are also ill-defined. Therefore, the quotient $1/3^n$does not exist and, naturally, the product in (1) does not exist. Ill-defined concepts appear many times in the proof.

(2) On page 8, the paragraph after fact (3), part of the sentence that begins in line 2 says:

"...and there are uncountably many lines through the origin in R^3"

which is, again, vacuous since those lines do not exist?

Only one of items (1) and (2) suffices to show that the Banach-Tarski paradox does not exist.

The ambiguity of infinity has far reaching implications for mathematics. It affects theorems in infinite mathematical spaces. It limits the domain of the real number system to terminating decimals, i.e., a nonterminating decimal is ill-defined (ambiguous). Thus, the concept irrational as nonperiodic nonterminating decimal is ill- defined and ambiguous since it is unverifiable. This is the reason it is impossible to add $\sqrt{2}$ and $\sqrt{3}$ since both terms are ambiguous. The rationals as quotients are limited to terminating decimals because division by a prime number other than 2 and 5 is ill-defined the quotient being nonterminating.

Moreover, ambiguity or uncertainty applies to small and large numbers. For example, changing a small divisor less than 1 (<1) leaves a wide margin of error. Furthermore, the digits of both small and large numbers (depending on context) cannot be entered into the calculator or computer screen and this is an ambiguity. In dealing with a large number the scientific notation is used which approximates the number by the largest order of magnitude and discarding the digits and terms of lower orders of magnitude. For a small number the nonzero (decimal) digits are too far to the right in the decimal expansion that it is difficult to distinguish its location relative to 0. This uncertainty applies to tiny object at great speed, e.g., electron at speed close to that of light which accounts for the Heisenberg's uncertainty principle of physics [26] . This is an example of motion in the real world affected by number not because of the numbers involved but the difficulty in determining the position of a tiny object and measuring large numbers. Furthermore, large integers have different behavior described by the law of large numbers in statistics. For instance, it is noted in [27] that the behavior of an integer N abruptly changes when its value goes past 10^k, where $k = 10^{70}$. Thus, both extremes of the norm of a real number suffer from ambiguity.

2.1.3. Russell Paradoxes

In 1901 Bertrand Russell who was then working on the Principles of Mathematics, rocked set theorists and logicians with this discovery named after him (Russell antimony) [28] :

Let M be the set of all sets where each element does not belong to itself, i.e., M = {m: m Ï m}. Either M Î M or M Ï M. If M Ï M, its defining conditions hold; therefore M Î M. On the other hand, if M Î M, then M also satisfies its defining condition; therefore M Î M and M Ï M.

Russell sent it in a letter to G. Frege in 1902 [28] just as Frege had submitted Grundlagen der Arithmetik that, like the Principles of Mathematics, attempted to build arithmetic on set theory, the foundation of logicism.

The letter invalidated much of the rigor of the work and Frege was forced to add a note at the end stating, "A scientist can hardly meet with anything more undesirable than to have the foundation give way just as the work is finished. I was put in this position by a letter from Mr. Bertrand Russell when the work was nearly through the press" [28] .

A contradiction arises whenever the expression "the set of all sets" is involved for it leads to the concept of a set being a member of itself and, therefore, not a member of itself.

In other words, the universal set and power set do not exist, they are vacuous. This, again, is an example, of thought creating nonsense. There are, in fact, numerous Russell paradoxes that can be reduced to this type of paradox [29] ; we give another example, the barber paradox [30] :

The barber of Seville shaves those who do not shave themselves. Who shaves the barber?

If the barber shaves himself then he does not shave himself; if he does not then he shaves himself. Obviously, there is no correct answer here, i.e., the barber of Seville does not exist. Thus, a Russell paradox reduces to a vacuous proposition.

2.1.4. Self-Reference

If we replace the relation "Î" by other relations that lead to nonsense we get a much larger class of contradictions called self-reference. For example, suppose we replace "Î" by "the ball half its size", can anyone comprehend what that ball is? Belonging to this type of paradox are these relations: "the snake that swallowed itself" (let it swallow the tail first) and "the basket inside itself". The indirect proof is self-referent because the conclusion of the theorem being proved rests on the negation of its hypothesis (a theorem, by the way, includes its proof). The indirect proof is based on the principle of the excluded middle of formal logic which is what the Russell antimony is and rejected by intuitionism and the broader philosophy of constructivism that extends to mathematics education where the teacher does not lecture but sets the right conditions for the student to create the appropriate concepts for the situation [31] -[35] .

Gödel's second incompleteness theorem states that no consistent axiomatic system which includes Peano arithmetic (Peano's axioms) can prove its own consistency. This can be stated simply and in more general terms: No consistent axiomatic system (mathematical space) can prove its own consistency. This theorem is true because a proof would be self-referent.

3. THE STATUS OF SET THEORY

The creator of set theory was Georg Cantor [20] [21] who introduced, among others, the concepts infinity, cardinal number and the axiom of choice. The subject was revisited by Russell and Whitehead who developed set theory on the axioms of formal logic in their joint work, Principia Mathematica [36] -[38] . This work attempts to reduce the foundations of mathematics to logic [39] [40] . Although the theory of types in Principia did resolve the Russell paradoxes, it was impossible to say that a class was or was not a member of itself. We have resolved this problem by putting the set of all sets where each element does not belong to itself and the set of all sets where each element belongs to itself in the category of vacuous set that must be avoided. Today, the most advanced development of set theory is Zermelo-Fraenkel with the axiom of choice or ZFC [20] . However, both its axioms of infinity and choice bring in uncontained ambiguity to it.

We identify two aspects of set theory: its role (1) as a fundamental language and repository of the basic principles of modern mathematics and (2) as an independent branch of formal or mathematical logic (propositional calculus). The first role does not apply to any other mathematical spaces since every mathematical space is defined solely by its axioms including its rules of inference for proving theorems. In other words, universal rules of inference such as formal logic do not apply to any other mathematical space since they have nothing to do with its axioms. As an independent mathematical space it is also flawed by the inclusion of the axiom of infinity and the axiom of choice. Therefore, the program of logicism to develop a universal language for mathematics failed and whatever concepts of set is needed in mathematics can take refuge in the appropriate mathematical spaces where they are defined.

4. THE PARTICIPANTS IN THE DEBATE

We present briefly the three philosophies of or schools of thought in mathematics.

4.1. Logicism

What is the status of logicism? It aims to build mathematics on the axioms and propositions (theorems) of formal logic which is cast in the language of set theory. The axioms of formal logic consists of the laws of classical logic [40] plus type theory [41] that provides the remedy for the Russell paradoxes and, essentially, avoids vacuous concepts such as the set of all sets where each element does not belong to itself from which follows that the universal and power sets of a set are vacuous and do not exist and the relation "Î" ("is a member of") is not reflexive. Moreover, we require that the relation "Î" is neither reflexive nor transitive, i.e., A Î B and B Î C does not imply A Î C; in other words, there is no conclusion that can be drawn from A Î B and B Î C. In addition, an axiom schema [39] is introduced that generates additional axioms. The rules of inference are defined by these axioms. A proposition is true by virtue of its form derived from the axioms [1] . The aim of logicism is to establish set theory as the universal language of mathematics, i.e., as the language of all mathematical spaces. First, the concepts of a mathematical space are defined in terms of the language of set theory. Then its axioms are derived from set theory in accordance with the rules of inference of formal logic or propositional calculus. Both, of course, are subject to the limitations of set theory. All the three philosophies of mathematics consider the axioms self-evident and true. However, what is self-evident to one may not be so to another and, as we have seen, thought can create contradictory and vacuous concepts, e.g., the complex number i. Consequently, there exist Euclidean and non-Euclidean geometry, Abelian and non-Abelian group and Hausdorff and non-Hausdorff mathematical spaces all of which created by thought. This nature of thought is explained in [13] [42] [43] .

The natural numbers are defined in the language of set theory by the axioms of Peano [20] :

1) Zero is a number.

2) If a is a number, the successor of a is a number.

3) Zero is not the successor of a number.

4) Two numbers of which the successors are equal are themselves equal.

5) (Induction Axiom) if a set S of numbers contains zero and also the successor of every number in S, then every number is in S.

Logicism failed to prove the axioms of arithmetic and its goal of deriving arithmetic from formal logic. Recall that it is the representation of thought in the real world that constitutes a mathematical space and not the concepts of individual thought. Therefore, writing the concepts of a mathematical space in the language of set theory and proving its theorems by the rules of inference of formal logic negates its axioms as the sole determinant of its structure, properties and behavior.

4.2. Intuitionism

The rejection of the indirect proof by intuitionism [2] is the foundation of constructivism which has been extended to a philosophy of mathematics-science education now called creative mathematics-science education [31] -[35] . Consequently, Brouwer had to reject his earlier important contribution to Analysis and Topology named after him, the Brouwer fixed point theorem [44] , because the proof rests on the indirect proof. However, constructivism as a philosophy is more than simply the rejection of the indirect proof and extension to mathematics-science-education since as we have seen it has engendered a new philosophy of education. Like logicism and formalism, intuitionism correctly views concepts as creation of thought but incorrectly assumes that human thought is invariant or the same among individuals. This is not the case according to [13] [42] [43] . In fact, individual thought is conditioned or shaped by the cultural milieu and one's training and experience. Moreover, we recall that individual thought is capable of creating ambiguous, vacuous and contradictory concepts. Constructivism is a modification of the Socratic method of teaching where the student states his opinion on a subject or an answer to a question and the teacher asks a series of leading questions towards the correct answer [33] -[35] . In constructivist teaching the teacher provides the condition in the classroom so that the student creates the appropriate concepts for the situation. Moreover, intuitionism correctly rejects concept or solution that cannot be constructed. Therefore, the solution must be determined by actual construction but the latter preceded by proof of its existence.

4.3. Formalism

Formalism recognizes the ambiguity of the concepts of individual thought since it is not accessible to others and cannot be discussed collectively [3] . Therefore, it requires concepts to be objects in the real world that can be collectively observed such as words, symbols, figures and chess pieces. In a chess game the axioms are the rules of the game. In other words, the subject matter of mathematics cannot be the concepts of individual thought but objects in the real world (also called concepts) subject to the axioms of the given mathematical space. Both logicism and intuitionism hold that an axiom is self-evident by virtue of its form regardless of content [1] [2] ; intuitionism says that a concept or an axiom is "evident from profound intuitionist reflection" [2] . At the same time, intuitionism admits correct ideas based on intuition but limits them to those that can be constructed or actualized which is consistent with the rejection of the principle of excluded middle, the basis of the indirect proof and a contradiction. Formalism does not challenge the logicist view but retains the indirect proof [3] . This contradicts the formalist requirement that the concepts and propositions of a mathematical space be subject to consistent axioms. Among the implications of this requirement are: 1) mathematical truth is not absolute but relative to the axioms of a mathematical space. In other words, when we say that a theorem is true it is relative to the mathematical space where the theorem belongs. A theorem in our sense is a sequence of arguments or propositions from the hypothesis to the conclusion, where each proposition except the hypothesis follows from the preceding proposition; 2) to avoid ambiguity every concept of a mathematical space must be defined by its axioms and the choice of the axioms is not complete until every concept is defined, i.e., undefined terms are inadmissible. 3) The rules of inference must follow from the axioms. 4) Therefore, as noted earlier, universalrulesofinferenceincludingformallogicdonotapplytoanyothermathe maticalspace. 5) Consequently, there cannot be a single set of axioms that defines traditional mathematics (which is the goal of logicism) or even geometry since there are contradictory mathematical spaces, e.g., Euclidean and Non-Euclidean geometry and Archimedean and non-Archimedean number systems. 6) The goal of logicism which is to derive arithmetic from formal logic is not achievable because formal logic has its own independent set of axioms.

Intuition still plays a role in mathematics based on experience but needs verification by the axioms or experience. For example, a skillful chess player takes the next move based on intuition (that comes with experience) and the position of the pieces; in this case, however, he cannot prove the correctness of his next move from the rules of the game due to time limitation. In this regard, we pose an open problem: Is there a position of a player's pieces that insures a win when attained in the course of a game?

The main contribution of logicism is the discovery of the Russell paradoxes and the remedy by type theory but left unresolved the flaw in set theory and contradiction in self-reference. It also failed to derive arithmetic from set theory. The main contribution of intuitionism is constructivist mathematics with the rejection of the indirect proof and the requirement that concepts and propositions be constructible, that of formalism the representation of thought by objects in the real world subject to consistent axioms. All three schools of thought agree that concepts are created by thought but do not distinguish between abstract concept, i.e., concept that has no referent in the real world such as mathematical concepts, and physical concept, i.e., concept that has referent in the real world such as atom, star and our universe. Moreover, formalism retains the indirect proof and like intuitionism does not openly challenge the logicist proof of theorems. However, in practice the formalist proves theorems directly from the axioms. Technical mathematics, especially, those with applications to science is constructivist. Non-constructivist applications of mathematics are unreliable and could be dangerous as what happened to the disastrous final flight of the Columbia Space Shuttle in 2004 [45] [46] .

5. THE NEW FOUNDATIONS OF MATHEMATICS

The resolution of the great debate which includes the critique-rectification of the foundations of traditional mathematics plus new requirements on every mathematical space to avoid uncontained ambiguity constitutes the new foundations of mathematics. The requirements include what axioms to choose that define a mathematical space. Our aim is to choose axioms that serve science and its applications, e.g., natural science, engineering, medicine and social science. The new requirements are:

1) Constructivism which requires the new mathematical methodology, qualitative mathematics. Among the earliest constructivist mathematical space is the rectified calculus of variations together with functional analysis and optimal control theory spurred by the discovery of the generalized curves and surfaces by L. C. Young that solved the calculus of variations and optimal control problems on curves and surfaces starting in 1937 [47] -[52] . Incidentally, this is also the constructivist solution of Problem 23 of Hilbert's 23 problems [20] posed a century ago. The most recent constructivist mathematics is the mathematics of GUT [53] that includes the new real number system and complex vector plane, nested generalized physical fractal and generalized integral and derivative [16] [17] [54] -[57] .

2) Admission of concepts only when defined solely by the axioms; this is a departure from traditional mathematics that admits undefined terms or

concepts other than grammatical connectives which have no bearing on thought. The choice of the axioms of a mathematical space is not complete until every concept is defined; when this phase is completed the mathematical space becomes deductive and every theorem follows from the axioms.

3) Avoidance of uncontained ambiguity.

4) Anchoring the rules of inference on the axioms.

5) Enhancing accuracy, clarity and simplicity by defining the relation "=" to mean "the same as" or "equivalent to" so that it is an equivalence relation and satisfies the identity, reflexivity, and transitivity properties which need not be taken as axioms.

At this point we focus on the real number system which lies at the foundations of mathematics. We have already noted the inconsistency arising from the ambiguity of the axiom of choice one of its defining field axioms. Then in the early 1920s Brouwer constructed a counterexample to the trichotomy axiom, another Field axiom [58] . In 1992 the author constructed a counterexample to it ([16]) independently which showed at the same time that the real number system is not linearly ordered; the latter collapses not only the real number system but also real and complex analysis. The counterexample is a theorem in [16] [17] [30] [57] . These inconsistences of the real number system are the rationale for its reconstruction into the consistent new real number system which provided the counterexample to FLT that proved it false [16] . The counterexample has since then been extended to countably infinite counterexamples in [16] -[18] [30] [57] [59] .

The axioms of the new real number system R^* are: Axiom 1: 0, 1 Î R^* where 0 and 1 are the additive and multiplicative identities defined by the addition and multiplication tables, respectively; Axiom 2: the addition table, and Axiom 3: the multiplication table. The identity elements 0 and 1, are defined by the addition and multiplication tables of Axioms 2 and 3; so are the rest of the integers which are the successors of 0, 1, 2, ···, (by integers we mean their representation in the real world since the concepts of thought are ambiguous; at present we use the Hindu-Arabic numerals for the integers and the metric system for the real numbers). Then in the definition of the real numbers (decimals) in [16] the integers are formally defined as the integral parts of the decimals. Then for the first time the isomorphism between the natural numbers and the integers was established defining at the same time that the natural numbers are the integers. Moreover, the addition to the identity, reflexivity and transitivity properties of "=", the commutativity and associativity of addition and multiplication and the distributivity of multiplication with respect to addition follow from the addition and multiplication tables and need not be taken as additional axioms for the real number system. How do we interpret 2 + 3 = 5? Addition is a binary mapping or operation f_+ on R^* so that 2 + 3 = $f_+(2, 3)$ = 5, the image of (2, 3) under f_+ in accordance with the addition table. This shuts off the controversy in cyberspace around the equation 1 = 0.99··· which raged for over a decade since 1998. The equation is wrong because 1 and 0.99··· are distinct objects; it is akin to the equation orange = apple. In fact, in the new real number system, 0.99··· < 1 and 1 – 0.99··· = the dark number d^* which is not a decimal but an element of R^*.

Consider the paradoxes (contradictions) that we have identified. The Russell paradoxes involve relations that are reducible to the relation Î = "is a member of" or Ï = "is not a member of" which have been resolved by Russell's theory of types that orders propositions into a linear hierarchy of orders [41] . Then the power set of a nonempty set A is inadmissible for it would lead to the Russell antimony m Î m; so is the universal set. The resolution of self-reference is simply to avoid it. If we eliminate self-referent propositions the indirect proof collapses since its conclusion rests on the negation of the hypothesis, a case of self-reference. Thus, the resolution of the Great Debate is constructivist and meets the other requirements we have set to avoid uncontained ambiguity (Constructivist mathematics is the subject of a sequel to this paper which will be illustrated in the constructivist development of R^*, the g-closure of R [16] [17]).

Finally, we include in the new foundations of mathematics the rectification work on the real number system R and complex number system C which is the replacement of the field axioms by the axioms of the new real number system and the replacement of the complex number i by the plane vector operator j:

Axiom 1. R^* contains the elements 0, 1.

Axiom 2. The addition table (Table 1).

Axiom 3. The multiplication table (Table 2).

The partial Table 1 and Table 2 below are axioms because they insure the existence of the integers, define addition and multiplication as binary operations on them and specify their properties. We first construct the digits or basic integers: 2 = 1 + 1, 3 = 2 + 1, 4 = 3 + 1, 5 = 4 + 1, 6 = 5 + 1, 7 = 6 + 1, 8 = 7 + 1, 9 = 8 + 1 which apply only to the integers which are sums and products. It is also the metric system of numeration or scientific notation for large numbers. To this end, we define the base integer 10 = 1 + 9.

Their extension to the full tables using the Hindu Arabic or base 10 numerals is quite familiar and will not concern us here.

Table 1. The addition table.

+	0	1	2	3	4	5	6	7	8	9
0	0	1	2	3	4	5	6	7	8	9
1	1	2	3	4	5	6	7	8	9	
2	2	3	4	5	6	7	8	9		
3	3	4	5	6	7	8	9			
4	4	5	6	7	8	9				
5	5	6	7	8	9					
6	6	7	8	9						
7	7	8	9							
8	8	9								
9	9									

Table 2. The multiplication table.

×	0	1	2	3	4	5	6	7	8	9
0	0	0	0	0	0	0	0	0	0	0
1	0	1	2	3	4	5	6	7	8	9
2	0	2	4	6	8					
3	0	3	6	9						
4	0	4	8							
5	0	5								
6	0	6								
7	0	7								
8	0	8								
9	0	9								

Since the tables define finite sums and products the laws of addition and multiplication of arithmetic can be verified from them. In either case, finite mathematical induction can be applied if needed. The system of Hindu-Arabic numerals quantitatively models the metric system as a system of measurement.

Our proposed remedy for the vacuous concept i of the complex number system C is its replacement by the operator j defined as left-right operator on or mapping of a plane vector by positive (counterclockwise) rotation about the origin by p/2. Then the coordinate axes are: the x-axis (x, 0), application of j on the x-axis (rotation of (x, 0) by π/2), to obtain the y-axis, i.e., j(x, 0) = jy; application of j on y-axis (rotation of the y-axis by π/2) to obtain the −x-axis, i.e., jj(x, 0) = −(x, 0) = (−x, 0); application of j on the −x-axis (rotation by π/2) to obtain the −y-axis; , i.e., j(−x, 0) = −jy and application of j on the −y-axis (rotation by π/2) to obtain the x-axis, i.e., j(−y, 0) = (x, 0). Thus, we have the cyclic values of the composites of j with itself:

$$\mathbf{j}(\mathbf{1}) = \mathbf{j},\ \mathbf{jj}(\mathbf{1}) = \mathbf{j}^2 = -\mathbf{1},\ \mathbf{jj}^2(\mathbf{1}) = \mathbf{j}^3 = -\mathbf{j},\ \mathbf{j}(\mathbf{j}^3) = \mathbf{j}^4 = \mathbf{1} \quad (2)$$

where 1 is the unit vector along the x-axis.

They are the vector basis of the complex vector plane [10] . We define the operator −j as inverse operator of j, i.e., when applied on a vector v we rotate it clockwise through π/2. Note: −j(1) = j(−1). Applying composite mappings on the unit vector 1 along the x-axis successively, we have the four cyclic images of 1 in (2) in reverse order. For n > 4, the cycle is repeated and we obtain $j^n = j(j^{n-1})$, n = 1, 2, ⋯, where we define $j^0 = 1$. Then by replacing both the x-axis and yi axis by the new real number system and yj, where y ranges through the new real numbers, we obtain the complex vector plane C^*, the full rectification of the complex number system [17] . The constructivist development of the new real number system and complex vector plane will be the content of a sequel to this paper.

6. STATUS OF HILBERT'S PROBLEMS AND OTHER LONG-STANDING PROBLEMS OF MATHEMATICS

Hilbert's problems originally consist of unsolved problems in mathematics proposed by Hilbert. Ten of them were presented at the Second International Congress of Mathematicians in Paris on August 8, 1900. They were problems 1, 2, 6, 7, 8, 13, 16, 19, 21, and 22. The complete list is in [9] . They were designed to serve as examples of the kinds of problems the solutions of which would advance mathematics. As such, some were areas for investigation not strictly "problems". Reference [60] lists over a hundred problems including some of Hilbert's problems that have been solved but we focus on Hilbert's problems.

1) "Cantor's problem of the cardinal number of the continuum." The question of whether there is a transfinite number between that of a denumerable set and the numbers of the continuum was answered by Gödel and Cohen in their solution to the continuum hypothesis to the effect that the answer depends on the particular version of set theory assumed. The question of whether the continuum of numbers should be considered a well ordered set is related to Zermelo's axiom of choice. In 1963, the axiom of choice was demonstrated to be independent of all other axioms of set theory. There is universal consensus on whether these results give a solution to this problem. (Of course, mathematics is not a matter of consensus)

2) "The compatibility of the arithmetical axioms." Gödel's second incompleteness theorem indicated that it cannot be proven that the

axioms of logic are consistent in the sense that any formal system interesting enough to formulate its own consistency can prove its own consistency if it is inconsistent. There is no consensus if the results of Gödel and Gentzen provide a solution.

3) Give two tetrahedra that cannot be decomposed into congruent tetrahedra directly or by adjoining congruent tetrahedra. Dehn (1900, 1902) showed that a regular tetrahedron cannot be decomposed into a finite number of congruent tetrahedra (directly or by joining congruent tetrahedra) which can be reassembled to make a cube. It follows immediately from this result that two tetrahedra cannot be decomposed, as Hilbert proposed.

4) Find geometries whose axioms are closest to those of Euclidean geometry if the ordering and incidence axioms are retained, the congruence axioms weakened, and the equivalent of the parallel postulate omitted. This problem was solved by G. Hamel.

5) Can the assumption of differentiability for functions defining a continuous transformation group be avoided? This is a generalization of the Cauchy functional equation solved by John von Neumann in 1930 for bicompact groups. Also solved for the Abelian case, and for the solvable case in 1952 with complementary results by Montgomery and Zippin (subsequently combined by Yamabe in 1953). Andrew Gleason showed in 1952 that the answer is also "yes" for all locally bicompact groups.

6) Can physics be axiomatized?

7) Let α ^{1} 1 ^{1} 0 be algebraic and β irrational. Is α^{β} transcendental? In particular, are the Gelfond-Schneider constant $2^{\sqrt{2}}$ and Gelfond's constant e^{p} transcendental (Wells, 1986)? A β is known to be transcendental for the special case of an irrational algebraic number, as proved in 1934 by Aleksander Gelfond in a result now known as Gelfond's theorem (Courant and Robins, 1996). However, the case of nonalgebraic irrational β has not been resolved, with solutions known only for degenerate constructions such as $\alpha = 2$, $\beta = \ln 3/\ln 2$.

8) Prove the Riemann hypothesis. The conjecture has been neither proved nor disproved.

9) Construct generalizations of the reciprocity theorem of number theory.

10) Does there exist a universal algorithm for solving Diophantine equations? The impossibility of obtaining a general solution was proven by Yuri Matiyasevich in 1970 (Matiyasevich, 1970; Davis, 1973; Davis and Hersh, 1973; Davis, 1982; Matiyasevich, 1993) by showing that the relation $n = F_{2m}$ (where is the $(2\ m)^{Th}$ Fibonacci number) is

Diophantine. More specifically, Matiyasevich showed that there is a polynomial Pin n, m, and a number of other variables x, y, z, ··· having the property that n = F_{2m} iff there exist integers x, y, z, ··· such that P(n, m, x, y, z) = 0.

11) Extend the results obtained for quadratic fields to arbitrary integer algebraic fields.

12) Extend a theorem of Kronecker to arbitrary algebraic fields by explicitly constructing Hilbert class fields using special values. This calls for the construction of holomorphic functions in several variables which have properties analogous to the exponential function and elliptic modular functions (Holzapfel, 1995).

13) Show the impossibility of solving the general seventh degree equation by functions of two variables.

14) Show the finiteness of systems of relatively integral functions.

15) Justify Schubert's enumerative geometry (Bell, 1945).

16) Study the topology of real algebraic curves and surfaces. See Gudkov and Utkin (1978), Ilyashenko and Yakovenko (1995), and Smale (2000) for additional details.

17) Find a representation of definite form by squares.

18) Build spaces with congruent polyhedra.

19) Analyze the analytic character of solutions to variational problems.

20) Solve general boundary value problems.

21) Solve differential equations given a monodromy group. More technically, prove that there always exists a Fuchsian system with given singularities and a given monodromy group. Several special cases had been solved, but a negative solution was found in 1989 by B. Bolibruch (Anasov and Bolibruch, 1994).

22) Uniformization.

23) Extend the methods of calculus of variations.

6.2 Assessment of Hilbert's Problems

We mark by an asterisk (*) when the solution of a problem has been challenged or replaced by the author, by double asterisk (**) when the solution of a problem has not been reviewed by the author and by triple asterisk (***) when a problem has been solved but not listed among the solved problems. An unmarked problem is open.

1)* Problem 1 was supposedly solved by Gödel and Cohen in their proof of the continuum hypothesis [23] [24] according to [9] . There are two reasons why the proof is flawed: (1) Cantor's proof that the reals are nondenumerable is flawed because it involves the application of the axiom of choice on infinite set and (2) the power set S of a denumerable set does not exist being a Russell antimony type paradox that leads to the conclusion S Î S and S Ï S. Therefore, the problem is vacuous and has no solution. Moreover, the discussion of the solution confirms our position that the language of set theory cannot be a universal language for mathematics because the arguments and conclusions vary with the kind of or axioms for the set theory being applied.

2)* Problem 2 is incorrectly formulated and we replace it by a general theorem: Any mathematical space (consistent axiom system) cannot prove its own consistency or inconsistency for such a proof would be self-referent and flawed. Consensus is inadmissible in constructivist mathematics being a deductive system; a theorem or solution to a problem follows from the axioms.

3)** It is reported in [9] that the solution follows from a result by Dehn (1900, 1902).

4)** It is reported in [9] that Problem 4 has been solved by G. Hamel.

5)** The problem has been solved for special cases [9] .

6)*** The answer to problem 6 is yes and the author provided the solution in 2008, namely, the Grand Unified Theory (GUT) [10] . The solution is constructivist.

7)* Problem 7 has been partially solved according to [9] but based on the new foundations of mathematics the problem itself is ambiguous because it involves the ill-defined irrational and, naturally, the supposed partial solution is flawed.

8) Problems 8 and 9 are open.

10)** We have seen that the real number system is not well defined because two of its axioms, the axiom of choice or its variants—the Dedekind cut and completeness—are ambiguous and false on infinite set. At the same time, the claim of impossibility of obtaining a general solution by Yuri Matiyasevich, Davis and Hershis are not true for the same reason.

11) Problems 11 to 15 are open.

16)** Gudkov and Utkin (1978), Ilyashenko and Yakovenko (1995) and Smale (2000) have done what Problem 16 asks for according to [9] .

17) Problems 17 to 20 are open.

21)** Several special cases of Problem 21 had been solved, but a negative solution was found in 1989 by B. Bolibruch (Anasov and Bolibruch, 1994) according to [9] .

22) Candidly, the author does not understand Problem 22.

23)*** L. C. Young not only extended the methods of the calculus of variations asked in Problem 23 but also introduced a new method that he applied to functional analysis and optimal control theory in R^n, n ³ 3, and solved the general calculus of variations and optimal control problems in the mathematical spaces of curves and surfaces [47] -[52] . The method entails enrichment of the domains of curves and surfaces with the introduction of generalized curves and surfaces, respectively, and a suitable norm called Young measure to achieve a complete space in the norm for each space so that the optimal solution is calculated as the limit in this norm of a sequence of simplicial curves or surfaces which is a generalized curve or surface in either case. The solution is constructivist. There is no indication at all in Young's work that he was aware he solve done of Hilbert's problems. His method of enrichment in a suitable norm has become a standard method that the authored rediscovered and used to develop the new real number system. Although Brouwer introduced the idea of constructivist mathematics and proved theorems that are constructivist, it was Young who first developed a whole body of constructivist mathematics in the 20th Century.

The list in [60] erred in not reporting this major development in mathematics from which sprung modern control theory.

***The 378-year-old conjecture by Fermat (popularly known as Fermat's last theorem (FLT)) [11] that says, "The Diophantine equation $x^n + y^n = z^n$ has no integer solutions for n > 2, xyz ¹ 0", was supposedly solved by Wiles and Taylor [61] [62] according to [60] . Unfortunately, the solution was not informed by the counterexample to the trichotomy axiom constructed by Brouwer in 1985 [58] and independently by the author in 1998 [16] . The trichotomy axiom is one of the defining axioms of the real number system [9] . The counter example to it is another proof that the real number system R is inconsistent. Since FLT is formulated in R, it is also ambiguous and, naturally, has no solution. Consequently, the proof by Wiles and Taylor is false. To resolve FLT it was necessary to reconstruct the real number system R into the consistent constructivist new real number system $R^*$and reformulate FLT in it. The construction was started in 1998 [16] with the introduction of the new integers N.99×××, and the dark number $d^* = N - (N - 1).99\cdots, 1 - 0.99\cdots, N = 1, \cdots$, where N is an integer. The element d^* is a new real number but not a real number; $0 < d^* < x$, x Î R. It is both the qualitative and quantitative model of the superstring, fundamental building block of matter [63] . The first countable counterexamples to FLT (1998 [18] is the solution of the above Diophantine equation with (x, y, z) = $((0.99\cdots)10^T, d^*,$

10^T), T = 1, 2, ⋯, that clearly satisfies it. The countably infinite solutions are the triples (kx, ky, kz), k = 1, 2, ⋯, which also satisfy Fermat's equation. The existence of such countably infinite solutions is a theorem in [16] -[18] [30] [57] [59] . The full construction of R^* will be published in a sequel, The Constructivist Development of the New Real Number System.

***The 273-year old strong Goldbach's conjecture [12] that says, All positive even integers ^{3} 4 can be expressed as the sum of two primes, was proved in 2003 [16] . It is a theorem in [17] [30] [57] [59] [63] . The conjecture and its proof are incorrectly excluded from the list of solved problems listed in [60] . The proof is likewise constructivist.

7. CONCLUSIONS

One thing stands out in this paper: to pursue constructivist mathematics a new mathematical methodology is needed: the introduction of qualitative mathematics, the qualitative model of rational thought [13] and the complement of computation and measurement and a suitable norm, in this case, the g-norm [16] [17] . In the work of L. C. Young this translates into the method of enrichment and the Young Measure.

Examples of qualitative mathematics are abstract mathematical spaces and the search for the laws of nature. In science, the appropriate methodology is qualitative mathematics and modeling [13] that explains how nature works its tool qualitative mathematics. It is the main contribution of [14] and was applied to physics for the first time to solve the gravitational n-body problem [64] . Qualitative mathematics and modeling shifts the subject matter of science from the appearances of nature (natural phenomena) to nature itself and how it works and lifts traditional science to the new science [65] articulated by GUT and its theoretical applications [13] [30] [43] [44] [46] [66] -[68] . Through this methodology GUT unites the natural sciences on the natural laws, the common thread being the superstring. It is clear that natural science and mathematics are distinct disciplines: natural science is the study of nature its language mathematics. On the philosophical side: at the super micro scale, nothing is smooth in nature because of the uncertainty of small numbers that manifests itself in the Heisenberg uncertainty principle. Moreover, while at the micro scale the second law of thermodynamics says that a closed system tends towards greater disorder or chaos or zero entropy [69] , the opposite is true at the macro scale. For example, from the chaos of the big bang [70] and the cosmic burst [71] , our universe has evolved to its present order in accordance with natural laws [66] .

REFERENCES

1. Speaks, J. (2007) Russell's Logicism. http://www3.nd.edu/~jspeaks/courses/2007-8/43904/_HANDOUTS/Russell-logicism.pdf
2. van Stigt, W.P. (1990) Brouwer's Intuitionism. http://www.mscs.mu.edu/~wimr/publica/120521_stigt2.pdf
3. Philosophy of Mathematics: Formalism. http://www.mathematics-in-europe.eu/home/43-information/history-philosophy/84-philosophy-of- mathematics-formalism
4. Irvine, A.D. Bertrand A. D. Russell. http://www-history.mcs.st-and.ac.uk/ Biographies/Russell.html
5. Whitehead, A.N. http://www.philosophybasics.com/philosophers_whitehead.html
6. O'Connor, J.J. and Robertson, E.F. Luitzen Egbertus Jan Brouwer.http://www-history.mcs.st-and.ac.uk/Biographies/Brouwer.html
7. Weir, A. (2011) Formalism in the Philosophy of Mathematics. http://plato.stanford.edu/entries/formalism-mathematics/
8. Snapper, E. (1979) The Three Crises in Mathematics: Logicism, Intuitionism and Formalism. http://www.maa.org/sites/default/files/images/upload_ library/22/Allendoerfer/1980/0025570x.di021111.02 p0048m.pdf
9. Weisstein, E.W. Hilbert's Problems, MathWorld—A Wolfram Web Resource. http://mathworld.wolfram.com/HilbertsProblems.html
10. Escultura, E E. (2008) The Grand Unified Theory. Nonlinear Analysis, A-Series: TMA, 69, 823 – 831.
11. Young, L.C. (1980) Mathematicians and Their Times. North-Holland, Amsterdam. [Citation Time(s):4]
12. Davies, P.J. and Hersch, R. (1981) The Mathematical Experience. Birkhäuser, Boston.
13. Escultura, E.E. (2015) Rational Thought, Cognition and Knowledge. International Education Research (Science and Education Centre of North America), 3, 21 – 37. http://www.todayscience.org/IER/article/ier.v3i1p21.pdf
14. Escultura, E.E. (1970) The Trajectories, Reachable Set, Minimal Levels and Chains of Trajectories of a Control System. PhD Thesis, University of Wisconsin, Wisconsin.
15. Lakatos, I., Worral, J. and Zahar, E., Eds. (1976) Proofs and Refutations. Cambridge University Press, Cambridge.
16. Escultura, E.E. (2009) The New Real Number System and Discrete Computation and Calculus. Neural, Parallel and Scientific Computations, 17, 59-84. [Citation Time(s):15]
17. Escultura, E.E. (2013) Critique-Rectification of Mathematics. In: Escultura, E.E., Ed., Qualitative Mathematics and Modeling: Theoretical and Practical Applications, LAP LAMBERT Academic Publishing, Saarbrücken, 77-129.
18. Escultura, E.E. (1998) Exact Solutions of Fermat's Equation (Definitive Resolution of Fermat's Last Theorem). Journal of Nonlinear Studies, 5, 227-254.
19. Royden, H.L. (1983) Real Analysis. 3rd Edition, MacMillan, New York.
20. Weisstein, E.W. Zermelo-Fraenkel Axioms, MathWorld—A Wolfram Web Resource. http://mathworld.wolfram.com/Zermelo-FraenkelAxioms.html
21. Weisstein, E.W. Cantor Diagonal Method; MathWorld—A Wolfram Web Resource. http://mathworld.wolfram.com/CantorDiagonalMethod.html
22. 19th Century Mathematics—Cantor. http://www.storyofmathematics. com/19th_cantor.html
23. Cohen, P.J. (1966) Set Theory and the Continuum Hypothesis. W. A. Benjamin, Amsterdam.
24. Szudzik, M. and Weisstein, E.W. Continuum Hypothesis, from Math World—A Wolfram Web Resource.. http://mathworld.wolfram.com/ContinuumHypothesis.html
25. Weston, T. Banach-Tarski Paradox. http://people.math.umass.edu/~weston/ oldpapers/banach.pdf
26. Merchbacher, E. (1970) Quantum Mechanics. 2nd Edition, John Wiley & Sons, Inc., New York.

27. Horgan, H. (1993) The Death of Proof. Scientific American, 269, 74-82.
28. Russell Antimony, MathWorld—A Wolfram Web Resource. http://mathworld.wolfram.com/RussellsAntinomy.html
29. Lakshmikantham, V., Escultura, E.E. and Leela, S. (2009) The Hybrid Grand Unified Theory. Atlantis Press, Paris.
30. Escultura, E.E. (2012) Creative Mathematics Education. Creative Education, 3, 45-54.http://www.scirp.org/journal/PaperInformation.aspx?PaperID=17266#.VNMJ6S6um44
31. Escultura, E.E. (2013) Creative Science Education. In: Escultura, E.E., Ed., Qualitative Mathematics and Modeling: Theoretical and Practical Applications, LAP LAMBERT Academic Publishing, Saarbrücken, 318-322.
32. Concept to Classroom. http://www.thirteen.org/edonline/concept2class/ constructivism/
33. Mathematics Education, Math World—A Wolfram Web Resource. http://mathforum.org/mathed/constructivism.html
34. Sandy Chapman. The Socratic Method: Fostering Critical Thinking. http://teaching.colostate.edu/tips/tip.cfm?tipid=53
35. 20th Century Mathematics—Russell and Whitehead. http://www.storyofmathematics.com/20th_russell.html
36. Irvine, A. D. Principia Mathematica.http://plato.stanford.edu/entries/principia-mathematica/
37. Principia Mathematica. http://plato.stanford.edu/entries/principia-mathematica/
38. Sakharov, A. and Weisstein, E.W. Propositional Calculus, MathWorld—A Wolfram Web Resource.http://mathworld.wolfram.com/PropositionalCalculus.html
39. Three Laws of Logic. https://school.carm.org/amember/files/demo3/2_logic/3logic.htm
40. Type Theory, Stanford Encyclopedia of Philosophy. http://plato.stanford.edu/ entries/type-theory/#1
41. Escultura, E.E. (2012) The Physics of Intelligence. Journal of Education and Learning, 1, 51-64.http://www.ccsenet.org/journal/index.php/jel/article/view/20095/13268 http://dx.doi.org/10.5539/jel.v1n2p51
42. Escultura, E.E. (2014) Physics of the Mind. The Journal of the Science of Healing Outcomes, 6, 6-15.
43. Weisstein, E.W. Brouwer Fixed Point Theorem, from Math World—A Wolfram Web Resource.http://mathworld.wolfram.com/BrouwerFixedPointTheorem.html
44. Escultura, E.E. (2007) The Pillars of the New Physics and Some Updates. Nonlinear Studies, 14, 241-260.
45. Escultura, E.E. (2013) Columbia's Disastrous Final Flight. In: Escultura, E.E., Ed., Qualitative Mathematics and Modeling: Theoretical and Practical Applications, LAP LAMBERT Academic Publishing, Saarbrücken, 281-293.
46. Young, L.C. (1933) Approximations by Polygons. Proceedings of the Royal Society of London A, 141, 325-341.
47. Young, L.C. (1937) Generalized Curves and the Existence of an Attained Absolute Minimum in the Calculus of Variations. Comptes Rendus des Séances de la Société des Sciences et des Lettres de Varsovie, 30, 211-234.
48. Young, L.C. (1969) Lectures on the Calculus of Variations and Optimal Control Theory. W. B. Saunders, Philadelphia.
49. Young, L.C. (1955) On Generalized Surfaces of Finite Topological Types. Memoirs of the AMS, Philadelphia.
50. Young, L.C. (1942) Generalized Surfaces in the Calculus of Variations: I. Generalized Lipschitzian Surfaces. Annals of Mathematics, 43, 84-103.
51. Young, L.C. (1942) Generalized Surfaces in the Calculus of Variations: II. Mean Surfaces and the Theory of the Problem $\iint f(x,y,p,q)dxdy$ – Min. Annals of Mathematics, 43, 530-544.
52. Escultura, E.E. (2009) The Mathematics of the Grand Unified Theory. Journal of Nonlinear Analysis, A-Series: Theory: Method and Applications, 71, e420-e431.

53. Escultura, E. (2013) The Logic and Fundamental Concepts of the Grand Unified Theory. Journal of Modern Physics, 4, 213-222.
54. Escultura, E.E. (2011) Extended Geometrical and Generalized Fractals, Chaos and Applications. In: Brennan, K.J., Ed., Handbook on the Classification and Application of Fractals, Nova Publishers, Hauppauge, 1-39.https://www.novapublishers.com/ catalog/ product_info.php?products_id=23231
55. Escultura, E.E. (2011) The Generalized Integral and Derivative. In: Escultura, E.E., Ed., Scientific Natural Philosophy, Bentham Science Publishers, Sharjah, 54-140. http://www.benthamscience.com/ebooks/9781608051786/index.htm
56. Escultura, E.E. (2008) Extending the Reach of Computation. Journal of Applied Mathematics Letters, 21, 1074-1081.
57. Benacerraf, P. and Putnam, H. (1985) Philosophy of Mathematics. Cambridge University Press, Cambridge.
58. Escultura, E.E. (2011) Scientific Natural Philosophy. Bentham Science Publishers.
59. Solved Problems, MathWorld—A Wolfram Web Resource. http://mathworld.wolfram.com/topics/SolvedProblems.html
60. Wiles, A. and Taylor, R. (1995) Modular Elliptic Curves and Fermat's Last Theorem. Annals of Mathematics, 141, 443-551.
61. Wiles, A. and Taylor, R. (1995) Ring-Theoretic Properties of Certain Hecke Algebras. Annals of Mathematics, 141, 553-572.
62. Escultura, E. E. (2003) The Flux Theory of Gravitation XVII. The New Mathematics and Physics. Applied Mathematics and Computation, 138, 127-149.
63. Escultura, E. E. (1997) The Solution of the Gravitational n-Body Problem. Nonlinear Analysis, Series A: Theory, Methods and Applications, 30, 5021-5032.
64. Escultura, E.E. (2014) The Leap from Traditional to the New Science. Nonlinear Studies, 21, 283-291.
65. Escultura, E.E. (2005) Dynamic Modeling of Chaos and Turbulence. Nonlinear Analysis, A-Series: TMA, 63, e519-e532.
66. Escultura, E.E. (2013) Chaos, Turbulence and Fractal: Theory and Applications, Open Access. International Journal of Modern Nonlinear Theory and Application, 2, 176-185.
Escultura, E.E. (2001) Turbulence: Theory, Verification and Applications. Journal of Nonlinear Analysis, A-Series: TMA, 47, 5955-5966.
67. 69. What Is the Second Law of Thermodynamics. http://www.livescience.com/50941-second-law-thermodynamics.html
68. Escultura, E.E. (2012) The Big Bang and What It Was. In: O'Connell, J.R. and Hale, A.L., Eds., The Big Bang: Theory, Assumptions and Problems, Nova Science Publishers, 61-102. https://www.novapublishers.com/catalog/productinfo.php?productsid=21109
69. Schilling, G. (1999) Science, Watching the Universe's Second Biggest Bang. Science, 283, 2013-2014.

CHAPTER 9

Recent Advances in Global Optimization for Combinatorial Discrete Problems

Adel R. Awad[1], Samia O. Chiban[2]

[1]*Department of Environmental Engineering, Tishreen University, Lattakia, Syria*

[2]*Higher Institute of Environmental Research, Tishreen University, Lattakia, Syria*

ABSTRACT

The optimization of discrete problems is largely encountered in engineering and information domains. Solving these problems with continuous-variables approach then convert the continuous variables to discrete ones does not guarantee the optimal global solution. Evolutionary Algorithms (EAs) have been applied successfully in combinatorial discrete optimization. Here, the mathematical basics of real-coding Genetic Algorithm are presented in addition to three other Evolutionary Algorithms: Particle Swarm Optimization (PSO), Ant Colony Algorithms (ACOA) and Harmony Search (HS). The EAs are presented in as unifying notations as possible in order to facilitate understanding and comparison. Our combinatorial discrete problem example is the famous benchmark case of New-York Water Supply System WSS network. The mathematical construction in addition to the obtained results of Real-coding GA applied to this case study (authors), are compared with those of the three other algorithms available in literature. The real representation of GA, with its two operators: mutation and crossover, functions significantly faster than binary and other coding and illustrates its potential as a substitute to the traditional optimization methods for water systems design and planning. The real (actual) representation is very effective and provides two near-optimal feasible solutions to the New York tunnels problem. We found that the four EAs are capable to afford hydraulically-feasible solutions with reasonable cost but our real-coding GA takes more evaluations to reach the optimal or near-optimal solutions compared to other EAs namely the HS. HS approach discovers efficiently the research space because of the random generation of solutions in every iteration, and the ability of choosing neighbor values of solution elements "changing the diameter of the pipe to the next greater or smaller commercial diameter" beside keeping good current solutions. Our proposed promising point to improve the performance of GA is by introducing completely new individuals in every generation in GA using a new "immigration" operator beside "mutation" and "crossover".

Keywords: Evolutionary Algorithms, Meta-Heuristic Algorithms, Real-Coding Genetic Algorithms, Water Supply System, New-York Tunnels, Optimal Design

1. INTRODUCTION

Global optimization is the task of finding the absolutely best set of admissible conditions to achieve an objective under given constraints, assuming that both are formulated in mathematical terms [1] . There exist many global optimization problems with discrete variables in engineering design, information science, and related fields. In such practical applications, some or all of the variables must be selected from a list of integer or discrete values. For example, structural members may have to be designed using sections available in standard sizes, member thicknesses may have to be selected from the commercially available ones, and the number of bolts for a connection must be an integer. The variables of optimization in these cases are not continuous [2] . One of the interesting discrete-variables global optimization problems is the optimization of Water Supply System WSS design. The complexity, nonlinearity and discrete nature of WSS optimization problems seriously limit the local search algorithms methodologies as they can be trapped in a local optima in the solution space "the set of all possible solutions" [3] . Evolutionary Algorithms EA have been widely popular nowadays in the global optimized design of WSS because it can deal easier with discrete problems, which permits, opposed to other optimization methods, the direct inclusion of commercial pipe diameters in the design, in addition to the fact that EA allow for a family of solutions, one can obtain various optimal solutions or solutions close to the optimal case [4] -[6] .

We present in this paper the WSS optimization as an example of combinatorial discrete optimization problems. The optimization algorithms used to solve this type of problem are divided into two categories the classical algorithms and the Evolutionary Algorithms EAs which are more popular. We highlight the main differences between the two categories, and then we concentrate on the last category. We demonstrate the mathematical basics of four EAs that have been used successfully to solve WSS optimization problems (Genetic Algorithms GAs, Particle Swarm Optimization PSO, Ant Colony Algorithm ACOA and Harmony Search HS). We adopt the real Genetic Algorithm, as it is one of the fundamental evolutionary algorithms and the most used nowadays. We compare the real GA as mathematical construction, results and efficiency with the three more-recent algorithms, taking the New-York City WSS as a case study.

2. PARTICULARITY OF WATER SUPPLY SYSTEM OPTIMIZATION

The complexity of water supply systems comes from multiple pumps, valves and reservoirs, head losses, large variations in pressure values, several demand loads, etc. In such systems, finding the global optimal solution is a difficult task using the conventional trial and error methods, while the incomplete mathematics precisely, the innovative optimization algorithms are more widely explored in optimization process of the water supply systems [7] .

Design optimization problems in WSS are based on searching the system characteristics which minimise the total system cost without affecting the proper operation of the hydraulic system and the consumers supply. This means that the system must be economic and reliable. However, increase in reliability can imply higher costs [8] . The problem formulation required for the optimization of WSS is different from that used for other combinatorial optimization problems, such as the traveling salesman problem, in the way the problem is constrained. For example, in the traveling salesman problem, the only constraints are that each city must be visited once only and that the finishing point must be the same as the starting point. In this situation, tabu lists are used to ensure that only feasible solutions are generated [9] . However, the constraints that need to be satisfied in the optimal design of WSS are of a different nature. The feasibility of a particular trial solution e.g., whether minimum pressure constraints have been satisfied can only be assessed after it has been constructed in its entirety, and consequently, the constraints cannot be taken into account explicitly during the construction of trial solutions. In this way, the feasibility of the solutions can only be assessed after their full construction.

In most works dealing with optimization design of WSS, some benchmark networks have been constantly used for the comparison between distinct developed methodologies, from which we mention the New York City Tunnels firstly presented by Schaake and Lai 1969 [10] , and the Hanoi network in Vietnam [11] : 1) The New York City Tunnels problem consists essentially of a single source in Hill View and two main city tunnels. The objective of this problem is to determine the need of laying a new pipe parallel to the existing ones and also to determine their diameters. A unique demand case is considered. 2) The Hanoi network contains three loops and also ramifications. The objective of this problem is to find the least-cost diameters for all pipes while respecting the minimum value for the head pressure at each node. In this work, we take The New York City Tunnels problem as case study.

2.1 Problem Statement

As a common practice, due to the nature of WSS, one uses an objective function which only takes into account the pipeline costs and penalty costs, if not complying with minimal values for the pressure at each node of the network and the objective function can be written as follows [5] [6] [12] [13]

$$F_{obj} = \sum_{i=1}^{\#pipes} c_{i,j} \cdot l_i + Pe \Rightarrow \min \tag{1}$$

where i is the pipe number (1 ® i ® S, S is the number of all pipes i.e. the number of variables), $c_{i,j}$ is the cost per meter depending on the diameter j of pipe i (1 ® j ® J, J is the number of available commercial diameters i.e. the number of options), l_i is the length of pipe i.

The penalty Pe only applies when the pressure in any node is less than a predetermined minimal value. In order to reveal the feasible solutions only, the hydraulic simulations is required to check pressure constraints and to obtain the flow velocity in each pipe, and if the resulting velocity in one pipe is larger than the determined threshold velocity, this pipe diameter is increased to the next larger commercial size. Hydraulic simulators are coupled successfully with the optimization algorithms to evaluate each solution.

3. CLASSIFICATION OF OPTIMIZATION ALGORITHMS

Global optimization algorithms can be distinguished according to two general classifications: 1) classical algorithms, based essentially on the computation of the objective function gradient and/or function evaluations and 2) Evolutionary /Meta-heuristic algorithms, consisting essentially on exploratory search and generally based on a phenomena that occur in nature or even based on artificial intelligence.

The classical algorithms used in Water Supply Systems as classified in [7] include: Linear Programming (LP), Nonlinear Programming (NLP), Integer Nonlinear Programming including MINLP (Mixed Integer NonLinear Programming) [14] and Dynamic Programming. These kinds of algorithms enable finding the position of an optimal solution with certain tolerance. They usually converge to local optimal solutions which could not be the global optimum. In addition, the need of derivative evaluations can, in some cases, complicate the optimization process. These classical methods of optimization involve the use of gradients or higher-order derivatives of the fitness function. However, they are not well-suited for many real-world problems since they are not able to process inaccurate, noisy, discrete and complex data [6] [15] [16] .

On the other hand, there are the incomplete mathematics presented by the Evolutionary /Meta-heuristic algorithms. These innovative techniques have shown great potential and good perspective for the solution of various optimization problems. Due to the exploratory nature of the heuristic algorithms, the probability of finding global optimal solutions using these advanced techniques is higher and they provide the advantages of not requiring derivatives calculations and not relying on the initial decision variables; but their main disadvantage is related to the higher computational effort [17] . However, Metamodels have proven be very useful when it comes to reduc- ing the computational requirements of Evolutionary Algorithm-based optimization by acting as quick-solving surrogates for slow-solving fitness functions [18] . As a common example of these metamodels, we cite the Artificial Neural Networks (ANNs) due to their demonstrated ability to model water quality variables in distribution systems [19] -[21] . In addition, many researches have focused recently on the amelioration of computational efficiency of evolutionary algorithms, as the work of Montalvo et al., 2010 [6] on Particle Swarm Optimization PSO with self-adaptive parameters, the work of Bi et al. 2015 [22] on Improved genetic algorithm optimization by incorporating domain knowledge and the work of Beheshti and Shamsuddin 2015 [23] on Non-parametric PSO that enhances the global exploration and the local exploitation in PSO without tuning any algorithmic parameter while solving global optimization problems.

Likewise, evolutionary algorithms are usually the most used for solving multi-objectives problems. They deal with a set of solutions during the search procedure, allowing to obtain a set of optimal solutions in a single run, while the classic complete methods only lead to a single solution and cannot guarantee the generation of different points based on the optimal global solution.

4. MATHEMATICAL BASICS OF SOME (EAS) USED IN WSS DISCRETE PROBLEM

Evolutionary algorithms (EAs) have shown considerable success in solving global optimization problems; thus, it has attracted much attention in the past few years. Recently, a large number of EAs, such as genetic algorithms (GAs), particle swarm optimization (PSO), Ant Colony Algorithm (ACOA) and Harmony Search (HS) have been proposed.

Here in, we state the most promising incomplete mathematics algorithms used in solving global optimization problems such as the water supply systems:

4.1. Genetic Algorithms (GA)

From the group of heuristic algorithms, it is usual to find works mostly applying the (GA). The theory behind GAs was proposed by Holland (1975) [24] and further developed by Goldberg (1989) [25] and others in the 1980s and 1990s.

There are many variations of GAs, but the following general description encompasses most of the important features. The analogy with nature requires creation by computer of a set of solutions called a population. Each individual in a population is represented by a set of parameter values that completely propose a solution. These are encoded into chromosomes, which are originally sets of character strings, analogous to the chromosome found in DNA. The GA search, sometimes with modification, has been proven to perform efficiently in a large number of applications. The selection of an appropriate chromosome representation of candidate solutions to the problem at hand is the foundation for the application of genetic algorithms to a specific problem. With a sufficiently large population of chromosomes, adequate representation will be achieved.

The typical size of the population for a GA can range from 30 to 1000. In addition, GAs use probabilistic (not deterministic) transition rules, and they also use objective function information, not derivatives or other auxiliary knowledge [25] .

In the WSS optimization problem, every pipe diameter is considered like a gene that has as many options as the available commercial diameters, while the whole network is considered as the chromosome "individual" that should "fit" to the environment presented by the objective function and the survival chances for the possible solutions or individuals in the population are bigger for those who propose the least cost (Figure 1). Detailed explanations of the notations of the reproduction, crossover and mutation operators using real representation can be seen in [4] .

Returning to the objective function in Equation (1), the penalty is applied when infeasible solutions, failed to meet the minimum nodal pressure requirements. This pressure is 30 m above ground level in the case of New-York city WSS. However, the infeasible solutions are not removed from the population. Instead they are allowed to join the population and help guide the search, but with penalty term incorporated in objective function. The penalty activation reduces the strength of the infeasible solution relative to other solutions in the population consequently leading to the degradation of its fitness. The penalty is a function of the distance from feasibility. The bigger the difference between the nodal pressure and the required nodal pressure, the greater the penalty is. The maximum violation of the pressure constraint is multiplied by the penalty multiplier Pm, which is a function of the generation number, which allows a gradual increase in the penalty term:

$$Pm = \delta\left(\frac{\text{current\_generation\_number}}{\text{Determined\_number\_of\_generations}}\right)^{\kappa} \quad (2)$$

The penalty multiplier Pm should be a value that does not allow the best infeasible solution to be better than any feasible solution in the population. As we note from Equation (2), when the generation number approaches the maximum determined number of generations Gmax, the penalty value increases. The selected appropriate values of κ and δ in Equation (2) are 0.8 and 100,000 respectively.

The application of real GA on the New-York network, hereafter, gives more details about the method with comparison to other evolutionary algorithms.

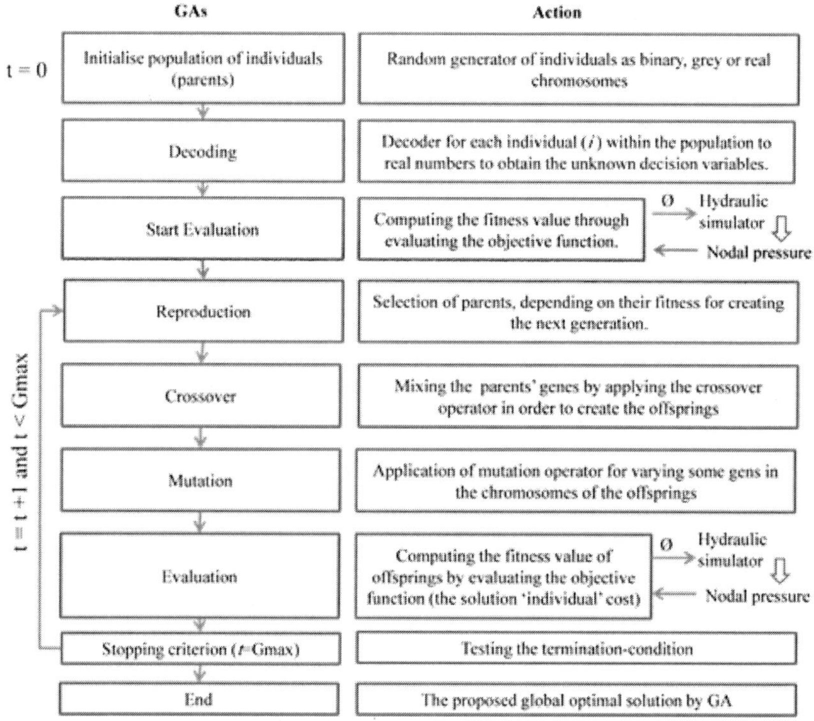

Figure 1. Flow-chart for the GA implementation with the related action.

4.2. Particle Swarm Optimization (PSO)

PSO is a relatively new category of stochastic, population-based optimization algorithms that draw inspiration from the collective behavior and emergent intelligence that arise in socially organized populations. Within the PSO algorithm, every solution is a bird of the flock and is referred to as a particle: in this framework the birds, besides having individual intelligence, also develop some social behavior and coordinate their movement towards a destination [26] .

For nodes with pressures larger than this minimal value, the associated individual penalties vanish, and one uses the usual Heaviside step function H in the explicit expression for Pe:

$$Pe = \sum_{i=1}^{\#nodes} H\left(P_{\min} - P_i\right) \cdot p \cdot \left(P_{\min} - P_i\right) \tag{3}$$

The factor p represents a fixed value that becomes effective when $\Delta P_i = (P_{min} - P_i) > 0$ which means that the minimal pressure requirement is not met.

Initially, the process of finding the optimal solution starts from a swarm of particles (we can take a number of particles 1 ® n ® N, N = 100 for example), in which each of them contains a solution needs to be improved by iteration. The n-th particle is associated with a position in an S-dimensional space, where S is the number of variables involved in the problem (the number of pipes in the case of WSS). The values taken by each variable of the S variables (the diameter j of each pipe in the case of WSS) represent a possible solution of the optimization problem. Each particle n is completely determined by three vectors: its current position, in iteration t, X_n^t, its best position reached in previous iterations Y_n^t, and its velocity in the current iteration V_n^t:

Current position

$$X_n^t = \{x_{i,j}^{n,t}\} \tag{4}$$

which reveals the diameter j of each pipe i, presented by the particle n in the iteration t. So, in every iteration we have N particles and thus N solution.

Best previous position

$$Y_n^t = \{y_{i,j}^{n,t}\} \tag{5}$$

Y_n^t is the best solution presented by particle n before the iteration t.

Flight velocity

$$V_n^t = \{v_{i,j}^{n,t}\} \tag{6}$$

The flight velocity depends on the position of the current "solution" comparing to the best solution of all the particles in the same iteration and the best previous solution as explained in Equation (7).

This algorithm simulates a flock of birds which communicate during flight. Each bird examines the search space from its current local position, and this process repeats until the bird possibly reaches the desired position. The birds learn through their own experience (local search) and the experience of their peers (global search).

In every iteration, one identifies the particle which has the best instantaneous solution to the problem; the position of this particle subsequently enters into the computation of the new position for other particles in the flock. This heuristic calculation is carried out according to:

$$X_n^{t+1} = X_n^t + V_n^{t+1} \tag{7}$$

where the primes denote new values for the variables, and the new velocity is given by:

$$V_n^{t+1} = \omega \cdot V_n^t + c_1 \cdot rand(\ ) \cdot \left(Y_n^t - X_n^t\right) + c_2 \cdot rand(\ ) \cdot \left(Y^* - X_n^t\right) \tag{8}$$

Here, c_1 and c_2 are two positive constants called learning factors; $c_1 = 3$, $c_2 = 2$, in [5] and they are bound to vary between 0 and 5 in [6] .

$rand(\ )$ represents a function which creates random numbers between 0 and 1.

w is a factor of inertia suggested by Shi and Eberhart [26] , it controls the balance between global and local search; it was suggested to have it decrease linearly with time, usually in a way to first emphasize global search and then, with each cycle of the iteration, to prioritize local search.

$$\omega = 0.5\left[1 + \frac{1}{\ln(t) + 1}\right] \tag{9}$$

where t is the iteration number.

Finally, Y^* is the best present solution of all Yi.

Once the current position is calculated, the particle directs itself towards a new position.

To control any change in the particle velocities, we introduce the respective upper and lower limits:

$$Vinf \le V_n \le Vsup \forall n. \tag{10}$$

where Vsup = 50% of the total variable range and Vinf = −Vsup.

This algorithm can be considered as the standard PSO algorithm, which is applicable to continuous systems and cannot be used for discrete problems. In discrete problem, the initial position vectors are generated with integer values, so if the flying velocity vector components are integer values then the new position vectors will be integer values as well. According to this scheme, Equation (7) turns out to be:

$$V_n^{t+1} = fix\left(\omega \cdot V_n^t + c_1 \cdot rand(\ ) \cdot \left(Y_n^t - X_n^t\right) + c_2 \cdot rand(\ ) \cdot \left(Y^* - X_n^t\right)\right) \tag{11}$$

where $fix(\ )$ implies that we only take the integer part of the result.

Thus, The PSO algorithm after some modifications can adapt to problems with discrete variables. The approaches which were presented in the works of Montalvo et al. 2008, 2010 [5] [6] have shown satisfactory performance when applied to water supply systems, namely the optimal design of the networks provided by the Hanoi and New York Cities.

However, the choice of the PSO algorithm's parameters (such as the group's inertia and acceleration coefficients) seems to be very important for the speed, convergence and efficiency of the algorithm [6] [27] .

4.3. Ant Colony Algorithm (ACOA)

Ant colony optimization algorithms are inspired by the fact that ants are able to find the shortest route between their nest and a food source. Ants deposit pheromone trails whenever they travel. The path taken by individual ants from the nest in search for a food source is essentially random [9] . As more ants travel on paths with higher pheromone intensities, the pheromone on these paths builds up further, making it more likely to be chosen by other ants and thus the shortest path between the nest and a food source is chosen. The main purpose of artificial ant systems is to find solutions to combinatorial optimization problems; mapped on a graph; they also incorporate features that are not found in their natural counterparts [13] .

In order to apply ACOAs to WSS optimization problems, the combinatorial optimization problem under consideration has to be able to be mapped onto a graph $G = (\mathbf{D}, \mathbf{L}, \mathbf{C})$, where D is a set of points at which decisions have to be made, L is the set of options j available at each decision point i, and C is the set of costs associated with options L. A set of finite constraints may be assigned over the elements of D and L. A feasible path over G is called a solution X and a minimum cost path is an optimal solution (X^*). The cost of a solution is denoted by $f(X)$ and the cost of the optimal solution by $F_{obj} = f(X^*)$ as Equation (1) indicates [28] .

Once the problem has been defined in the terms set out above, the ACOA can be applied. The main steps in the algorithm are as following:

1) Trial solutions are constructed incrementally as artificial ants move from one decision point to the next until all decision points have been covered.
2) The cost of the trial solution generated is calculated. The generation of a complete trial solution and calculation of the corresponding cost is referred to as a cycle n. Every ant draws one cycle in iteration, this is why we keep the same abbreviations, n for particle's number, and N for total particles number in PSO and in ACOA.
3) Pheromone is updated after the completion of one iteration t, which consists of N cycles "N is the number of ants used". In other words, the construction of N trial solutions means that we have N ants that made N tours or cycles, each one of them presents a solution.

$$pr_{i,j}^{n,t} = \frac{\left[\tau_{i,j}^{t}\right]^{\alpha} \cdot \left[\eta_{i,j}\right]^{\beta}}{\sum_{i,j}\left[\tau_{i,j}^{t}\right]^{\alpha} \cdot \left[\eta_{i,j}\right]^{\beta}} \tag{12}$$

where $pr_{i,j}^{n,k}$ is the probability that option $l_{i,j}$ is chosen by ant n in iteration t; $\tau_{i,j}^{t}$ is the concentration of pheromone associated with option $l_{i,j}$ at iteration t; $\eta_{i,j} = 1/c_{i,j}$ is a heuristic factor favoring options that have smaller "local" costs; α and β are exponent parameters that control the relative importance of pheromone and the local heuristic factor, respectively.

When one iteration is completed, then N trial solutions have been constructed, the pheromone trails are updated in a way that reinforces good solutions. The general form of the pheromone update equation is as following [28] :

$$\tau_{i,j}^{t+1} = \rho \cdot \tau_{i,j}^{t} + \Delta\tau_{i,j} \tag{13}$$

where $\tau_{i,j}^{t+1}$ is the concentration of pheromone associated with option $l_{i,j}$ at iteration t + 1; $\tau_{i,j}^{t}$ is the concentration of pheromone associated with option

$l_{i,j}$ at iteration t; ρ is a coefficient representing pheromone persistence; and $\Delta\tau_{i,j}$ is the change in pheromone concentration associated with option $l_{i,j}$ as a function of the trial solutions found at iteration t. The pheromone persistence coefficient has to be less than one and simulates pheromone evaporation. This enables greater exploration of the search space and avoids premature convergence to suboptimal solutions. The pheromone evaporation process is analogue to the mutation process in GA, with the difference that the first is deterministic, whereas mutation in GAs is stochastic.

$\Delta\tau_{i,j}$ is given by Stützle and Hoos 2000 [29] :

$$\Delta\tau_{i,j} = \Delta\tau_{i,j}^{n^*} \tag{14}$$

where n* is the cycle number that results in the best solution during iteration t.

$$\Delta\tau_{i,j}^{n} = \begin{cases} \dfrac{R}{f(X)^{n}} - P_{pher} \cdot \Delta H_{\max} & \text{If option } l_{i,j} \text{ is chosen by ant } n \\ 0 & \text{Otherwise} \end{cases} \tag{15}$$

where R is pheromone reward factor; $f(X)^n$ is the cost of the trial solution generated at cycle n, P_{pher} is pheromone penalty factor and ΔH_{max} is maximum pressure deficit in the WSS, which is obtained using a hydraulic simulator for each trial solution. Equation (14) clarifies that the amount of pheromone added to each of the options chosen during a cycle is a function of the cost of the trial solution obtained; the better the trial solution, and hence the lower the cost, the larger the amount of pheromone added, and thus the more likely the concerned solution options will be chosen by other ants in next iterations [29] . The steps of generating trial solutions, calculating the costs of the chosen solutions and updating the pheromone concentrations are repeated until certain stopping criteria are met, usually a specified number of iterations.

4.4. Harmony Search (HS)

The harmony search algorithm was conceptualized from the musical process of searching for a "perfect state" of harmony, such as jazz improvisation, just as the optimization algorithm seeks a best state (global optimum) determined by evaluating the objective function. The harmony quality is enhanced practice after practice, just as the solution quality is enhanced iteration by iteration [12] .

As in other optimization processes in the water network design, the objective function is the pipe cost function; the pipe diameter is the decision variable; the number of decision variables is the number of pipes in the network; the set of decision variable values is the range of possible candidate diameters.

The steps in the procedure of harmony search are as follows [30] :

Step 1. Initialize the problem and algorithm parameters. The HS algorithm parameters are also specified in this step. These are the harmony memory size (HMS), or the number of solution vectors in the harmony memory which is equivalent to the number of particles N in previous algorithms; harmony memory considering rate (HMCR); pitch adjusting rate (PAR); and the number of improvisations (NI), or stopping criterion. The harmony memory (HM), is a memory location where all the solution vectors (sets of decision variables) and corresponding objective function values are stored. The function values are used to evaluate the quality of solution vectors. This HM is similar to the genetic pool in the GA. In the study of Geem 2006 [12] , an HMS of 30 - 100, an HMCR of 0.7 - 0.95, and a PAR of 0.05 - 0.7 were used based on the frequently used value ranges in other HS applications [30] - [32] .

Step 2. Initialize the harmony memory.

$$HM = \begin{bmatrix} x_{1,1} & x_{1,2} & \cdots & x_{1,S-1} & x_{1,S} \\ x_{2,1} & x_{2,2} & \cdots & x_{2,S-1} & x_{2,S} \\ \vdots & \cdots & \vdots & \cdots & \cdots \\ x_{HMS-1,1} & x_{HMS-1,2} & \cdots & x_{HMS-1,S-1} & x_{HMS-1,S} \\ x_{HMS,1} & x_{HMS,2} & \cdots & x_{HMS,S-1} & x_{HMS,S} \end{bmatrix} \begin{matrix} \rightarrow f(X_1) \\ \rightarrow f(X_2) \\ \vdots \\ \rightarrow f(X_{HMS-1}) \\ \rightarrow f(X_{HMS}) \end{matrix} \tag{16}$$

In Step 2, the HM matrix is filled with randomly generated solution vectors. Each solution undergoes a hydraulic analysis to verify that it satisfies the minimum pressure requirements at each node. As in other Evolutionary algorithms, the penalty in the objective function will cause that the infeasible solutions that violate the minimum requirements, become uneconomic but still have a chance to be included in the HM in the hope of forcing the search towards the feasible solution area.

Step 3. Improvise a new harmony. A new harmony vector, $X' = (x'_1, x'_2, \cdots, x'_S)$ is generated based on three rules: 1) memory consideration, 2) pitch adjustment, and 3) random selection. Generating a new harmony is called "improvisation". The HMCR, which varies between 0 and 1, is the rate of choosing one value from the historical values stored in the HM, while (1 ? HMCR) is the rate of randomly selecting one value from the possible range of values (see Figure 2). The value of HMCR is taken generally between 0.8 and 0.95 for diver optimization problems, according to the survey made by Geem et al. 2006 [12] during their research.

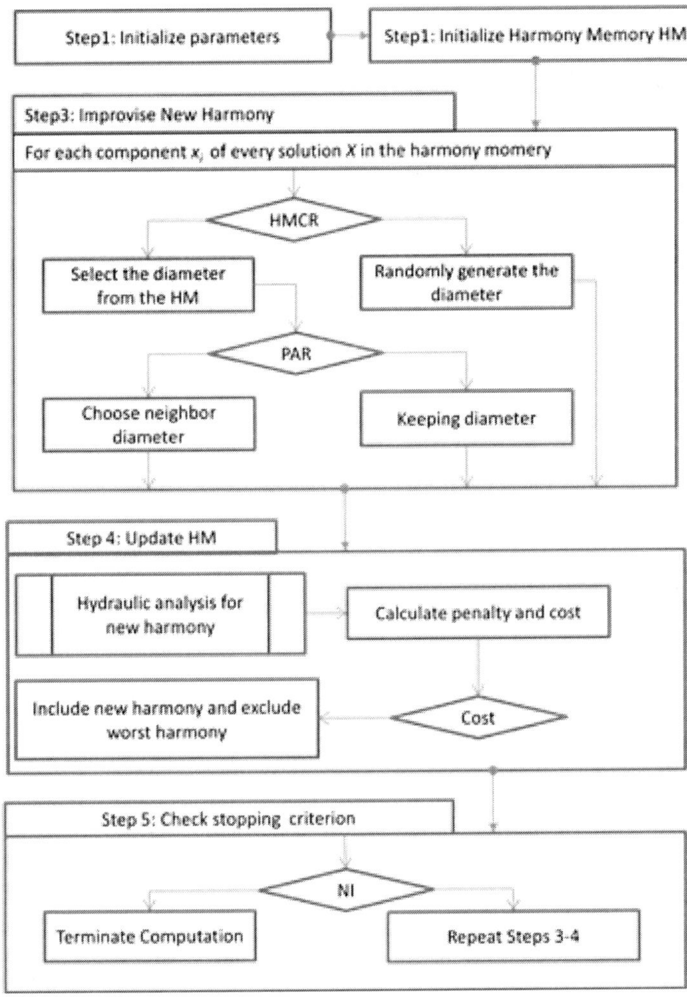

Figure 2. Harmony Search procedure for water supply system design as an example of combinatorial discrete optimization. Adapted from Geem et al. 2006. HMCR: Harmony Memory Consideration Rate, PAR: Patch Adjustment Rate, NI: Number of Improvisations.

Every component obtained by the memory consideration is then examined to determine whether it should be pitch-adjusted. This operation uses the PAR parameter. If the pitch adjustment decision for x_i' is Yes, is replaced with neighboring value, i.e. the next greater or next smaller diameter with equal chances. The PAR value between 0 and 1, varies considerably from problem to another. Geem 2006 [12] tried the HS optimization on five WSS and the PAR varies in each case from 0.05 to 0.7.

In this step, the application of HM consideration, pitch adjustment, or random selection to each variable of the new harmony vector in turn, help the algorithm find globally and locally improved solutions, respectively.

Step 4. Update the harmony memory. If the new harmony vector $X' = (x_1', x_2', \cdots, x_S')$ is better than the worst harmony vector in the HM, judged in terms of the objective function value, and no identical harmony vector is stored in the HM, the new harmony is included in the HM and the existing worst harmony is excluded from the HM. For the water network optimization problem, the new solution vector also undergoes a hydraulic analysis, just the same as in Step 2.

Step 5. Steps 3 and 4 are repeated until the number of improvisations (NI) is reached.

However, other techniques have also been investigated in this context, such as: Tabu Search (TS), Simulated Annealing (SA) and Shuffled Complex Evolution (SCE).

5. DISCUSSION ON THE NEW YORK CITY TUNNELS OPTIMIZATION

Given an existing network like the New-York City network shown in Figure 3, the objective is to “renovate” it by considering the increase of the water demand due to the population growth. If the existing network is no longer adequate for the increased demand, that can be translated in pressure violations at several junctions. Thus, the network must be modified by duplicating some of the pipes, i.e., putting new pipes in parallel with the existing ones, at a minimum cost. The decisions one has to take are: 1) Select the pipes that need to be duplicated; 2)

For each of these pipes, choose a diameter within the available diameter set [14].

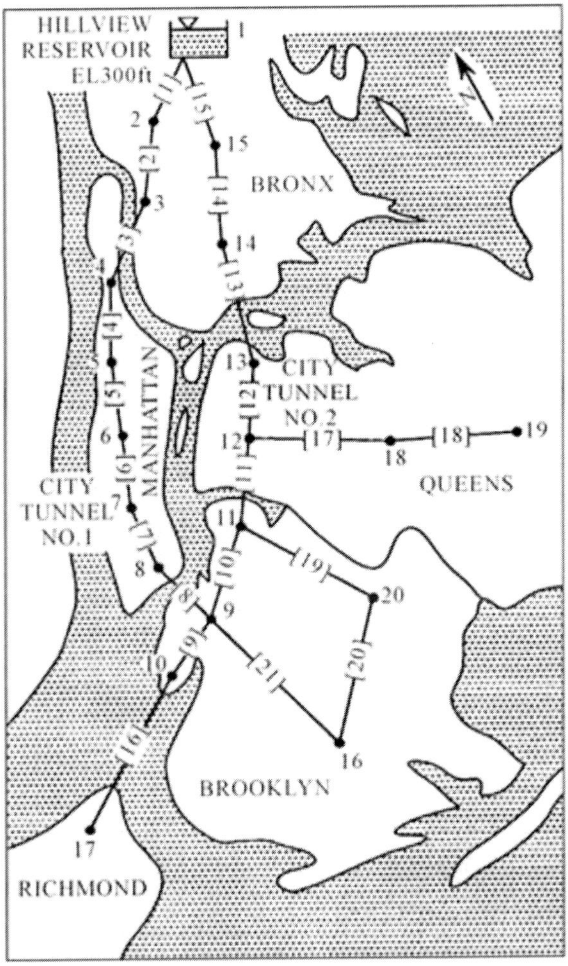

Figure 3. New-York City tunnels [10] .

Because of age and increased demands, the existing gravity flow tunnels in New-York City have been found to be inadequate to meet the pressure requirements in some nodes (nodes 16, 17, 18, 19 and 20 shown in Figure 3) for the projected consumption level. There are 21 pipelines that may be duplicated with one of the 15 available commercial diameter pipes or zero-diameter pipe, that is, "do nothing". The total solution space is $16^{21} = 1.93 \times 10^{25}$ possible network designs.

In order to solve this optimization problem in real-coding GA algorithm, a real string made of 21 substrings (genes) is used for coding the 21 decision variables "21 pipes that can take one of the 16 discrete diameters".

Table 1. Adjusting different evolutionary algorithms to be used in the New-York tunnels optimization.

Method and ref.	Research space (N)	Parameters	Notations in the text	Iterations (t)
Real GA	100 "individual"	The probability of crossover = 0.8 The probability of mutation = 0.1, $\kappa = 0.8$, $\delta = 100{,}000$	Figure 1 and Equation (2)	Number of generations allowed = 2500
PSO [5]	100 "particle/bird"	$c_1 = 3$, $c_2 = 2$.	Equations (8) and (11)	Number of flights = 100
ACOA [13]	100 "ant"	$\alpha = 3.5$, $\beta = 0.5$, $\rho = 0.95$, $P_{pher} = 0.1$, $R = 15{,}000{,}000$	Equations (12) and (15)	Number of iterations = 500
HS [12]	50 "musical harmony"	HMCR = 0.9, PAR = 0.1	Figure 2	Number of improvisations = 20,000

GA: Genetic Algorithms, PSO: Particle Swarm Optimization, ACOA: Ant Colony Algorithm, HS: Harmony Search.

The fitness of the chromosome that consists of 21 genes is computed through the objective or evaluation function which determines the cost of a solution by summing the cost of the pipes making up the network. Simulation of the network flows and pressure heads is then carried out through EPANET computer program to assess the feasibility of solutions [33].

Many optimization works were done in the literature for the case of New York City Tunnels optimization. Basing on the literature review made by Coelho and Andrade-Campos [7] in 2014, there are 26 published research works on this exact title, began from the work of Schaake et al., 1969 [10]. For obvious reasons, we choose to compare our Real GA results with recent researches applying equivalent evolutionary algorithms from 2003 and go on. The parameters chosen to run real GA to find the optimal design of New-York WSS in addition to the parameters used by the three other presented evolutionary algorithms to run the same case, as stated in their original references, are illustrated in Table 1.

Table 2 shows the details of four global optimal designs found by the different evolutionary algorithms. All the proposed designs indicate that pipes 16, 17, 18, 19 and 21 require duplication, with the addition of pipe 15 in our viewpoint. The Four presented designs are all feasible hydraulically, i.e. the pressure in the critical nodes is equal or slightly greater than the minimum required pressure as Table 3 shows. It worth's to be mentioned that there are other evolutionary algorithms, in the available literature, that have proposed the same cost as Design I (38.64 M$): PSO [5] and NSGA-II "Non-Dominated Sorting Genetic Algorithm" [3]. However, they did not indicate the design details in their published works such as the proposed pipes diameters, the feasibility of their solutions, nor the number of evaluations to reach this cost; this is why we could not place them for comparison in Table 2 and Table 3.

Regarding the four designs proposed by the evolutionary algorithms, the real GA affords two feasible designs "the first after 45,904 evaluations and the second after 116,805 evaluations" with costs comparable to the costs of designs I and II. The availability of alternatives with nearly equivalent costs has a big advantage in practice, because that enables the engineers to take into account other terrain factors to clearly make a choice on the adopted design.

Comparing the performance of the three algorithms in Table 2, we find that the Harmony search presents the least cost with the minimum number of evaluations compared to ACOA and real GA. However, associating these results with the theoretical basics of each algorithm, and with the corresponding parameters in Table 1, reveals that the ACOA needs five parameters to be calibrated, compared with four parameters in GA, two parameters in HS.

Table 2. The detailed four designs of New-York City WSS network proposed by EAs.

Pipe number	Diameters of parallel pipes, inches			
	ACOA [13] Design I	HS [12] Design II	Real GA 1 Design III	Real GA 2 Design IV
1	0	0	0	0
2	0	0	0	0
3	0	0	0	0
4	0	0	0	0
5	0	0	0	0
6	0	0	0	0
7	144	96	0	0
8	0	0	0	0
9	0	0	0	0
10	0	0	0	0
11	0	0	0	0
12	0	0	0	0
13	0	0	0	0
14	0	0	0	0
15	0	0	120	96
16	96	96	84	96
17	96	96	96	96
18	84	84	96	84
19	72	72	84	72
20	0	0	72	0
21	72	72	72	72
Cost $ M	38.64	36.66	38.8	38.5
Diameter design	Discrete	Discrete	Discrete	Discrete
Number of Evaluations	13,928	6000	45,904	116,805

Table 3. Total head at critical nodes for New-York City WSS in the EAs' four proposed designs.

Critical node	Minimum Allowable Head (ft)	Head difference from minimum (ft)			
		ACOA	HS	Real GA1	Real GA2
16	260.00	0.08	0.15	0.59	0.02
17	272.80	0.07	0.00	0.11	0.00
19	255.00	0.05	0.22	0.78	0.17
Used Hydraulic Simulator		WADISO/EPANET	EPANET	EPANET	
Feasibility		F	F	F	F

So, the success of ACOA in obtaining feasible solution with reasonable cost, can be referred to the adopted values of parameters, even if the ACOA handled unrealistically the zero pipe option by assigning a "virtual" unit cost of $70/m for the purpose of calculating the heuristic factor $\eta_{i,j}$ which is the inverse of the pipe cost used in Equation (11) [13] . The main drawback of GA is that it needs greater number of evaluations than other proposed evolutionary algorithms to reach equivalent solutions. Objectively, we can say that the search process in the harmony search, illustrated in Figure 2, enables the algorithm to discover the research space more efficiently, where it generates solutions randomly, choose neighbor solutions and keep current solutions; while in the Genetic Algorithms, the evolution of a solution depends only on the mutation and crossover process applied on its two parents only. The mutation is a limited operator that considers changing some gens in the solution, so the offspring in GA are more influenced by the initial population than the offspring in other evolutionary algorithms, this is the reason why GA takes longer time to discover the research space and reach the optimal solutions. The process that we propose to improve the performance of real-coding GA, and GAs in general, is to incorporate in every generation, some completely new solutions "chromosomes" that are generated randomly. This process is analogue to the "immigration" process in real life, where these new individuals "solutions" give a new blood to the population, and they are evaluated like other solutions, and couple with existing solutions to find new ones.

6. SUMMARY

Evolutionary algorithms (EAs) have shown considerable success in solving global optimization of discrete problems and have proved to be a valuable tool in the search for alternatives, and not only a unique solution.

The real-coding Genetic Algorithm GA is based on the phenomena of natural selection, the possible solutions are seen as individuals "chromosome" and they are evaluated basing on their fit to the objective function. The conventional GA uses two main operators applied on the initially generated population and the next generations (mutation and crossover). The real GA approach has considerable flexibility in application to nonlinear problems and to complex systems. In addition, a real GA will generate several near-optimal solutions.

The second presented evolutionary algorithm is the Particle Swarm Algorithm (PSO). Within the PSO algorithm, every solution is a bird of the flock "a particle". Birds develop some social behavior and coordinate their movement towards a destination which is the objective function. The flight velocity depends on two terms with random portions: the position of the current "solution" comparing to the best solution of all the particles in the same iteration, and the best previous solution.

The third presented evolutionary algorithm is the Ant Colony Algorithm (ACOA).This algorithm is inspired by the fact that ants are able to find the shortest route between their nest and a food source. The ants have tendency to choose paths where other ants have crossed. They can recognize these paths by pheromone intensities deposed by the traveling ants. In the ACOA, the pheromone concentration is an auxiliary parameter used to help artificial ants to make their options.

The fourth presented evolutionary algorithm is Harmony Search (HS). (HS) conceptualized from the musical process of searching for a "perfect state" of harmony, such as jazz improvisation. The harmony memory (HM), is first filled by solution vectors. Improvise a new harmony is generated based on three rules: 1) memory consideration, 2) pitch adjustment, and 3) random selection.

A considerable amount of work is done in this paper for unifying the symbols and notations in the mathematical description of the four mentioned EAs in order to be comparable and easy to understand.

The New-York City WSS tunnels network for WSS, is taken as an example to compare the real GA with PSO, ACOA and HS. The results of the three last EAs are stated in literature. Primarily, the comparison while preparing each algorithm to be applied to a specific case study, reveals that the ACOA needs the biggest number of parameters to be calibrated (5 parameters), compared with 4 parameters in GA, 2 parameters in PSO and HS. Four distinguished designs are offered by the EAs. However, the detailed results of PSO are not included in the published work of Montalvo et al. 2008 [5] , they mentioned only the cost of their solution of 38.64 M$. The same cost is found using ACOA [13] . Our real GA offer two near-optimal designs, the first with a cost of 38.8 M$ after 45,904 evaluations, while the second with a cost of 38.5 M$ after 116,805 evaluations. Harmony search found the least-cost design of 36.66 M$ in the minimum number of evaluations (6000 evaluations).

The superiority of Harmony search in results and performance is probably due to the competent search process illustrated in Figure 2, which enables the algorithm to discover the research space more efficiently, as it generates solutions randomly, choose neighbor solutions and keep current solutions. On the other hand, the Genetic Algorithms takes greater number of evaluations to reach equivalent solutions. The reason of the reduced efficiency of GA compared with HS or ACOA, can be explained as follows: the evolution of solution in GA, depends only on the mutation and crossover process applied on its two parents only. The mutation is a limited operator that considers changing some gens in the solution, so the offspring in GA are more influenced by the initial population than the offspring in other evolutionary algorithms, this is the reason why GA takes longer time to discover the research space and reach the optimal solutions. The process that we propose to improve the performance of real GA, and GAs in general is to incorporate in every generation, some completely new solutions "chromosomes" that are generated randomly. This process is analogue to the "immigration" process in real life, where these new individuals "solutions" give a new blood to the population, and they are evaluated like other solutions, and couple with existing solutions to find new ones.

Finally, we conclude that the four proposed EAs were able to find four near-optimal designs for the same discrete optimization problem with close costs. The existence of these alternatives is a major advantage in practice, where other factors can be taken into account. The time needed by each EAs to reach the optimal or near-op- timal global solutions varies considerably depending on its mathematical construction and the choice of parameters involving. The development of EAs nowadays is moving towards improving their performance and calculation time, since their robustness has been proven.

REFERENCES

1. Neumaier, A. (2004) Complete Search in Continuous Global Optimization and Constraint Satisfaction. In: Iserles, A., Ed., A Chapter for Acta Numerica, Cambridge University Press, Cambridge.
2. Arora, J.S., Huang, M.W. and Hsieh, C.C. (1994) Methods for Optimization Variables of Nonlinear Problems with Discrete: A Review. Structural Optimization, 8, 69-85.
3. Olsson, R.J., Kapelan, Z. and Savic, D.A. (2009) Probabilistic Building Block Identification for the Optimal Design and Rehabilitation of Water Distribution Systems. Journal of Hydroinformatics, 11, 89-105.
4. Awad, A. and Poser, I. (2005) Genetic Algorithm Optimization of Water Supply Networks. Water Intelligence Online Journal (WIO), International Water Association (IWA) Publishing, 8, 1-25.
5. Montalvo, I., Izquierdo, J., Perez-Garcia, R. and Tung, M.M. (2008) Particle Swarm Optimization Applied to the Design of Water Supply Systems. Computers and Mathematics with Applications, 56, 769-776.
6. Montalvo, I., Izquierdo, J., Perez-Garcia, R. and Herrera, M. (2010) Improved Performance of PSO with Self-Adaptive Parameters for Computing the Optimal Design of Water Supply Systems. Engineering Applications of Artificial Intelligence, 23, 727-735.
7. Coelho, B. and Andrade-Campos, A. (2014) Efficiency Achievement in Water Supply Systems—A Review. Renewable and Sustainable Energy Reviews, 30, 59-84.
8. Swamee, P.K. and Sharma, A.K. (2008) Design of Water Supply Pipe Networks. John Wiley & Sons, Inc., New Jersey.
9. Dorigo, M., Maniezzo, V. and Colorni, A. (1996) The Ant System: Optimization by a Colony of Cooperating Ants. IEEE Transactions on Systems, Man, and Cybernetics Society, 26, 29-42.

10. Schaake, J.C. and Lai, D. (1969) Linear Programming and Dynamic Programming Applications to Water Distribution Network Design. Department of Civil Engineering, Cambridge.
11. Fujiwara, O. and Khang, D.B. (1990) A Two-Phase Decomposition Method for Optimal Design of Looped Water Distribution Networks. Water Resources Research, 26, 539-549.
12. Geem, Z.W. (2006) Optimal Cost Design of Water Distribution Networks Using Harmony Search. Engineering Optimization, 38, 259-280.
13. Maier, H.R., Simpson, A.R., Zecchin, A.C., Foong, W.K., Phang, K.Y., Seah, H.Y. and Tan, C.L. (2003) Ant Colony Optimization for Design of Water Distribution Systems. Journal of Water Resources Planning and Management, 129, 200-209.
14. Bragalli, C., D'Ambrosio, C., Lee, J., Lodi, A. and Toth, P. (2012) On the Optimal Design of Water Distribution Networks: A Practical MINLP Approach. Optimization and Engineering, 13, 219-246.
15. Bonabeau, E., Dorigo, M. and Theraulaz, G. (1999) From Natural to Artificial Swarm Intelligence. Oxford University Press, New York.
16. Kennedy, J., Eberhart, R.C. and Shi, Y. (2001) Swarm Intelligence. Morgan Kaufmann Publishers, Burlington.
17. Coelho, B. and Andrade-Campos, A. (2012) Using Different Strategies for Improving Efficiency in Water Supply Systems. Proceedings of 1st ECCOMAS Young Investigators Conference, Aveiro, 24-27 April 2012.
18. Broad, D.R., Dandy, G.C. and Maier, H.R. (2015) A Systematic Approach to Determining Metamodel Scope for Risk Based Optimization and Its Application to Water Distribution System Design. Environmental Modelling & Software Volume, 69, 382-395.
19. May, R.J., Dandy, G.C., Maier, H.R. and Nixon, J.B. (2008) Application of Partial Mutual Information Variable Selection to ANN Forecasting of Water Quality in Water Distribution Systems. Environmental Modelling & Software, 23, 1289-1299.
20. Bowden, G.J., Nixon, J.B., Dandy, G.C., Maier, H.R. and Holmes, M. (2006) Forecasting Chlorine Residuals in a Water Distribution System Using a General Regression Neural Network. Mathematical and Computer Modelling, 44, 469-484.
21. Gibbs, M.S., Morgan, N., Maier, H.R., Dandy, G.C., Nixon, J.B. and Holmes, M. (2006) Investigation into the Relationship between Chlorine Decay and Water Distribution Parameters Using Data Driven Methods. Mathematical and Computer Modelling, 44, 485-498.
22. Bi, W., Dandy, G.C. and Maier, H.R. (2015) Improved Genetic Algorithm Optimization of Water Distribution System Design by Incorporating Domain Knowledge. Environmental Modelling & Software, 69, 370-381.
23. Beheshti, Z. and Shamsuddin, S.M. (2015) Non-Parametric Particle Swarm Optimization for Global Optimization. Applied Soft Computing, 28, 345-359.
24. Holland, J.H. (1975) Adaptation in Natural and Artificial Systems. University of Michigan Press, Ann Arbor.
25. Goldberg, D.E. (1989) Genetic Algorithms in Search, Optimization and Machine Learning. Addison-Wesley, Reading.
26. Shi, Y.H. and Eberhart, R. (1998) A Modified Particle Swarm Optimizer. The 1998 IEEE International Conference on Evolutionary Computation Proceedings, IEEE World Congress on Computational Intelligence, Anchorage, 4-9 May 1998, 69-73.
27. Yao, X., Liu, Y. and Lin, G.M. (1999) Evolutionary Programming Made Faster. IEEE Transactions on Evolutionary Computation, 3, 82-102.
28. Dorigo, M. and Di Caro, G. (1999) The Ant Colony Optimization Metaheuristic. In: Corne, D., Dorigo, M. and Glover, F., Eds., New Ideas in Optimization, McGraw-Hill, London, 11-32.
29. Stützle, T. and Hoos, H.H. (2000) MAX-MIN Ant System. Future Generation Computer Systems, 16, 889-914.
30. Geem, Z.W., Kim, J.H. and Loganathan, G.V. (2001) A New Heuristic Optimization Algorithm: Harmony Search. Simulation, 76, 60-68.

31. Lee, K.S. and Geem, Z.W. (2004) A New Structural Optimization Method Based on the Harmony Search Algorithm. Computers and Structures, 82, 781-798.
32. Kim, J.H., Geem, Z.W. and Kim, E.S. (2001) Parameter Estimation of the Nonlinear Muskingum Model Using Harmony Search. Journal of the American Water Resources Association, 37, 1131-1138.
33. Rossman, L.A. (1993) EPANET, Users Manual. US Environmental Protection Agency, Cincinnati.

CHAPTER 10

The Pestalozzi Method: Mathematics as a Way to the Truth

Peri Mesquida, Fabio Inácio Pereira, Maurício Eduardo Bernz

Pontifical Catholic University of Paraná (PUCPR), Curitiba, Brazil

ABSTRACT

Johann Heinrich Pestalozzi (1746-1827), the father of modern pedagogy (Soëtard, 1985), dedicated his life to the pursuit of truth for mankind. This quest considered the education and teaching of children as a whole hence, his famous formula of "head, heart, and hands" was conceived not just as a slogan but, particularly, as a concept of comprehensive education. But how do you get to the truth behind education and teaching? How do you teach and educate children? What method ought to be used for this task? Pestalozzi wrote five volumes on his own Method (in Germany), which were later translated into French, and in these, he aimed to address the three questions mentioned above. Ever since the first attempt to explain the Pestalozzi Method, by Marc Antoine JULLIEN in his work Spirit of the Method of Pestalozzi education (Esprit de la Méthode d'Éducation de Pestalozzi) (1812), followed by Daniel Alexandre CHAVANNES in his Elemental Method of H. Pestalozzi (Méthode Élémentaire de H. Pestalozzi) (1819) and the works of Soëtard (1994) and Thröler (2008) , many of the Method's concepts have been discussed and defended. However, the various authors do not appear to have settled on a common ground. Nevertheless, this article aims to offer a reflection on the subject based on a reading and analysis of the texts by Pestalozzi that were translated into French. It will attempt to defend the argument that Pestalozzi appropriated the inductive Method of Locke and Comenios who, like himself, were protestants, and whose methodological resources for finding the "Truth" through both education and teaching involved geometry and mathematics; these were elements that Pestalozzi considered useful for approaching and identifying truth and accuracy.

Keywords: Pestalozzi Methods, Children's Education, Geometry, Mathematics, Truth

1. INTRODUCTION

Pestalozzi: An overview of his life

Johann Heinrich Pestalozzi was born in Zurich, Switzerland, on January 12, 1746, in a middle-class family; his father was a doctor and his mother was a housewife. In his youth, Pestalozzi became involved in politics, and his attempts to expose the abuses and corruption committed by officials led to his imprisonment. His militancy and political views led him to write about the topic in a work that reflects on the French revolution (Pestalozzi, 2007) . However, politics would also appear in other works by Pestalozzi, such as To the innocence, the gravity and the nobility of my time and my country (À l'Innocence, à la gravité et à la noblesse de mon époque et de ma patrie) (Lausanne: LEP Éditions Loisirs et Pédagogie SA, 2012). The importance he accorded to politics (not partisan politics but the politics considered by Aristotle), and its relationship with education, as well as the relationship between education and politics can be discerned in the following statement: "The beginning and end of my politics is education" (Pestalozzi, 2012: p. 34) .[1] Pestalozzi spent the final years of his life in Yverdon-Les-Bains, in "exile" from his home in the Canton of Zurich. He died in 1827.

The Swiss pedagogue took every step possible to convince the government to adopt his educational Method and to have it implemented in public schools throughout the Swiss Confederation. As such, he requested that the Swiss authorities give him their opinions regarding the content and practice of his Method and its feasibility at the national level. His Method was assessed on three occasions: in July 1802 (Report Ith), in March 1806 (Rapport Chavannes), and in May 1810 (Rapport Girard); however, Pestalozzi, despite being praised by the evaluators for his Method, never saw it adopted or made universal by the government, that is, by the Swiss Confederation. In fact, even he did not believe that his Method was finished and ready for universal application; rather, to him, the Method needed to be reworked on a case-by-case basis: "I have been seeking for quite some time now a world to simply yet accurately express what my Method in fact is. I have never found it, neither before nor after. The reason for this is clear: my Method is not yet complete" (Pestalozzi, 2008: p. 18) .[2]

2. PESTALOZZI: HEAD, HEART, AND HAND (KOPF, HERZ, HAND)

All of Pestalozzi´s educational work is based and performed on this trio. It is no accident that heart is placed at the center of the trio. It feeds and fulfills the wholeness of human beings: their minds, feelings, and actions or, even, their intellect, feelings, and actions. In the present day, it could be translated into the formulation of knowing how to think, feel, and act, as knowing how to think is nourished by feeling, just like the dynamic source of knowing how to act is the heart. Strictly speaking, if the emphasis is placed on the heart, it is because Pestalozzi's pedagogy is a pedagogy of love. In the first volume of Writings on the Method (Écrits sur la Méthode), Pestalozzi defends what we would call the anthropology of love. Love, in the articulation of moral education, religious education, and teaching/learning, makes the act of educating a practice in which the interrelated elements of trust, justice, respect, and affection express themselves, making the practice of pedagogy a vocation that is dynamic and joyful (as Snyders (1988) writes in Joy in School. São Paulo: Ed. Manole, 1988). The central role played by the pedagogy of love in Pestalozzi's educational thought led him to observe that "elementary education aims at educating, through the use of its own means and by virtue of its harmony, the mind, heart, and hands towards performing to the best of our abilities" (Pestalozzi, 2008: p. 99) .[3] It is important to stress the fact that by emphasizing the "heart," feeling, and love as points of reference in the practice of childhood education, which in turn illuminates intellectual and "artistic, aesthetic, and professional" training, Pestalozzi highlights that the results of this loving relationship are trust, respect, and justice, which are fundamental in the relationship between the educator and the educated. Therefore, Pestalozzi asserts that "an intellectual training that is accompanied by the education of the heart can bear nothing other than the fruit of love" (Pestalozzi, 2008: p. 89) .[4]

In her Letter from Stans (Lettre de Stans), (1985), Pestalozzi says that as long as "man" and, by extension, the educator, wishes "well" and demonstrates this to the child, the child will in turn:

Listen carefully; but, the child does not do this for you, master, nor because of you, educator, the child does so for himself/herself. The goodness that you want the child to achieve must not be the result of a whim of yours, or of your passion...this goodness must be good in itself because of the nature of the goodness itself, and must appear this way in the child's eyes (Pestalozzi, 1985: p. 25).[5]

Ultimately, to create this goodness, it is necessary that the educator desires that the child feels and follows the example he has set. This is not a discourse on goodness, but rather an unequivocal demonstration that educators are good, mean well, and love both their own profession and the child.

It is to this end that the meaning of the Latin verb educare can be understood. Etymologically, it means education that starts at the cuore, the heart, and which is expressed as love. Nevertheless, it is worth remembering that Pestalozzi's idea of love is closely related to consciousness, which in turn is related to human beings' rational ability inherent in their own nature, a fact that eliminates both soppiness and sentimentalism. With this in mind, autonomy, freedom, and the ability to maintain a dialogue might also be a part of this construct.

2.1 Plutarch's Water Pail, and Pestalozzi's Vase

However, this perspective on education is not originally Pestalozzi's. Plutarch (45?120), also known as Lucius Mestrius PLUTARCHIUS, had already come to this conclusion 1.700 years before the Swiss pedagogue. Plutarch's work, titled in English on the education of children (In French: Sur l'éducation des enfants) (Plutarch, 1927 , Moralia, Volume 1. Cambridge, MA:Loeb Classical Library, Harvard University Press), includes the following idea:

"The mind does not require to be filled like a bottle, but rather, like wood, it only requires kindling to create in it an impulse to think independently and an ardent desire for the truth."[6]

By providing us modern educators with this sentence, at a time when we are concerned about merely infusing students with knowledge without considering that they need "kindling" to spur their abilities and to encourage them to continue seeking the truth, Plutarch has clearly summarized what educating means.

3. PESTALOZZI AND HIS METHOD

Strictly speaking, what Pestalozzi calls "My Method" (always with a capital "M") concerns not only the use of teaching techniques or didactic-pedagogic instruments but, rather, the combination of pedagogic theory and practice, including philosophy, psychology, and education policy. From his very first pedagogic experiences with poor children in Neuhof and later in Stans, Pestalozzi sought to unite practice and theory. Authors differ as to the meaning of the Method, or as some say, "the spirit of the Method" (Soëtard, 1982) . However, it is clear, either through the Swiss educator's writings or especially in the expert reports that were produced upon Pestalozzi's request by educational institutions, that Pestalozzi's Method is a set of educational actions designed to be performed in schools as well as in the institutes he established. These educational actions involve teaching not only techniques and didactic-pedagogic resources, but also pragmatic content and, in particular, effecting a teacher-student relationship based on respect through dialogue, freedom, and what he calls "Moral.". It ranges progressively from concrete to the abstract, from the sensitive to the resilient, and from theory to experience: "This Method aims to develop as many of the children's intellectual abilities as possible, keeping pace with nature" (Pestalozzi, 2011: p. 126) .[7]

The Method progressively proceeds through "gradation from the simple to the complex" (Pestalozzi, 2013: p. 84) , leading the child to uncover the reality expressed by words. This is a development of education according to the children's ages ("development of education according to the children's ages") (Pestalozzi, 2009: p. 178 and 2013: p. 19) .

With this in mind, the free and independent man and education play a fundamental role in helping human beings forge themselves.

However, the Method, as we have seen above, is not, in essence, a group of techniques or didactic instruments, but rather a set consisting of techniques, didactic means, pedagogic action, content, and practice. Moreover, Pestalozzi's Method must be developed by the mature initiative of each pupil.

Therefore, he recommends that every educational institution organize itself in a suitable fashion in all branches of education, so that "every lesson can be tailored to the willingness and needs of the child."

The induction takes the opposite course by leading thought and action from the specific towards the universal, the sensitive to the resilient, from the empirical to the abstract, from practice to theory. This sub-chapter seeks to discern the theoretical-practical elements of what we call "method" in Pestalozzi's Methods, as well as the didactic-pedagogic means that he requires, temporarily setting aside the specific elements that constitute the method.

Pestalozzi says that "from the intellectual point of view, the idea of elementary education maintains the pedagogical principle: 'life is a factor of education" (Pestalozzi, 1947: p. 49) .[8] Bearing this assertion in mind, a fundamental element appears in this assertion made by the Swiss pedagogue: a child's education is based on nature (life) and, concurrently, must be an education for life.

Therefore, Pestalozzi claims that the foundation of both the method and the knowledge that the child will obtain, as well as the education to be offered, is "sensitive intuition" that is performed by the sensitive perception that the child will have, of concrete, empirical things.

The child has a special way of learning and communicating with the world; that is why it is the responsibility of educators, during the development of the pedagogical actions of teaching/learning, to make use of a method that requires "didactic-pedagogic" means fit for the child's reality.

Every educational action is performed in the sense of leading "the child to learn to know himself/herself, that is, to learn to feel, observe, to know" (Pestalozzi, 2008: p. 150-151) [9] and to become autonomous or independent. Pestalozzi explains:

The fate of the child is to become independent in all fields, by means through which his/her strengths are developed. He/she must necessarily receive care that is the root of his/her entire education, but that contains the seed of independence that will lead him/her—that must lead him/her—to free himself/herself of it (Pestalozzi, 2008: p. 116-117) .[10]

Pestalozzi is even more emphatic in his view on childhood education when he explains that the pedagogical activity of the educator is performed and achieved when he/she manages to make the child begin

Developing the forces and strengths in himself/herself that can make him/her no longer require the help of others. He/she is now ready to move forward at a steady pace following the path to perfection and to complete his/her education on his/her own. Had it not been this way, my home would not exist and my initiative would have failed (Pestalozzi, 2008: p. 87) .[11]

However, the Swiss pedagogue, consistent with his worldview, defends the idea that the first steps of a child's education must be conducted within the context in which the child lives.

Therefore, "instructing a child...must be done according to the child's external situation, the context in which he/she lives, considering the totality of his/her truth" (Pestalozzi, 2009: p. 39) .[12] By proceeding in this way, Pestalozzi wanted to teach man to forge himself through his own work ("as work of myself the world becomes world for me") (Pestalozzi, 2012: p. 226) because "through my own work, the world becomes meaningful for me" (Pestalozzi, 1994: p. 154, 161) .

3.1 Strategies of the Way—Method—For Achieving Objectives

To Pestalozzi, the first foundation of all human knowledge is "sensitive impression, learned and fixed through attention, precisely expressed by language and made noble by its purpose"(Pestalozzi, 2011: p. 36) ,[13] and its

Raw material is provided by sensitive impressions, particularly those of sight; these sensations that attract through novelty stir great interest in the child and indelibly mark him/her. This sensory representation of objects as they appear in nature may be called natural sensitive perception or intuition; for the school, it is empirical perception (Pestalozzi, 2011: p. 37) .[14]

Therefore, it is not possible to propose "anything distant, anything strange to the child's conscience" (Pestalozzi, 2009: p. 29) ,[15] since there is "a natural bond that must be established between the knowledge of words and of letters" (Pestalozzi, 2013: p. 35) .[16] It is precisely due to this natural "bond" that child literacy in children's elementary education is developed based on the "connections" created between the sight of things in nature, objects, the signs that represent them, images, and the words that express them. This is why the course of children's education (bildung) goes from the thing, the object, the image of the object, and information about the thing, to the word that the child learns to write and communicate; this enables the child to learn to name things, or to master them through the word (Pestalozzi, 2013: p. 92, 100, 146) . This constitutes a process involving "knowing and thinking, naming, and acting" (connaître et penser, nommer et agir)"(Pestalozzi, 2013: p. 163) .

In fact, Pestalozzi, consistent with his worldview and with his vision for human education (bildung), intended to awaken in children and in human beings the ability to think for themselves, to dare to think, and dare to know: "Dare to know... Am I allowed to wish that men speak these terms whose words will have more importance than mine" (Pestalozzi, 1947: p. 304-305) .[17] Therefore, "the dare to know" by Immanuel Kant, is employed by Pestalozzi to justify his Method, which must encourage the child (the man) to discover his/her "wisdom" (as did Rousseau's Emilio and Sofia), forge himself/herself, and pronounce her/ his word.

Pestalozzi maintains the importance of "touch" (toucher) in relation to the objects, as then the child can "feel" them concretely and make connections between the words; he also believes that the image of the objects placed alongside the words helps the child to learn to read and write, following Comenius' recommendations. As Pestalozzi himself explains:

Yet, the circle of objects that surround man is limited; and Comenius ... felt the need to resort to images, thus creating his world through painting ... To the child, the images represent not only what he/she could not see in the objects around him/her, but also what he/she could see every day in nature itself. (Pestalozzi, 2009: p. 178) [18]

Therefore, the images transmit true "sensitive impressions" to children, i.e., they are "imprinted," through the sight of them and the words that express them, with their own meaning, since the images represent and need to represent things that relate to the child's world (Song fo the swan: Chant du cygne, 1947, p. 130). As long as the images (pictures) of things are also artistic expressions, they contribute to developing the child's creative ability (Song of the swan: Chant du cygne, 1947, p. 100 and 102) and to an "education through art and for the art of creating".(Pestalozzi, 2011: p. 45)

Another important element that is prominent among the methodological "instruments" proposed and used by Pestalozzi is the progressive teaching and learning process that, in his view, must be gradual, respecting the child's age and the development of their ability to understand what Piaget would call in the twentieth century "cognitive development" according to age (Psychology of the child: Psychologie de l'enfant. Paris: Presses Universitaires de France, 1966; The birth of intelligence in children: La naissance de l'intelligence chez l'enfant. Neuchâtel: Délachaux et Niestlé, 1936).

Pestalozzi's dream was to concern himself with education, the poor, the unhappy, he believed that educating the poor, the "wretched of the earth," was difficult not because they are poor, but because

We do not make enough means available to be able to educate them so that they can beneficially use the strengths and abilities that their situations desperately require...i.e., means that can encourage their thinking and creativity (Pestalozzi, 2009: p. 224, 225) .[19]

Being aware of the fact that "education is the road to salvation for the poor," he sought to fulfill his life-long dream by founding a school for educating the poor. This opened in Clendy in 1818 but, unfortunately, it did not prosper, even after it was converted into an annex or the Pestalozzi Institute in Yverdon. This school was intended for poor children and for training teachers who would work in rural areas, teachers whom Pestalozzi believed he could train in no more than three months(Pestalozzi, 2009: p. 234) .

In his writings, Pestalozzi sought to clarify what the relationship between the educator and the educated would be. In his view, the educator, in performing pedagogical practice, must conduct "a dynamic unity of faith, love, and truth" (Song of the swan: Chant du cygne, Pestalozzi, 1947: p. 145, 152) .[20] That means that the educator cannot settle for superficial knowledge of the subject; it is essential for the educator to master it. Yet, this does not make him/her a gatekeeper of knowledge—proud, yet humble in the performance of the "mission." He is a "true" defender of the truth. In his practice, the educator leads the child to "notice things and knowledge itself as a live and organic whole whose elements overlap harmoniously with each other" (Pestalozzi, 2008: p. 151) .[21] Bearing this in mind, the educator "cannot be a tool, an inanimate distributor of a dead substance, similar to a machine, but must put himself/herself in the child's shoes" (Pestalozzi, 2008: p. 151) .[22] This requires the educator to assume an energetic stance and, concurrently, give lots of love—not love as "theory, but rather as love in action" (théorie, mais l'amour en acte)(Pestalozzi, 2009: p. 27) , since "the person who does not love people is not worthy of them"(Pestalozzi, 2008: p. 61) .[23]

4. THE TRUTH AS A PURPOSE OF THE PESTALOZZI METHOD AND MATHEMATICS AS A METHODOLOGY

Pestalozzi attributes the purpose of leading child and man to the truth of his teaching/learning Method by claiming that all of the Method's didactic-peda- gogic actions lead to a sole possible objective, namely, the truth. According to him, these actions:

...lead to a sole possible target: the truth...it is thus that they lead to nothing other than approaching the truth, do not lead to anything but the truth, are not interested in anything but the truth. They cannot offer anything else, both due to their nature as well as due to their form; they cannot lead to anything more than the truth, nor interest anything other than the truth (Pestalozzi, 2008: p. 19) .[24]

By way of conclusion, he said that "the only end of my teaching and of my Method is the truth. I wholeheartedly search for it" (2009, III, 154). This statement, as we have seen, can still be found on pages 150 and 170 of volume III of the Writings on the Method (Ecrits sur la Méthode), as mentioned above. I believe that searching for the truth as a purpose for the Method is related to, firstly, the fact that Pestalozzi, as a Christian and pietist, had the Gospel according to John (chapter 4, verse 16) in mind; specifically, the passage where the evangelist reproduces the statement by Jesus Christ that he is "the Way, the Truth, and the Life." Thus, the Method would be the path that would lead one to the truth and to life. Secondly, we can attribute the intransigent search for the truth with Pestalozzi's tenacity in valuing accuracy, numbers, mathematics, and figures with exact shapes, since, according to Pestalozzi, "every truth aims at certainty. Certainty is necessary; but we approach the truth and move in its direction, but what is arbitrary is erased; the greatest perfection of understanding is infallibility"

(Pestalozzi, 2011: p. 58, 59) .[25] Hence, the "elementary number and form of exercises lead not only to recognizing, but also to discovering the truth"(Pestalozzi, 2009: p. 174) . This is how Pestalozzi can conclude that his Method "is not the teaching of truths, but rather, the teaching about truth, since it guided me to the certainty of my judgment"(Pestalozzi, 2013: p. 158) [26]. For to him truth is mistaken for accuracy, and mathematics reflects precisely the exact.

5. CONCLUSION

Overall, according to the numbers and the forms,

"My method does nothing other than reproducing the simple course of nature. Therefore, following the path of nature in the development of the human genus, my method, due to the effect it has, and based on an intuitive vision it produces, forms in the mind of the child's capacity, beginning with observations of squares and rectangles and their divisions, to become aware, in their reality, of extremely complicated fractions and...become aware of the surface of a triangle, etc."(Pestalozzi, 2008: p. 20) [27]

Therefore, from the observation of the forms that nature presents and of the reproduction of objects found in nature, the child learns to write, to use numbers, reason, and the truth, and to read words as well as discover the world and its expressions. Pestalozzi explains his manner of seeing and working for the child's education and learning based on sensitive perception (i.e., induction):

"Every sensitive perception deeply imprinted in the human mind triggers a series of secondary notions that come more or less close to this perception...thus bringing together objects whose essence is the same; your understanding of the inner truth of these objects will be expanded, sharpened, and strengthened" (Pestalozzi, 2013: p. 86) .[28]

Further on, he says:

"Yet if the form bears in itself, successively, and in various ways, the seed of error and illusion, the numbers do not... Only the number carries the infallible results. And, if the art of measuring demands the same priority, it might not intend this other than for the purpose of its alliance with the art of calculation and the help that it receives: it is infallible because it calculates" (Pestalozzi, 2013: p. 131) .[29]

That is, when he conducts an entire reflection about arithmetic: the numbers are infallible; arithmetic is based on them, and this leads to accurate results. As long as they are accurate, they are true. Therefore, he believes that by using geometry, numbers, calculation, arithmetic, and mathematics, his Method leads to the purpose of the educational action: the truth: "I brought together sensitive intuition, reason, the sensitive mechanism and the purely intellectual action and, when finishing in the midst of this method with the confusing chaos of the many individual truths, I brought back teaching about the truth" (Pestalozzi, 2013: p. 158) .[30]

Due to the emphasis on forms, numbers, geometry, and mathematics as paths to the truth, people said that he wanted to "mechanize education." Nothing could be further from the truth, Pestalozzi believed: "I, like Rousseau, simply want to give the child himself/herself back again and lead him/her through teaching to a learning that follows the process of nature" (Pestalozzi, 2013: p. 36, 37) .[31]

REFERENCES

1. Pestalozzi, J. H. (1947). The Song of the Swang Followed by My Destinies (Le chant du cygnesuivi des mesdestinées). Boudry-Neuchatel: Editions de la Baconnière. [Paper reference 3]
2. Pestalozzi, J. H. (1985). Letters from Stans (Lettres de Stans). Uverdonles-Bains: Ed. CDRP. [Paper reference 1]
3. Pestalozzi, J. H. (1994). My Research on the Progress of Nature in the Evolution of Mankind (Mes recherches sur la marche de la nature dans l'évolution du genre humain. Lausanne: Ed. Payot. [Paper reference 1]
4. Pestalozzi, J. H. (2007). Yes or Not? Considerations on the Political Humour of European Humanity in the Upper and Lower Classes by a Free Man (Oui ou non? Considérations sur l'humeur politique de l'humanité européenne dans les hautes et basses classes par un homme libre). Lausanne: LEP éditions Loisirs et Pédagogie SA. [Paper reference 1]
5. Pestalozzi, J. H. (2008). Writings on the Method I(écrits sur la Méthode I). Lausanne: LEP éditions Loisirs et Pédagogie SA. [Paper reference 11]
6. Pestalozzi, J. H. (2009). Writings on the Method III (écrits sur la Méthode, III). Le Montsur-Lausanne: Ed. Loisirs et Pédagogie S/A. [Paper reference 8]
7. Pestalozzi, J. H. (2011). Writings on the Method IV (écrits sur la Méthode IV). Le Montsur-Lausanne: Ed. Loisirs et Pédagogie S/A. [Paper reference 5]
8. Pestalozzi, J. H. (2012). To the Innocence, the Gravity and the Nobility of My Time and My Country (à l'Innocence, à la gravité et à la noblesse de mon époque et de ma patrie). Lausanne: LEP éditions Loisirs et Pédagogie SA. [Paper reference 2]
9. Pestalozzi, J. H. (2013). Writings on the Method V(écrits de la Méthode V). Le Montsur-Lausanne: ed. Loisirs et Pédagogie S/A. [Paper reference 9]
10. 10. Plutarch. (1927). Moralia (Volume 1). Cambridge, MA: Loeb Classical Library, Harvard University Press. [Paper reference 1]
11. Snyders. G. (1988). Reflexions on Joy in School (A alegria na escola). Sao Paulo: Ed. Manole. [Paper reference 1]
12. Soetard, G. (1985). Introduction (Introduction). In J. H. Pestalozzi, Letters from Stans (Lettres de Stans). Uverdon-les-Bains: Ed. CDRP. [Paper reference 1]
13. Soetard, G. (1994). Commentary (Commentaire). In J. H Pestalozzi, My research on the progress of nature in the evolution of mankind (Mes recherches sur la marche de la nature dans l'évolution du genre humain. Lausanne: éd. Payot. [Paper reference 1]
14. Soetard, M. (1982). Pestalozzi or the Birth of the Educator (Pestalozzi ou la naissance de l'éducateur). Berne, Francfort, Las Vegas: Peter Lang. [Paper reference 1]

15. Throler, D. (2008). Preface (Préface). In J. H. Pestalozzi (Ed.), Writings on the Method I (écrits de la Méthode I). Lausanne: LEP Editions Loisirs et Pédagogie S/A. [Paper reference 1]

NOTES

[1]Le commencement et la fin de ma politique est l'éducation.

[2]"Il y a longtemps déjà, j'ai cherché un mot pour exprimer de manière simple mais exacte ce qu'est en fait ma Méthode. Je ne l'ai pas trouvé, ni alors, ni depuis. La raison en est évidente: c'est que ma Méthode n'est pas achevée..."

[3]La formation élémentaire ne se proporse pas moins que d'élever, à travers chacun des moyens qu'elle utilize et grace à leur harmonie, le coeur, l'esprit et la main vers le meilleur de ce dont notre nature est capable.

[4]Une formation intellectuelle qui ne vient pas accompagnée d'une formation du coeur, ne peut pas produire les fruits de l'amour.

[5]Lui prête si volontiers une orielle attentive; mais il ne veut pas pour toi, maître, il ne veut pas pour toi, éducateur, il le veut pour il même. Le bien auuel tu veux le faire accéder ne doit pas être le fruit dun caprice de ton humeur ou de la passion...il doit être bon en soi, conformément à la nature de la chose, il doit apparaître comme bon aux yeux de l'enfant.

[6]L'esprit n'est pas comme un jarre qu'il ne faille que remplir. Semblable aux matières combustibles, il a plutôt besoin d'un aliment qui l'échauffe, qui donne l'essor à ses facultés et l'enflamme pour la recherche de la vérité.

[7]Cette Méthode a pour but de chercher à donner à plus grand développement possible aus facultés intellectuelles des enfants, et cela, en suivant la marche de la nature.

[8]"Du pont de vue intellectuel l'idée de formation élémentaire vient également à l'appui du príncipe pédagogique: 'la vie est un facteur d'éducation".

[9]"l'enfant à apprendre à se connaître lui-même, c'est-à-dire qu'il apprend à sentir, observer, connaître".

[10]le destin de l'enfant est de se rendre indépendant, dans tous les domaines, des moyens grâce auxquels se forces se sont développées. Il doit nécessairement recevoir des soins qui sont les racines de toute son éducation, mais que contiennent le germe de l'indépendance qui le conduira-qui doit le conduire-à s'en détacher.

[11]développer en elle-même des forces et des moyens tels qu'elle peut se passer désormais de l'aide et de l'assistance des autres. Elle est alors en mesure de s'avancer d'un pas sûr dans la voie du perfectionnement et d'achever par elle-même sa formation. S'il n'en était ainsi, ma maison n'existerait pas et mon enterprise aurait échoué.

[12]l'instruction de l'enfant.doit être en accord avec la situation extérieure de l'enfant, avec le contexte dans lequel il vit, pris dans toute sa vérité.

[13]"le premier fondement de tout savoir humain est l'impression sensible, saisie et fixée par l'attention, précisée para la langue et ennoblie par la finalité".

[14]Matière première est fournie par les impressions sensibles, et tout particulièrement para celle de la vue, ces sensations qui ont l'attrait de la nouveauté, suscitent un grand intérêt chez l'enfant; elles le marquent de manière indélibile. On peut appeler cette représentation sensorielle des objets tels qu'ils se presentment dans la nature perception ou intuition sensible naturelle; pour l'école il s'agit de la perception empirique.

[15]"à sa conscience-de l'enfant-rien de lointain, rien d'étranger".

[16]Lien naturel qui doit s'établir entre la connaissance du monde réel et la connaissance des mots et des lettres.

[17]"Aude sapere, incipe ... Il m'est permis de souhaiter que des hommes pronnoncent ces mots dont la parole aura plus d'importance et de poids que la mienne".

[18]Mais le cercle des objets qui entourent l'homme est borné; et Coménius ... sentit besoin de recourir aux images et créa son monde péint... Des images représetent à l'enfant non seulement ce qu'il ne pouvait pas voir dans les objets qui l'entouraient, mais aussi ce qu'il pouvait voir chaque jour en réalité dans la nature même.

[19]"nous n'avons pas assez mis à disposition les moyens aptes à les former à une utilisation bénéfique des forces et capaités dont Il a un urgent besoin dans sa situation...c'est-à-dire, les moyens aptes à stimuler leur réflexion et leur criativité".

[20]une unité dynamique de la foi, de l'amour et de la vérité.

[21]"percevoir les choses et la connaissance elle-même comme un tout vivant et organique, dont les éléments s'imbriquent harmonieusement".

[22]"ne peut pas être un pur outil, distributeur inanimé d'une matière morte, semblable à une machine, mais Il doit se placer du point du vue de l'enfant".

[23]car celui qui n'aime pas le peuple n'est digne de lui.

[24]...conduisent à un seul but possible: la vérité...c'est ainsi qu'elle parvient à n'approcher que de la vérité, à ne conduire qu'à la vérité, à n'intéresser qu'à la vérité... Elle ne peut donner rien d'autre, tant à cause de sa nature qu'à cause de sa forme, elle ne peut conduire à rien d'autre ni intéresser à rien d'autre.

[25]"toute vérité vise la certitude. La certitude est nécessaire; plus nous nous rapprochons de la vérité et avançons vers elle, plus l'arbitraire s'efface; la plus haute perfection de l'entendement est l'infaillibilité".

[26]n'est pas l'enseignement des vérités, mais elle est l'enseignement de la vérité, car il me conduit à la certude de mon jugement.

[27]Ma méthode ne fait donc rien d'autre que de reproduire le simple cours de la nature. Donc en suivant strictement la voie de la nature dans le développement du genre humain, ma méthode, grâce aux effets qu'elle produit ayant pour base une vision intuitive forme dans le sprit de l'enfant la capacité en partant des observations des carrés et des retangles avec leurs divisions, de prendre conscience, dans sa vérité, de fractions extrémement compliquées e... prendre conscience de la super- ficie du triangle, etc.

[28]À chaque perception sensible profondément imprimée dans l'esprit humain et rendu ineffaçable, s'enchaîne très une série de notions secundaires plus ou moins provhes de cette perception...en réunissant ensemble des obejts dont l'essence est la même, ta connaissance de la vérité intérieure de ces objets s'entrouvera profondément élargie, aiguisée et affermie.

[29]et si la forme porte en soi, bien souvent et de bien des manières, le germe de l'erreur et de l'illusion, le nombre jamais, et si l'art de mesurer revendique le même privilège, il ne peut pretendre qu'une raison de son alliance avec l'art du calcul et du secours qu'il em reçoit. Il est infaillible parce qu'il calcule.

[30]J'ai mis en harmonie l'intuition sensible et le jugement, le mécanisme sensible et la démarche purement intelelectuelle, et qu'em mettant fin, par cette méthode, au chaós confus de la multitude des vérités particulières, j'ai ramené l'enseignement à la vérité.

[31]Méchanizer l'éducation. "c'est pas vraie. Je ne veux que, comme Rousseau, rendre l'enfant à lui même, et l'amener par l'enseignement vers un apprentissage qui suit le processus de la nature.

CHAPTER 11

From Fractal Geometry to Fractal Analysis*

Gabriele A. Losa[1#†], Dušan Ristanović[2†], Dejan Ristanović[3], Ivan Zaletel[4], Stefano Beltraminelli[1]

[1] *Institute of Scientific Interdisciplinary Studies, Locarno, Switzerland*

[2] *Department of Biophysics, Faculty of Medicine, Belgrade, Serbia*

[3] *PC Press, Belgrade, Serbia*

[4] *Department of Histology and Embriology, Faculty of Medicine, Belgrade, Serbia*

ABSTRACT

Euclidian geometry pertained only to the artificial realities of the first, second and third dimensions. Fractal geometry is a new branch of mathematics that proves useful in representing natural phenomena whose dimensions (fractal dimensions) are non-integer values. Fractal geometry was conceived in the 1970s, and mainly developed by Benoit Mandelbrot. In fractal geometry fractals are normally the results of an iterative or recursive construction using corresponding algorithm. Fractal analysis is a nontraditional mathematical and experimental method derived from Mandelbrot's Fractal Geometry of Nature, Euclidean geometry and calculus. The main aims of the present study are: 1) to address the dimensional imbalances in some texts on fractal geometry, proving that logarithm of a physical quantity (e.g. length of a segment) is senseless; 2) to define the modified capacity dimension, calculate its value for Koch fractal set and show that such definition satisfies basic demands of physics, before all the dimensional balance; and 3) to calculate theoretically the fractal dimension of a circle of unit radius. A quantitative determination of the similarity using the set of Koch fractals is carried out. An important result is the relationship between the modified capacity dimension and fractal dimension obtained using the log-log method. The text includes some important modifications and advances in fractal theory. It is important to notice that these modifications and quantifications do not affect already known facts in fractal geometry and fractal analysis.

Keywords: Capacity Dimension, Dimensional Balance, Fractal Dimension, Log-Log Method, Similarity

1. INTRODUCTION

Geometry is a branch of mathematics that concerns itself with questions of shape, size, relative position of figures and properties of space. The foundation of Euclidean geometry is a group of elementary notions and axioms. In Euclidean geometry, which has been posed in modern terms by David Hilbert [1] , the elementary notions are the point, straight line and plane. His 20 axioms logically connect the elementary notions. In fractal geometry the elementary notions and axioms are not introduced. Instead, some basic notions are posed, the most important being the fractal set. These basic notions can be defined by means of mathematics. Using the Hausdorff (or capacity) dimensionformula, for every fractal set a fractal dimension can be calculated. We show that this value for the fractal dimension can be obtained using log-log method on a specific manner. The same method can be used for calculating the fractal dimension of nonfractal sets and natural objects as a problem of fractal analysis. Euclidian geometry relates to the artificial realities of the first, second and third dimensions. Fractal geometry is a branch of mathematics concerned with irregular patterns made of parts that are in some way similar to the whole. Unlike conventional geometry which is concerned with regular shapes and whole-number dimensions, fractal geometry deals with shapes that have non-integer or fractal dimensions. This geometry has proven to be a useful tool in analyzing various phenomena [2] since its meaning received a great deal of attention in many natural sciences and engineering. Now it has become widespread and multidisciplinary, and its applications have evolved into many disciplines. Compared to Euclidean geometry, which has a long history for more than 2000 years, fractal geometry is very new. Mandelbrot's famous book "The Fractal Geometry of Nature" was published relatively recently, in 1982. Although conventional approaches based on Euclidean geometry can be well applied to describe regular geometries, it fails to describe the complex geometries. Fractal analysis consists of several methods (most of them being computer-based) useful to assign a fractal dimension and other fractal characteristics to a dataset in all areas of science. Although fractal geometry has been widely applied and used for quantitative morphometric studies of objects, mainly using fractal analysis and calculating the fractal dimensions of sets of objects [3] , there are still some unresolved issues that need to be addressed. Since some concepts of fractal geometry and fractal analysis are determined descriptively and/or qualitatively, this paper offers for some of them the more exact mathematical determination or explanation. Fractal geometry developed as a new geometry of nature [4] -[8] has been conceived with the aim to describe and evaluate the complexity and irregularity of shapes and processes in nature [3] . Up to now fractal geometry and fractal analysis are used in diverse research areas [9] -[12] . The goal of fractal geometry is to provide a platform for the demonstration of the importance of the fractal disciplines and to continuously explore the world of fractals not only in mathematics but also in sciences and technologies. One of the aims of the present paper is to show how fractal geometry gives rise to the fractal analysis confining our consideration to fractal sets and mathematics.

2. METHODS AND RESULTS

2.1. The Geometrical Fractal Set

In Fractal geometry the geometrical fractal set should be considered as an infinite ordered series of geometrical objects defined on a metric space. To determine a fractal set we need specify three things [2] [13] [14] : 1) the shape of a starting object; (the initiator), 2) the iterate algorithm enabling its iterative application on the initiator and then, repeatedly, on all obtained geometrical objects (the generators), and 3) the conditions which these generators should satisfy, before all the properties of geometrical similarity (see Subsection 2.3). In that case, such geometrical object (generator) is called a prefractal [2] [4] [6] . The final result of such infinite procedure is the limit fractal or simply fractal [5] [8] . The initiator, prefractals and fractal represent the geometrical fractal set [13] [15] .

2.2. Koch Fractal Set

Basic definitions and laws of fractal planimetry can be demonstrated on some classical fractal models [3] [13] such as Cantor dust, Sierpinski gasket, Koch triadic curves, Mandelbrot set, etc. For that purpose we chose the triadic (snowflake) Koch curve set. The sequential construction of the this set can begin with the initiator which is an equilateral triangle of the edges length r_0(Figure 1(a)). The iterate algorithm to generate the set of Koch curves (prefractals) consists of recursive reduction of the straight line segments (or the scales) by 1/3 exchanging repeatedly the middle third of each side of the initiator, or a preceding generator, with two sides of a smaller, equilateral triangle whose side is one-third the length of the previous side. The result after the first iteration (the stage of construction z = 1) is shown in Figure 1(b), and that after the second iteration (the stage of construction z = 2), in Figure 1(c). For the Koch prefractals, the length of a segment at the zth stage of construction (r_z) and the number of segments at the same stage (N_z) are respectively,

$$r_z = \frac{r_0}{3^z}; \quad N_z = 3 \cdot 4^z. \tag{1}$$

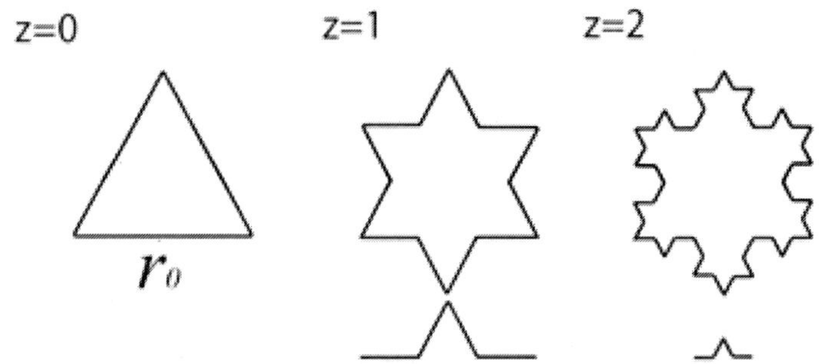

(a) (b) (c)

Figure 1. Koch fractal set. (a) Initiator-equilateral triangle, r_0 is its side length; (b) The first stage of construction (z = 1), detail beyond the drawing is the corresponding generating element; (c) The second stage of construction (z = 2), detail beyond the drawing is the corresponding generating element.

2.3. Geometrical Similarity

Similarity is a typical property of fractal sets. To define this concept we introduce a generating element of a generator. A generator is usually made up of straight-line segments (for example, see Figure 1). A logical concatenation of some segments of a generator can be thought of as the generating element [13] of a generator if the whole object can be completely built with such elements by their translations and/or rotations.

For example, the drawing in Figure 1(b) can be subdivided into 3 generating elements, that in Figure 1(c) into 12 generating elements, and so on. In our example shown in Figure 1 the generators of the Koch prefractals, at the first and second stages of construction, have the generating elements made of four equal segments each, as shown as details below the drawings in Figure 1(b) and Figure 1(c). Two successive generating elements of two generators of a set of objects can be geometrically similar or not. According to the definition of similarity in Euclidean planimetry, two generating elements of the generators at stages z and z + 1 (say, those in Figure 1(b) and Figure 1(c)) are similar to each other [16] if (1) the ratio of the measure of a segment of the generating element at stage z + 1 and the measure of the corresponding segment of the element at the stage z is constant for all pairs of corresponding segments (e.g., for the four pairs of segments of the two mentioned details in Figure 1) and for all z, and (2) the angles between the pairs of corresponding segments of the two generating elements are congruent.

2.4. The Capacity Dimension

Fractal dimension is a quantitative measure of morphological complexity of an object. There are many different definitions of fractal dimension [2] . Mandelbrot [8] thinks that the “plethora” of distinct definitions of the fractal dimension should be reduced to two: the similarity dimension and Hausdorff dimension. The similarity dimension is basic dimension related to all similar (fractal) sets but this dimension is not the subject of our study. The most important definitions are the Hausdorff and capacity dimensions [2] . These two dimensions are quite similar [2] , but the Hausdorff dimension is rather sophisticated, being a subject of mathematical measure theory. The capacity dimension can be given by [2]

$$D = \lim_{z \to \infty} \left[\frac{\log(N_z)}{\log\left(\frac{1}{r_z}\right)} \right]. \tag{2}$$

Since the logarithm is defined only for dimensionless values (numbers), not for physical quantities like 1/r (where r is a length), we define the modified capacity dimension as:

$$D = \lim_{z \to \infty} \left[\frac{\log(N_z)}{\log\left(\frac{r_0}{r_z}\right)} \right] \quad (3)$$

where r_0 is a reference scale [17] [18] (for example, r_0 can be the length r_0 of the scale of an initiator of the Koch set, Figure 1). The quantity r_0 (Equation (1)) could be included into the definition of the capacity dimension (Equation (3)) because the ratio r_0/r_z in Equation (3) is a dimensionless quantity (a number).

If we submit Equation (1) into the last definition (Equation (3)) and put z to tend to infinity, one can see that the expression in Equation (3) gives $\log(4)/\log(3) = 1.262$:

$$\lim_{z \to \infty} \frac{\log(N_z)}{\log\left[\frac{r_0}{r_z}\right]} = \lim_{z \to \infty} \left(\frac{1}{z} + \frac{\log 4}{\log 3} \right) = D = 1.262. \quad (4)$$

Fractal dimension D is the main quantifier to measure complexity of a set of geometrical and natural objects. The larger the D, the higher the complexity of the set is. For example, since the fractal dimension of the Sierpinski set is $\log(3)/\log(2) = 1.585$, it means that the last set is more complex than the Koch set (Equation (4)).

2.5. Log-Log Method

Considering Equation (3) it would be interesting to analyze a relation between N_z and r_z of the Koch fractal set in a log-log coordinate system. If along the horizontal coordinate axis we put the values $r_z = 1/3^z$ for $z = 0, 1, 2, 3, \cdots$ (the value of r_0 is set to 1for visual clarity) and along the vertical coordinate axis the values of $N_z = 3 \cdot 4^z$ also for $z = 0, 1, 2, 3, \cdots$ and fit the power function to these data using Microsoft Excel software, the decreasing straight line graph shown in Figure 2 is obtained. The fitting power function is:

$$N_z = 3 \cdot r_z^{-1.262} \quad (5)$$

with the coefficient of determination $R^2 = 1$. The fitting parameter is −1.262 which is equal to the negative fractal dimension of the Koch fractal set already obtained using the formula for modified capacity dimension (Equation (3)). This procedure can be thought of as the log-log method. This method is also used in many other fractal techniques, particularly in computer-based ones.

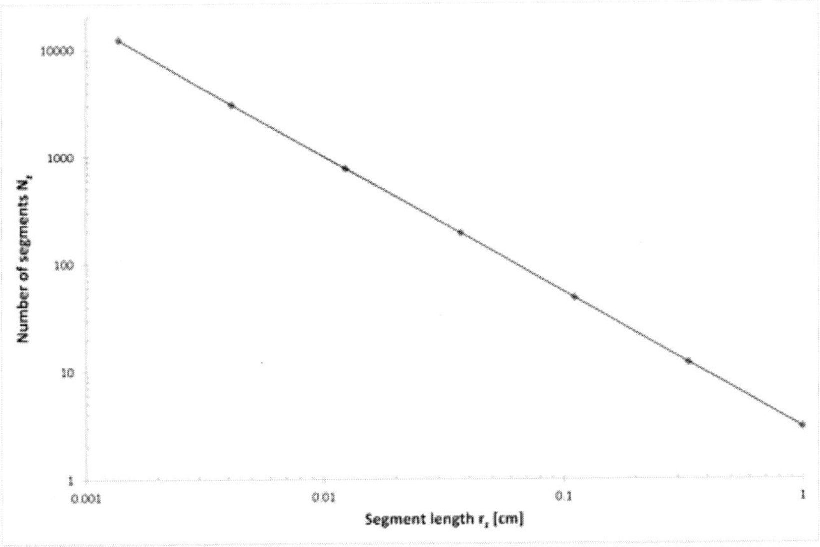

Figure 2. Log-log method applied to the Koch construction (the beginning of the construction is shown in Figure 1). The fitting graph of relationship between the number of segments and segment sizes (in cm) is a decreasing straight line with the high coefficient of determination R^2.

2.6. A Nonfractal Set

Following the fractal methodology, we inscribe an equilateral triangle (as the initiator) in a circle of unit radius (Figure 3(a)) and state the following iterate algorithm: from the middle of each triangle side the normal is erected to the sections with the circle. These three sections are connected to the nearest vertices of the triangle sides forming a hexagon inserted in the circle (Figure 3(b)). The result of the next iteration (Figure 3(c)) is a dodecagon. The next polygons can be constructed using the same algorithm. Inspecting the details below the drawings in Figure 3(b) and Figure 3(c), which are the generating elements for these polygons (generators), it is obvious that they are not mutually similar since the angles between the corresponding segments of the generating elements are different. The condition of side ratios constancy (geometrical similarity) also fails. Namely, it is known that for a circle of unit radius:

$$r_{2n} = \sqrt{2 - 2\sqrt{1 - 0.25 \cdot r_n^2}} \tag{6}$$

where r_n is the side length of a regular polygon inserted in the circle, n is the number of sides of the polygon and r_{2n} is the side length of the polygon with 2n sides. From this expression it is obvious that the ratio r_{2n}/r_n is not constant for every n (e.g., $r_6/r_3 = 0.577$, $r_{12}/r_6 = 0.518$ etc.), so that the class of such regular polygons inscribed in a circle cannot be considered as a set of prefractals. If n tends to infinity, the polygons tend to the circle, which is not a limit fractal. Therefore this set of regular polygons inscribed in a circle cannot be a fractal set.

On the other hand, if we apply the log-log method to the data of the polygonal set obtained, a straight-line graph is again obtained (Figure 4). The equation of this graph is

$$N_z = 5.814 \cdot r_z^{-1.027} \tag{7}$$

which is analogous to Equation (6). From this equation the fitting parameter −1.027 corresponds to the negative value of the fractal dimension D of the nonfractal set considered. This is the fractal dimension of a circle of unit radius (which is the limit object). Using the mass method Smith et al. [3] have measured the fractal dimension of a circle and found D = 1.02. Using mathematics (with no measurements) we found for a circle of unit radius a more precise value D = 1.027 (Equation (7)).

2.7. Properties of Fractal and Nonfractal Sets

The main properties of geometrical fractal sets are geometrical self-similarity and scale invariance. The object's property known as self-similarity was first coined by Mandelbrot [5] . It can be geometrical or statistical. A fractal pattern is said to be geometrically self-similar if every small piece of an object resembles the whole object. This definition of the concept "geometrical self-similarity" should be quantified since small pieces that constitute geometrical or natural objects are rarely copies of the whole object [14] . We have offered a more exact interpretation of this descriptive definition introducing a generating element of a generator as a "small piece". Therefore every fractal set (which must satisfy geometrical similarity) should satisfy the property of geometrical self-similarity since by definition every small generating element is always similar to any other large generating element.

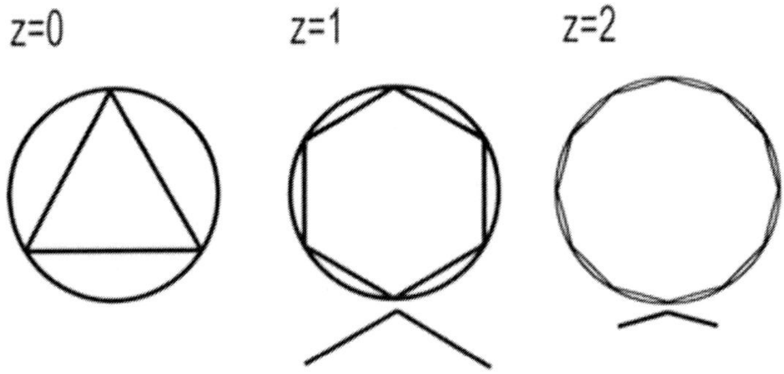

(a) (b) (c)

Figure 3. Nonfractal set of regular polygons. (a) Initiator- equilateral triangle inscribed in a circle of unit radius; (b) The first stage of construction (z = 1) is a hexagon; the detail beyond the drawing is its generating element; (c) The second stage of construction (z = 2) is a dodecagon with the corresponding generating element.

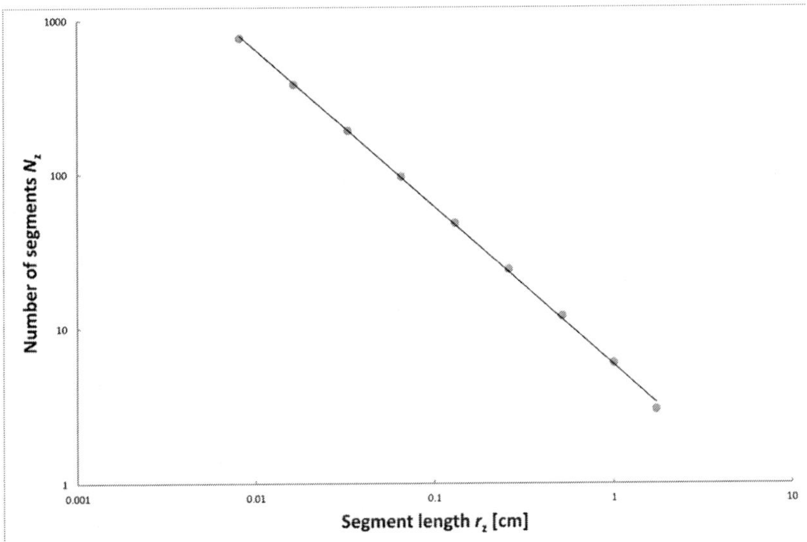

Figure 4. Log-log method applied to the set of polygons (Figure 3). The fitting graph is a decreasing straight line (see Figure 2 for comparison).

Scale invariance is a feature of objects or laws that they do not change if scales of length or other variables are multiplied by a common factor. For example, this common factor for the Koch fractal set is 1/3. If r_0 is 1 cm (Figure 1(a)), the length of the segment in Figure 1(b) is 1/3 cm, that in Figure 1(c) is 1/9 cm, etc. This means that if we multiply the length of each segment of the generating element in Figure 1(b) by 1/3, the same but reduced generating element of the prefractal shown in Figure 1(c) will be obtained. Nonfractal sets cannot satisfy geometrical similarity and properties of fractal sets.

2.8. A Fractal Set versus a Nonfractal Set

We showed that Koch set is a fractal set because it satisfies the geometrical similarity. It means that the ratio r_{z+1}/r_z is the same for every z. If z tends to infinity, we obtained the limit fractal which cannot be drawn or imagined. Nevertheless this fractal exists because we calculated the fractal dimensions of the limit fractal using both the formula for modified capacity dimension (Equation (3)) and log-log method. In both cases we found a real and finite value D = 1.262. We conducted the same procedure using the set of polygons inscribed in a circle. Although the polygonal set is not a fractal set we calculated the fractal dimension of this limit object using the log-log method. In fact, we calculated the fractal dimension of the circle which corresponds to the limit object in which the set of polygons is inscribed. Besides, the initiator in Fractal geometry is included into the fractal iterations as a part of prefractals while the starting object (the circle) in fractal analysis does not change its integrity during iterations.

The circle is a closed line, but the log-log method can be applied to any open line (Figure 5). We found that in Fractal geometry the values of segment lengths were strongly determined by the property of geometrical similarity, while in the fractal analysis values and shapes of open lines are not restricted (Figure 5 and Figure 6). Fractal analysis represents a formation of Fractal geometry consisting in using log-log and others methods [3] , and ignoring the property of geometrical similarity. Therefore we can use any set of regular polygons and always calculate the fractal dimension of the object in which these polygons are inscribed. For example, fractal analysis enables to calculate the fractal dimension of the border of any natural object (not of the set of objects) inscribing a set of straight lines of known and constant lengths along the object's border and count the number of segments. These data in a log-log coordinate system give the fractal dimension of the natural object (coastlines, borders between nations, dendrites of neurons etc.). This method has been the basis of the Richardson's method of fractal dimension measurement of natural objects [19] . It may be opportune to underline that an unlimited extension of log-log coordinate system is not valid when natural and biological objects are investigated. As previously reported, the fractal dimension could be reliable evaluated within a defined scale interval, called "fractal window" [17] [18] .

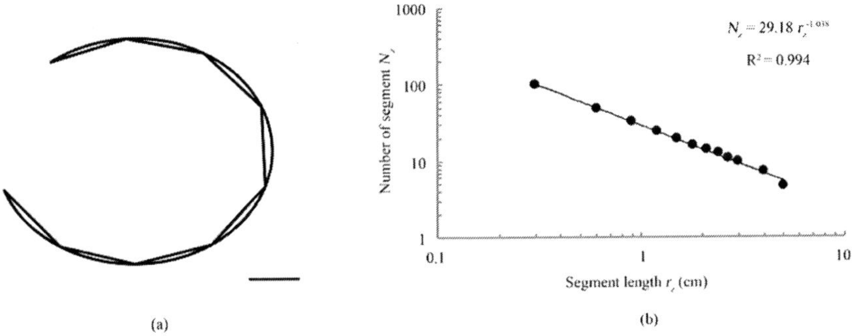

Figure 5. The fractal dimension of an open circle-like curved line. (a) The circle-like line considered, the length of a straight line segment is 1.5 cm, the bar is 1 cm; (b) Relation between N and r represented in a log-log coordinate system, the graph is a straight line whose exponent represents a negative value of the fractal dimension of the curve, R^3 is the coefficient of determination.

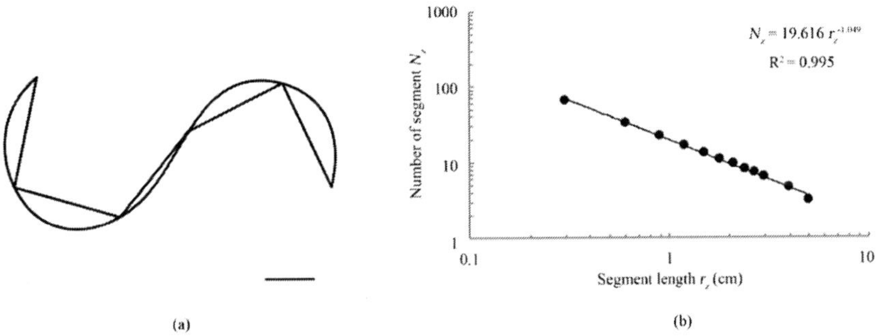

Figure 6. The fractal dimension of an open wavy line. (a) The curved line considered, the length of a straight line segment is 2.2 cm, the bar is 1 cm; (b) Relation between N and r, the graph is a result of application of the log-log method and re- presents a decreasing straight line whose exponent is a negative value of the fractal dimension of the curve in (a).

2.9. Richardson's Method

The polygonal method needs a group of about ten printed copies of a considered line and in each of them inscribes different regular polygon (Figure 5 and Figure 6). This method is rather tedious and time-consuming. A simpler but still time-consuming method is presented in Richardson's paper [20] . Richardson, as a physicist and Quaker, wondered if there are more wars between nations that share longer common border. Answering this question required measurement of the national borders lengths. To measure the length of the border drawn on a map of Atlas, he used as a ruler a divider having sharp points at the ends of both arms. For a given distance between the sharp points of the divider the total length of the border was given by the number of divider's steps along the border multiplied by the divider step size.

Since the length of a polygon's segment (Figure 5 and Figure 6) is the same as the corresponding divider's step size, all the measurements can be done on one printed line. We used this method when preparing Figure 2(b), Figure 4(b), Figure 5 and Figure 6.

The term fractal was introduced by Mandelbrot based on his paper on self-similarity [21] . In that paper, Mandelbrot cited previous work by Richardson [20] and noted that the diagrams presented in Richardson's paper lead to the conclusion that approximate length of the border between two nations can be given by the relationship [8]

$$L(r) \propto F \cdot r^{1-D} \tag{8}$$

where the value of the exponent D seems to depend upon the shape of the border chosen. Topology fails to discriminate between different borders [8] . Since Equation (9) can be rewritten as

$$L(r) = r \cdot Fr^{-D} \tag{9}$$

it is obvious that the factor Fr^{-D} corresponds to the number of divider's steps. For a given r the border of smaller length has smaller number of steps (smaller F) in relation to a longer border of the same shape (the same D) which has larger number of steps (larger F).

To Richardson, the D was a fitting parameter of no particular significance. Having in mind "unearthed" Richardson's work in which he claimed that his lines' slopes had no theoretical interpretation [8] , Mandelbrot claimed that the exponent D should be interpreted as a fractal dimension. We find that among many fractal analysis techniques, only Richardson's method enables correct calculation of the length of an object's border or irregular line.

3. DISCUSSION

When the physical or biological problem is stated in mathematical terms, dimensional balance should be a routine part of the solution of any problem. Exponents and logarithms must always be dimensionless [22] . When dimensioned quantities appear in exponents or in logarithms, they must be combined with the same dimensioned quantities so that their quotient or product is dimensionless [17] [18] . It seems that some of the authors do not take care of these facts. The problem is particularly distinct in defining the fractal dimensions in fractal geometry [2] [13] [23] [24] . One of the main aims of the present study is to define and discuss physically correct of modified capacity dimension. Bassingthwaighte et al. [2] noted that the capacity dimension tells us how much balls needed to cover the object when the size of the balls (r) decreases. In the analytical definition of capacity dimension there exists log (1/r). West and Deering [13] have defined the similarity dimension as the ratio of two logarithms log N/log (1/r). In defining the box-counting dimension Falconer [23] showed lower and upper box- counting dimension by the expression containing log δ where δ is the diameter of a corresponding set. Edger [24] also used log (1/r) in the definition of the upper box-counting dimension, where r is a diameter of the set. Richardson's method of measuring the fractal dimension is robust with very high correlation coefficients but it is, at the same time, tedious and time-consuming. Therefore, the need for more handsome methods for that measurement emerges. The conventional and most popular box-counting (computer-based) method is one suitable to measure the fractal dimensions of geometrical and natural objects. This method is commonly used among other fractal techniques [3] -[5] . Unfortunately, it seems that this method demonstrates some mistakes. Using this method we showed that the fractal dimension of a straight line segment is 0.927 not 1. The fractal dimension of a circle of the radius 1 cm obtained using purely mathematical method, is 1.027, while that measured using the box-counting method is larger than 1.15. The fractal dimension increases with the resolution of a figure and depends on its position on the screen. We showed that Richardson's method does not depend on these factors.

REFERENCES

1. Hilbert, D. (1899) Grundlagen der Geometrie. Teubner, Leipzig.
2. Bassingthwaighte, J.B., Liebovitch, L.S. and West, B.J. (1994) Fractal Physiology. Oxford University Press, New York. http://dx.doi.org/10.1007/978-1-4614-7572-9
3. Smith Jr., T.G., Lange, G.D. and Marks, W.B. (1996) Fractal Methods and Results in Cellular Morphology: Dimensions, Lacunarity and Multifractals. Journal of Neuroscience Methods, 69, 123-136. http://dx.doi.org/10.1016/S0165-0270(96)00080-5
4. Panico, J. and Sterling, P. (1995) Retinal Neurons and Vessels Are Not Fractal but Space-Filling. Journal of Comparative Neurology, 361, 479-490. http://dx.doi.org/ 10.1002/cne.903610311
5. Fernández, E. and Jelinek, H.F. (1998) Use of Fractal Theory in Neuroscience: Methods, Advantages and Potential Problems. Methods, 24, 9-18.
6. Ristanovic, D., Nedeljkov, V., Stefanovic, B.D., Milosevic, N.T., Grgurevic, N. and Stulic, V. (2002) Fractal and Nonfractal Analysis of Cell Images: Comparison and Application to Neuronal Dendritic Arborization. Biological Cybernetics, 87, 278-288. http://dx.doi.org/ 10.1007/s00422-002-0342-1
7. Losa, G.A., Di Ieva, A., Grizzi, F. and De Vico, G. (2011) On the Fractal Nature of Nervous Cell System. Frontiers in Neuroanatomy, 5, 51-62. http://dx.doi.org/10.3389/ fnana. 2011.00045
8. Mandelbrot, B.B. (1982) The Fractal Geometry of Nature. W Freeman & Co, New York. [Citation Time(s):6]
9. Bizzarri, M., Giuliani, M., Cucina, A., Anselmi, F.D., Soto, M.A. and Sonnenschein, C. (2011) Fractal Analysis in a System Biology Approach to Cancer. Seminars in Cancer Biology, 21, 175-182.http://dx.doi.org/10.1016/j.semcancer.2011.04.002
10. Thamrin, C., Stern, G. and Frey, U. (2010) Fractals for Physicians. Paediatric Respiratory Reviews, 11, 123-231.http://dx.doi.org/10.1016/j.prrv.2010.02.001
11. Perrotti, G., Aprile, G., Degidi, M. and Piattelli, A. (2011) Fractal Analysis: A Novel Method to Assess Roughness Organization of Implant Surface Topography. The International Journal of Periodontics and Restorative Dentistry, 3, 633-639.
12. Talu, S. (2011) Fractal Analysis of Normal Retinal Vascular Network. Oftalmologia, 5, 11-16.
13. West, B.J. and Deering, W. (1994) Fractal Physiology for Physicists Levi Statistics. Physics Reports, 246, 1-100.
14. Grizzi, G., Grimaldi, G. and Dioguardi, N. (2001) Fractal Geometry: A Useful Tool for Quantifying Irregular Lesions in Human Liver Biopsy Specimens. Italian Journal of Anatomy and Embryology, 106, 337-346.
15. Feder, J. (1998) Fractals. Plenum Press, New York.
16. Milosevic, N.T. and Ristanovic, D. (2007) Fractal and Nonfractal Properties of Triadic Koch Curve. Chaos, Solitons and Fractals, 34, 1050-1059. http://dx.doi.org/10. 1016/j.chaos.2006.03.117
17. Losa, G.A. and Nonnenmacher, T.E. (1996) Self-Similarity and Fractal Irregularity in Pathologic Tissues. Modern Pathology, 9, 174-182.
18. Losa, G.A. (2002) Fractal Morphology of Cell Complexity. Biological Forum, 95, 239-258.
19. Ristanovic, D. and Milosevic, N.T. (2012) Fractal Analysis Methodologies for Biomedical Researchers. Theoretical Biology Forum, 105, 99-108.
20. Richardson, L.F. (1961) The Problem of Contiguity: An Appendix to Statistics of Deadly Quarrels. General System Yearbook, 6, 139-187.
21. Mandelbrot, B.B. (1967) How Long Is the Coast of Britain? Statistical Self-Similarity and Fractional Dimension. Science, 165, 636-638. http://dx.doi.org/10.1126/ science. 156.3775.636
22. Simon, W. (1972) Mathematical Techniques for Physiology and Medicine. Academic Press, New York, London.

23. Falconer, K. (2003) Fractal Geometry, Mathematical Foundations and Applications. 2nd Edition, John Wiley and Sons, New York and London. http://dx.doi.org/ 10.1002/ 0470013850
24. Edger, G. (2008) Measure, Topology and Fractal Geometry. 2nd Edition, Springer, New York. http://dx.doi.org/10.1007/978-0-387-74749-1

NOTES

*The present paper is dedicated to the imperishable memory of Professor Benoît Mandelbrot, who passed away October 10, 2010.

CHAPTER 12

A History, the Main Mathematical Results and Applications for the Mathematics of Harmony

A. P. Stakhov[1,2]

[1]*International Higher Education Academy of Sciences, Moscow, Russia*

[2]*Institute of the Golden Section, Academy of Trinitarism, Moscow, Russia*

ABSTRACT

We give a survey on the history, the main mathematical results and applications of the Mathematics of Harmony as a new interdisciplinary direction of modern science. In its origins, this direction goes back to Euclid's "Ele- ments". According to "Proclus hypothesis", the main goal of Euclid was to create a full geometric theory of Pla- tonic solids, associated with the ancient conception of the "Universe Harmony". We consider the main periods in the development of the "Mathematics of Harmony" and its main mathematical results: algorithmic measurement theory, number systems with irrational bases and their applications in computer science, the hyperbolic Fibo- nacci functions, following from Binet's formulas, and the hyperbolic Fibonacci l-functions (λ = 1, 2, 3, ···), fol- lowing from Gazale's formulas, and their applications for hyperbolic geometry, in particular, for the solution of Hilbert's Fourth Problem.

Keywords: Golden Ratio; Pascal's Triangle and Fibonacci Numbers; Binet's Formulas; Gazale's Formulas; Hyperbolic Fibonacci Functions; Number Systems with Irrational Bases; Fibonacci l-Numbers; "Metallic Proportions"; Hilbert's Fourth Problem

1. INTRODUCTION

In 2009, the International Publishing House "World Scientific" has published the book of Alexey Stakhov: "The Mathematics of Harmony—From Euclid to Contemporary Mathematics and Computer Science" [1] http://www.worldscientific.com/worldscibooks/ 10.1142/ 6635

The American philosopher Professor Scott Olsen, author of the excellent book: "The Golden Section: Na- ture's Greatest Secret" [2] made a huge assistance in writing the book [1] and its preparation for the publication.

The publication of these books (see Figure 1) is a reflection of one of the most important trends in the devel- opment of modern science. The essence of this trend is very simple: a return to the "harmonic ideas" of Pytha- goras and Plato (the "golden ratio" and Platonic solids), embodied in Euclid's "Elements" [3].

The newest discoveries in chemistry and crystallography: fullerenes, based on the "truncated icosahedron" (The Nobel Prize—1996), and quasi-crystals, based on the icosahedral or pentagonal symmetry (Nobel Prize—2011), are brilliant examples, which confirm this trend.

The amount of such discoveries is increasing continuously. These include: "the law of structural harmony of systems" by Edward Soroko [4], based on the golden p-proportions, and the "law of spiral biosimmetry transformation" by Oleg Bodnar [5], based on the "golden" Fibonacci hyperbolic functions. It also includes a new theory of the genetic code, based on the "golden genomatrices" (author—Doctor of Physical and Mathematical Sciences Sergey Petoukhov, Moscow) [6]. These examples could go on.

Thus, the modern philosophy and theoretical natural sciences begun to use widely the "harmonic ideas" by Pythagoras, Plato and Euclid. And we have every right to talk about the "revival" of ancient Greeks' "harmonic ideas" in modern theoretical natural sciences. This fact puts forward a problem of the renaissance of these ancient harmonic ideas in modern mathematics. The publication of the book [1] is the answer of mathematics to this important trend.

Figure 1. Scott Olsen's book (2006) and Alexey Stakhov's book (2009).

The purpose of this article is to give a survey of the main stages, events and scientific findings, which led to the creation of "Mathematics of Harmony", the new interdisciplinary direction of modern science.

2. THE MATHEMATICS OF HARMONY: THE OPINION OF ACADEMICIAN YURI MITROPOLSKY

What is the Mathematics of Harmony"? What is its role in modern science and mathematics? The outstanding Ukrainian mathematician, the leader of the Ukrainian School of Mathematics, Honorary Director of the Institute of Mathematics of the Ukrainian Academy of Sciences and Editor-in Chief of the "Ukrainian Mathematical Journal" Academician Yuri Mitropolsky (Figure 2) wrote the following in his commentary [7]:

"One may wonder what place in the general theory of mathematics is occupied by Mathematics of Harmony created by Prof. Stakhov? It seems to me, that in the last centuries, as Nikolay Lobachevsky said, "mathematicians turned all their attention to the Advanced Parts of Analytics, neglecting the origins of Mathematics and not willing to dig the field that already been harvested by them and left behind." As a result, this created a gap between "Elementary Mathematics" - basis of modern mathematical education, and "Advanced Mathematics". In my opinion, the Mathematics of Harmony developed by Prof. Stakhov fills up that gap. I.e, "Mathematics of Harmony" is a big theoretical contribution, first of all to the development of "Elementary Mathematics" and as such should be considered of great importance for mathematical education."

Thus, according to Mitropolsky, the Mathematics of Harmony is a new mathematical discipline, which fills the gap between the "Elementary Mathematics" and "Higher Mathematics". That is, this new theory puts forward new challenges in the field of the "Elementary Mathematics". In its origins, this theory goes back to the ancient mathematical topics: "Measurement Theory", "Number Theory", "Numeral Systems", "Elementary Functions" and so on.

The main task of the "Mathematics of Harmony" is to find new mathematical results in the field of the "Elementary Mathematics" based on the "golden ratio" and Fibonacci numbers. It is proved a high efficiency of these results in such areas as theory of recurrence relations, theory of elementary functions, hyperbolic geometry, and finally, the computer and measuring engineering, coding theory [8-41].

3. ABOUT THE TERM OF "THE MATHEMATICS OF HARMONY"

For the first time the term "the Mathematics of Harmony" was used in the article "Harmony of spheres", placed in "The Oxford dictionary of philosophy" [42]:

"Harmony of spheres. A doctrine often traced to Pythagoras and fusing together mathematics, music, and astronomy. In essence the heavenly bodies being large objects in motion, must produce music. The perfection of the celestial world requires that this music be harmonious, it is hidden from our ears only because it is always present. The mathematics of harmony was a central discovery of immense significance to the Pythagoreans."

Figure 2. Academician Yuri Mitropolsky (1917-2008).

Thus, the concept of "the Mathematics of Harmony" is associated here with the "harmony of the spheres", which is also called the "harmony of the world" (Latin "harmonica mundi") or world music (Latin "musical mundane"). The harmony of the spheres is the ancient and medieval doctrine about the musical-mathematical construction of Cosmos, which goes back to the Pythagorean and Platonic philosophical tradition.

Another mention about "the Mathematics of Harmony", as the ancient Greek great discovery, we find in the book by Vladimir Dimitrov "A new kind of social science. Study of self-organization of human dynamics" [43]. Let us consider the quote from the book:

"Harmony was a key concept of the Greeks, a conjunction of three strands of meaning. Its root meaning was aro, join, so harmonia was what joined. Another meaning was proportion, the balance of things that allowed an easy fit. The quality of joining and proportion then came to be seen in music and other arts.

The precondition for harmony for the Greeks was expressed in the phrase "nothing too much". It also had a mysterious positive quality, which became the object of enquiry of their finest minds. Thinkers such as Pythago- ras sought to capture the mystery of harmony as something both inexpressible yet also illuminated by mathe- matics. The mathematics of harmony explored by the ancient Greeks is still an inspiring model for contempo-rary scientists. Crucial to it is their discovery of its quantitative expression in astonishing diversity and complexity of nature through the golden mean (golden ratio), Φ (phi): $\Phi = \frac{1+\sqrt{5}}{2}$, which is approximately equal to 1.618. It is described by Euclid in book five of his Elements: "A straight line is said to have been cut in extreme and mean ratio when, as the whole line is to greater, so is greater to the less".

Thus, in the book [43] the concept of "the Mathematics of Harmony" is directly associated with the "golden ratio"—the most important ancient mathematical discovery in the field of harmony, which at that time was called "the division of the segment in extreme and mean ratio".

Finally, it is pertinent to mention that this term was used in Stakhov's speech "The Golden Section and Modern Harmony Mathematics", made at the 7th International Conference "Fibonacci Numbers and Their Applications" (Austria, Graz, 1996) [11].

4. THE MOST IMPORTANT PERIODS IN THE DEVELOPMENT OF THE MATHEMATICS OF HARMONY

As mentioned above, very successful term of the "Mathematics of Harmony" was introduced to emphasize the most important feature of ancient Greek science (studying the Harmony of the Universe) [42,43]. The greatest interest in the "Harmony of Universe", that is, in the ideas of Pythagoras,

Plato and Euclid, always arose in the periods of greatest prosperity of the "human spirit". From this point of view, the studying the "Mathematics of Harmony" can be divided into the following critical periods.

4.1. Ancient Greek Period

Conventionally, it can be assumed that this period starts with the research of Pythagoras and Plato. Euclidean' "Elements" is a final event of this period. According to Proclus' hypothesis [3], Euclid created his "Elements" in order to give a complete geometric theory of the five "Platonic solids," which have been associated in the ancient Greek science with the "Harmony of the Universe". Fundamentals of the theory Platonic solids (Figure 3) were placed by Euclid in the final Book (Book XIII) of his "Elements".

In addition, Euclid simultaneously had introduced here some advanced achievements of ancient Greek mathematics, in particular, the "golden ratio" (Book II), which was used by Euclid for the creation of the geometric theory of "Platonic solids".

4.2. The Middle Ages

In the Middle Ages it was made very important mathematical discovery. The famous Italian mathematician Leonardo of Pisa (Fibonacci) wrote a book "Liber Abaci" (1202). In this book, he described "the task of rabbits reproduction".

In solving this problem he introduced the remarkable numerical sequence—the Fibonacci numbers F_n: 1, 1, 2, 3, 5, 8, 13, 21, 34, 55, 89, ··· which are given by the recurrence relation:

$$F_n = F_{n-1} + F_{n-2}; \quad F_1 = F_2 = 1 \tag{1}$$

The Fibonacci numbers can be extended to the negative values of the indices n (see Table 1).

In 17 c., the famous French astronomer Cassini had proved the following remarkable formula, which connects three neighboring Fibonacci numbers:

$$F_n^2 - F_{n-1}F_{n+1} = (-1)^{n+1} \tag{2}$$

4.3. The Renaissance

This period is connected with the names of the prominent figures of the Renaissance: Piero della Francesca (1412-1492), Leon Battista Alberti (1404-1472), Leonardo da Vinci (1452-1519), Luca Pacioli (1445-1517), Johannes Kepler (1571-1630). In that period two books, which were the best reflection of the idea of the "Universe Harmony," were published. The first of them is the book "Divina Proprtione" ("The Divine Proportion") (1509). This book had been written by the outstanding Italian mathematician and scholar monk Luca Pacioli under the direct influence of Leonardo da Vinci, who illustrated Pacioli's book. The brilliant astronomer of 17th century Johannes Kepler made an enormous contribution to the development of the "harmonic ideas" of Pythagoras, Plato and Euclid.

In his first book Mysterium Cosmographicum (1596), he built the so-called "Cosmic Cup" (Figure 4)—an- original model of the Solar system, based on the Platonic solids. The book Harmonice Mundi (Harmony of the Worlds) (1619) is the main Kepler's contribution into the Doctrine of the Universe Harmony. In Harmony, he attempted to explain the proportions of the natural world—particularly the astronomical and astrological aspects—in terms of music.

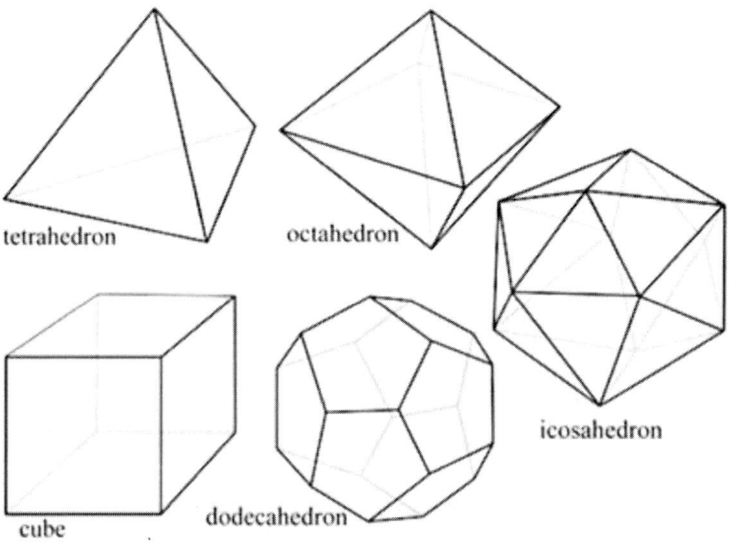

Figure 3. Platonic solids: tetrahedron, octahedron, cube, dodecahedron, icosahedrons.

The central set of "harmonies" was the musica universalis or "music of the spheres", which had been studied by Pythagoras, Ptolemy and many others before Kepler.

Table 1. The "extended" Fibonacci numbers.

n	0	1	2	3	4	5	6	7	8	9	10
F_n	0	1	1	2	3	5	8	13	21	34	55
F_{-n}	0	1	−1	2	−3	5	−8	13	−21	34	−55

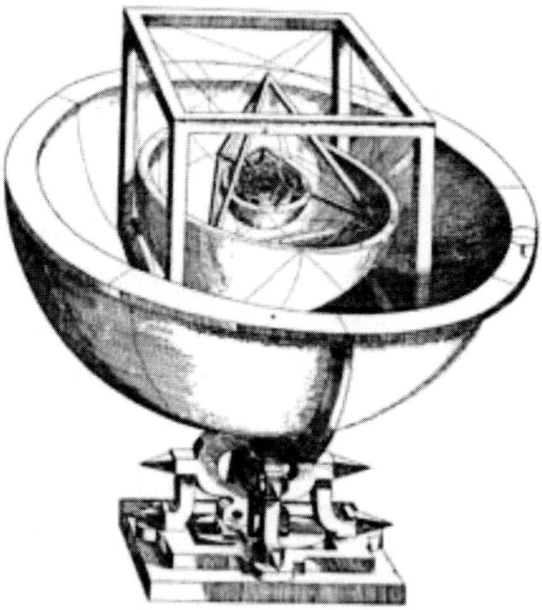

Figure 4. Kepler's Cosmic Cup.

4.4. The 19th Century

This period is connected with the works of the French mathematicians Jacques Philip Marie Binet (1786-1856), Francois Edouard Anatole Lucas (1842-1891), German poet and philosopher Adolf Zeising (born in 1810) and the German mathematician Felix Klein (1849-1925).

Jacques Philip Marie Binet derived a mathematical formula to expresses the "extended" Fibonacci numbers (see Table 1) through the "golden mean" $\Phi = \frac{1+\sqrt{5}}{2}$:

$$F_n = \begin{cases} \dfrac{\Phi^n + \Phi^{-n}}{\sqrt{5}} & \text{for } n = 2k+1; \\ \dfrac{\Phi^n - \Phi^{-n}}{\sqrt{5}} & \text{for } n = 2k \end{cases} \quad (3)$$

where $k = 0, \pm 1, \pm 2, \pm 3, \cdots$.

Francois Edouard Anatole Lucas introduced the Lucas numbers L_n, which are calculated by the same recurrence relation as the Fibonacci numbers (2), but with other seeds:

$$L_n = L_{n-1} + L_{n-2}; \quad L_1 = 1, L_2 = 3 \quad (4)$$

Recurrence relation (4) generates the following numerical sequence:

$$L_n : 1, 3, 4, 7, 11, 18, 29, \cdots \quad (5)$$

The Lucas numbers (5) can be extended to the negative values of the indices n (see Table 2).

The merit of Binet and Lukas is the fact that their researches became a launching pad for Fibonacci researches in the Soviet Union, the United States, Britain and other countries [44-46].

German poet Adolf Zeising in 1854 published the book "Neue Lehre von den Proportionen des menschlichen Körpers aus einem bisher unerkannt gebliebenen, die ganze Natur und Kunst durchdringenden morphologischen Grundgesetze entwickelt". The basic Zeising's idea is to formulate the Law of proportionality. He for- mulated this Law as follows:

"A division of the whole into unequal parts is proportional, when the ratio between the parts is the same as the ratio of the bigger part to the whole, this ratio is equal to the golden mean."

The famous German mathematician Felix Klein in 1984 published the book "Lectures on the icosahedron and the solution of equations of the fifth degree", dedicated to the geometric theory of the icosahedron and its role in the general theory of mathematics. Klein treats the icosahedron as a mathematical object, which is a source for the five mathematical theories:

geometry, Galois theory, group theory, invariant theory and differential equa- tions.

Table 2. The "extended" Lucas numbers.

n	0	1	2	3	4	5	6	7	8	9	10
L_n	2	1	3	4	7	11	18	29	47	76	123
L_{-n}	2	−1	3	−4	7	−11	18	−29	47	−76	123

What is the significance of the ideas of the outstanding mathematician from the point of view of the Mathe- matics of Harmony [1]? According to Klein, the Platonic icosahedron, based on the "golden ratio," is the main geometric figure of mathematics. It follows from this that the "golden ratio" is the main geometric object of mathematics, which, according to Klein, can unite all mathematics.

This idea of Felix Klein is consistent with the ideas of the article "Generalized Golden Sections and a new approach to geometric definition of a number," published by the author in the "Ukrainian Mathematical Journal" [24]. This article presents the concept of the "golden" number theory, which can be the basis for the "golden" mathematics, based on the "golden ratio" and its generalizations.

4.5. The First Half of the 20th Century

In the first half of the 20th century the development of the "golden" paradigm of the ancient Greeks is associated with the names of the Russian Professor of architecture Grimm (1865-1942) and the classic of the Russian religious philosophy Paul Florensky (1882-1937).

In the theory of architecture, it is well-known the book "Proportionality in Architecture", published by Prof. Grimm in 1935 [47]. The purpose of the book has been formulated in the "Introduction" as follows:

"In view of the exceptional significance of the Golden Section in the sense of the proportional division, which establishes a permanent connection between the whole and its parts and gives a constant ratio between them (which is unreachable by any other division), the scheme, based on it, is the main standard and is accepted by us in the future as a basis for checking the proportionality of historical monuments and modern buildings ··· Taking this general importance of the Golden Section in all aspects of architectural thought , the theory of proportio- nality, based on the proportional division of the whole into parts corresponding to the Golden Section, should be recognized as the architectural basis of proportionality at all."

In the 20th years of 20th century, Pavel Florensky wrote the work "At the watershed of a thought." Its third chapter is devoted to the "golden ratio". The Belorussian philosopher Edward Soroko in the book [4] evaluates Florensky's work as follows:

"The aim was to derive analytically the stability of the whole object, which is in the field of the effect of oppositely oriented forces. The project was conceived as an attempt to use the "golden ratio and its substantial basis, which manifests itself not only in a series of experimental observations of nature, but on the deeper levels of knowledge, for the case of penetration into the dialectic of movement, into the substance of things".

4.6. The Second Half of the 20th Century and the 21st Century

In the second half of 20th century, the interest in this area is increasing in all areas of science, including mathe- matics. The Soviet mathematician Nikolai Vorobyov (1925-1995) [44], the American mathematician Verner Hoggatt (1921-1981) [45], the English mathematician Stefan Vajda [46] and others became the most outstanding representatives of this trend in mathematics.

Reviving the idea of harmony in modern science is determined by new scientific realities. The penetration of the Platonic solids, the "golden ratio" and Fibonacci numbers in all areas of theoretical natural sciences (crystallography, chemistry, astronomy, earth science, quantum physics, botany, biology, geology, medicine, genetics, etc.), as well as in computer science and economics was the main reason for the renewed interest in the ancient idea of the Universe Harmony in modern science and the stimulus for the development of the "Mathematics of Harmony" [1-41].

4.7. The Article by Victor Shenyagin

Recently on the website "Academy of Trinitarizm, Institute of the Golden Section" has published the article by well-known Russian Fibonacci-scientist Victor Shenyagin [48] with the following appeal:

"Appeal to the international scientific community: On the appropriateness of nomination of Professor Stakhova AP to award the Abel Prize in Mathematics in 2014."

Of course, it is pleasant for the author to get a high appreciation of his research from his colleagues. Similar estimations were given also by other well-known scientists: the Ukrainian mathematician academician Yuri Mitropolsky [7], the American philosopher Scott Olsen [49], the Belarusian philosopher Edward Soroko [50], the Russian philosopher Sergei Abachiev [51], and many others. Their feedback is the highest award for the author.

Unfortunately, this scientific discipline, called the "Mathematics of Harmony" [1] until now is not a generally recognized part of the conventional mathematics. However, this scientific direction became widely known in modern science after the publication of the book [1], as well as the author's publications in the Ukrainian aca- demic and English scientific journals [3,8-41].

Let us consider the basic mathematical results and theories of the "Mathematics of Harmony," mentioned by Victor Shenyagin in the article [48].

5. ALGORITHMIC MEASUREMENT THEORY

The algorithmic measurement theory is the first mathematical theory, created by the author during 70th years of 20 c. The foundations of this theory are described in the Stakhov's books [12,13] (Figure 5).

Although this theory by its initial idea had purely applied character ("Synthesis of the optimal algorithms for analog-to-digital conversion", by the way, this was the title of author's doctoral dissertation, 1972), but in process of creating this theory, the author went out far for the application frameworks and touched the founda- tions of mathematics.

First of all, the author have excluded from consideration the abstraction of actual infinity as an internally con- tradictory concept (the "completed infinity"), because still Aristotle protested against this. As a result, the author has excluded from the algorithmic measurement theory Cantor's axiom, which is the basis of the classical mea- surement theory, based on the continuity axioms. The author found the contradiction between Cantor's axiom and the Archimedes' axiom [12]. This contradiction implies automatically the contradictions in the foundations of mathematics. The famous Russian mathematician and philosopher Alexander Zenkin had developed these ideas in the article [52], where he criticizes Cantor's theory of infinite sets.

Such a constructive approach to the algorithmic measurement theory, based on the abstraction of potential in- finity, allowed the author to solve the problem, which was never been considered in mathematics. We talk about the problem of the synthesis of optimal measurement algorithms, which are a generalization of well-known measurement algorithms: the "counting algorithm," which lies at the basis of Euclidean definition of natural numbers, and the "binary algorithm," which lies at the basis of the binary system, the basis of modern comput- ers.

Each measurement algorithm corresponds to some positional number system. From the algorithmic measure- ment theory, it follows all the known positional number systems. It is very important, that the new, unknown positional number systems, in particular, the "Fibonacci" and "binomial" number systems arose in the algorith- mic measurement theory [12].

Thus, the algorithmic measurement theory is a source for the new positional number systems. Note that mathematics was never seriously engaged in number systems. The mathematics moved forward not very far in this area compared with the period of its origin. And this is the mistake of mathematics. New positional number sys- tems (in particular, introduced by the author Fibonacci p-codes) can become an alternative to the binary system for many critical applications.

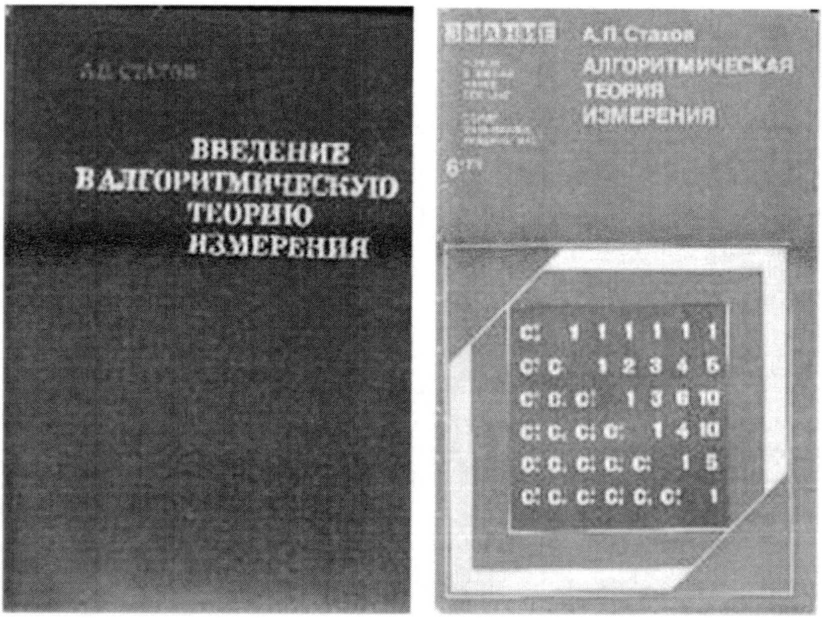

Figure 5. Stakhov books on the Algorithmic Measurement Theory (1977, 1979) (in Russian).

6. PASCAL'S TRIANGLE AND THE GENERALIZED FIBONACCI NUMBERS

Pascal's Triangle is so widely known and studied mathematical object that many mathematicians will be surprised what's new can be found in this triangle? The book "Mathematical Discovery" (Russian translation, 1970) [53] by the American mathematician Polya is well-known. In this book Polya showed an unexpected connection of Pascal's triangle with the Fibonacci numbers (see Figure 6).

The study of the so-called "diagonal sums" of Pascal's triangle, conducted by many mathematicians (includ- ing the author of this article), has led to the discovery of an infinite amount of the recurrent sequences, called the Fibonacci p-numbers [12]. These recurrent sequences for the given $p = 0, 1, 2, 3, \cdots$ are generated by the fol- lowing general recurrence relation:

$$F_p(n) = F_p(n-1) + F_p(n-p-1) \text{ for} \tag{6}$$

at the seeds:

$$F_p(1) = F_p(2) = \cdots = F_p(p+1) = 1. \tag{7}$$

7. A GENERALIZATION OF THE "GOLDEN RATIO"

By studying the Fibonacci p-numbers and considering the limit of the ratio of neighboring Fibonacci p-numbers $\underbrace{\lim}_{n\to\infty} \frac{F_p(n)}{F_p(n-1)} = x$, the author came to the following algebraic equation [12]:

$$x^{p+1} - x^p - 1 = 0 \ \left(p = 0, 1, 2, 3, \cdots\right). \tag{8}$$

The positive roots of the Equation (8) form a set of the new mathematical constants Φ_p, which describe some algebraic properties of Pascal's triangle. The classical "golden ratio" is a special case (p = 1) of the con- stants Φ_p. On this basis, the constants Φ_p have been called the "golden p-proportions".

This result has caused an admiration by academician Mitropolsky, who in his commentary [7] wrote the following:

"Let's ponder upon this result. Within several millennia, starting since Pythagoras and Plato, the mankind used the widely known classical Golden Proportion as some unique number.

And here in the end of the 20th century the Ukrainian scientist Stakhov generalized this result and proved the existence of infinite number of the Golden Proportions! And all of them have the same right to express Harmony, as well as the classical Golden Proportion. Moreover, Stakhov proved, that the golden p-proportions $\Phi_p (1 \le \Phi_p \le 2)$ represented a new class of irrational numbers, which express some unknown mathematical properties of Pascal triangle. Clearly, such mathematical result is of fundamental importance for the development of modern science and mathematics."

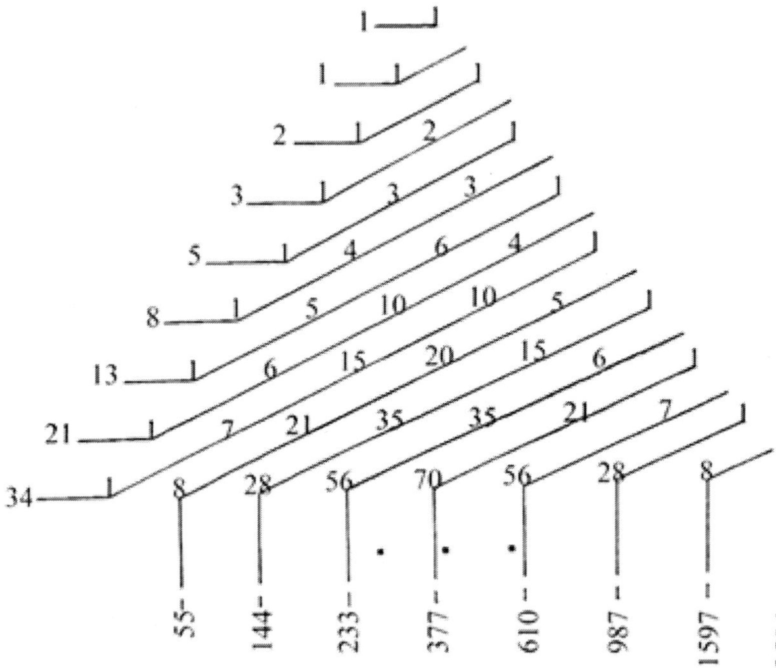

Figure 6. Pascal's triangle and fibonacci numbers.

8. THE REDUNDANT BINARY POSITIONAL NUMBER SYSTEMS (THE FIBONACCI P-CODES)

Fibonacci measurement algorithms, based on the Fibonacci p-numbers [12], led to the discovery of the new po- sitional number systems, called the Fibonacci p-codes:

$$N = a_n F_p(n) + a_{n-1} F_p(n-1) + \cdots + a_i F_p(i) + \cdots + a_n F_p(1), \qquad (9)$$

where $p = 0,1,2,3,\cdots$ is a given integer; $a_i \in \{0,1\}$ is the binary numeral of the ith digit; the Fibonacci p-number $F_p(i)\,(i = 1,2,3,\cdots,n)$ is the weight of the ith digit.

This expression is a generalization of the classical binary code (p = 0):

$$N = a_n 2^{n-1} + a_{n-1} 2^{n-2} + \cdots + a_i 2^{i-1} + \cdots + a_1 2^0, \qquad (10)$$

which underlies the modern computers. For the case p > 0 all the Fibonacci p-codes are redundant positional binary systems, which can be used for the design of new computers (Fibonacci computers), which have super- high informational reliability.

9. FOREIGN PATENTS ON THE FIBONACCI COMPUTERS

In 1976, the author worked during 2 months as Visiting-Professor of the Vienna University of Technology. At the final stage of the stay in Austria, the author made a speech "Algorithmic Measurement Theory and Foundations of Computer Arithmetic" at the joint meeting of the Austrian Computer and Cybernetics Societies. The success of the speech was stunning. It was recognized that this speech provides new informational and arithmetical foundation of computers.

This sudden interest of Austrian scientists in author's speech became a reason of the letter of the Soviet Ambassador in Austria Ivan Yefremov, who sent it to the USSR State Committee for Science and Technology. On the basis of this letter, the Soviet State Invention Committee took the decision about the urgent patenting author's inventions on "Fibonacci computers "abroad. 65 foreign patents of the U.S. Japan, England, France, Germany,

Canada and other countries are official legal documents, which confirmed a priority of the author in this area.

10. CODES OF THE GOLDEN PROPORTION

In 1957, the young (12 year old) American wunderkind George Bergman in one of the U.S. mathematical Journals had described the unique number system, called a number system with an irrational base [54]:

$$A = \sum_i a_i \Phi^i, \qquad (11)$$

where A is a some real number, $a_i \in \{0,1\}$ is a binary numeral of the ith digit $(i = 0, \pm1, \pm2, \pm3, \cdots)$, Φ^i is the weight of the ith digit, $\Phi = \frac{1+\sqrt{5}}{2}$ is a base of "Bergman's system (11). Unfortunately, the mathematicians of 20th century did not pay attention to this mathematical discovery, despite the fact that "Bergman's system" (11) is the greatest contemporary mathematical discovery in the field of positional number systems after the discov- ery of the Babylonian positional principle of number representation and decimal and binary number systems.

In 1980, the author had generalized the number system (11) and introduced in mathematics a wide class of the positional number systems with irrational bases, called "codes of the golden p-proportions" [10]:

$$A = \sum_i a_i \Phi_p^i, \qquad (12)$$

where $i = 0, \pm1, \pm2, \pm3, \cdots$, $p = 0, 1, 2, 3, \cdots$, $a_i \in \{0,1\}$ is a binary numeral of the ith digit, Φ_p is a base of the number system (12) (the golden p-proportion, following from Pascal's triangle), Φ_p^i is the weight of the ith di- git, connected with weights of the previous digits by the following identity:

$$\Phi_p^i = \Phi_p^{i-1} + \Phi_p^{i-p-1}. \qquad (13)$$

Theory of the codes of the golden p-proportions was set forth in the book [14]. This book attracted the attention of the Soviet scientific- popular Journal "Technology for young people". In 1985, this Journal had published author's article "Codes of the Golden Proportion, or the number systems for future computers?" [55]. This article was the "key" article of this Journal what had been emphasized by the collage "Codes of the Golden Proportion," placed on the back cover of this Journal (see Figure 7).

The publication of author's article in the famous scientific-popular Journal by circulation of 1.7 million copies caused a big interest of scientific community in this research direction, which has became well known not only in the USSR but also abroad.

Note that the expression (12) includes an infinite number of the positional binary number systems, because every $p(p=0,1,2,3,\cdots)$ generates its own number system. In particular, for the case $p=0$ the base $\Phi_p=\Phi_0=2$ and the code of the golden p-proportion (12) are reduced to the classical binary system, and for the case $p=1$ to Bergman's system (11).

11. THE "GOLDEN" NUMBER THEORY

Academician Yuri Mitropolsky, Chief-Editor of the "Ukrainian Mathematical Journal", has invited the author to submit the article on the codes of the golden p-proportions for the Journal. In 2004, according to Mitropolsky's recommendation, the article "The generalized golden proportions and a new approach to geometric definition of a number" was published in this prestigious mathematical Journal [24].

The main idea of the article is reduced to the following. The codes of the golden p-proportions, which are connected with Pascal's triangle, can be considered as the beginning of the new number theory, the "golden" number theory. Indeed, with the help of the codes of the golden p-proportions we can represent all real numbers, including natural, rational and irrational. The codes of the golden p-proportions change our ideas about the relationship between rational and irrational numbers, because the special irrational numbers (the golden p-proportions) are becoming the basis of all numbers and, therefore, of all mathematics.

Consider one unusual result of the "golden" number theory on the example of the Bergman's system (11). For this, we represent the natural number N in the Bergman's system (11):

$$N=\sum_i a_i\Phi^i \quad . \qquad (14)$$

It is proved [24] that the sum of (14) for any natural number N is finite always, that is, any natural number N can be represented as the finite sum of the powers of the "golden proportion". Since all powers of the "golden proportion" are irrational numbers (except for $\Phi^0 = 1$), then this assertion is far from trivial.

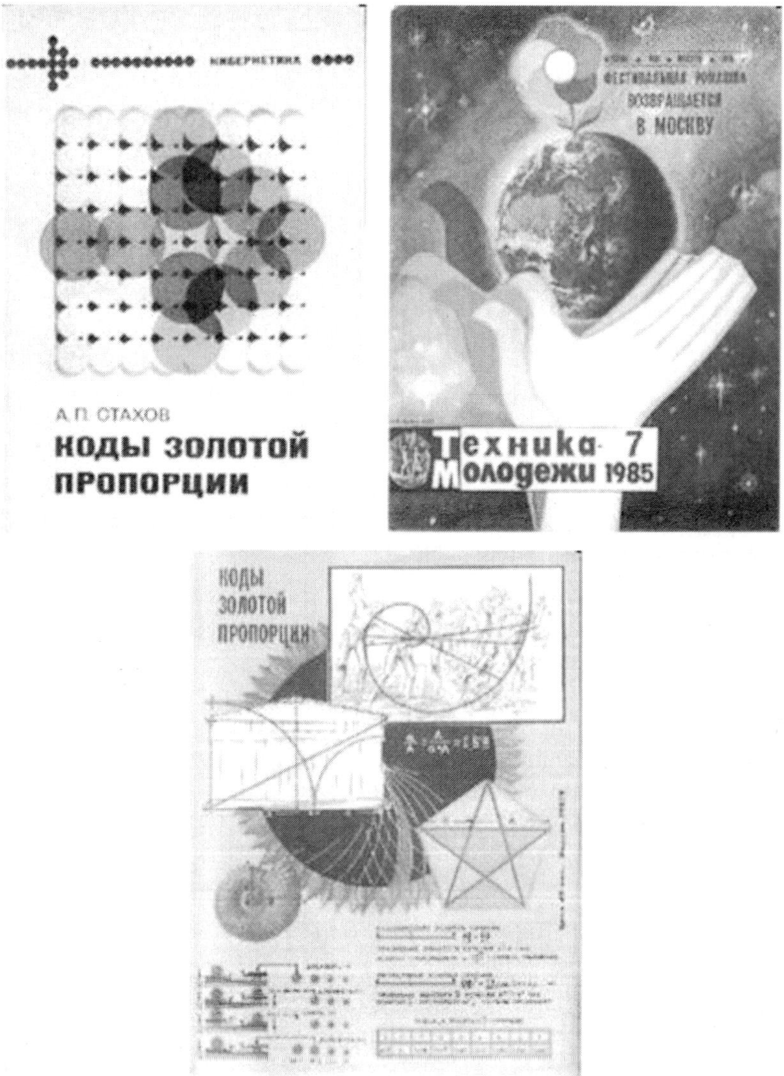

Figure 7. Stakhov's book "Codes of the Golden Proportion" (1984) and the Journal "Technology for Young People" (1985).

Let us consider the so-called "extended" Fibonacci numbers (see Table 1).

We substitute now the "extended" Fibonacci numbers $F_i\,(i = 0, \pm 1, \pm 2, \pm 3, \cdots)$ instead the powers Φ^i into the expression (14). To our surprise, we find [24] that this sum is equal to 0 for any natural number N, that is,

$$\sum_i a_i F_i = 0 \quad . \tag{15}$$

This property has been called the Z-property of natural numbers [24]. Since this property is valid only for natural numbers, this means that in [24] we found a new property of natural numbers, via 2.5 thousand years of their theoretical study (academician Mitropolsky was thrilled!).

12. PROCLUS HYPOTHESIS: A NEW VIEW ON EUCLID'S "ELEMENTS" AND THE HISTORY OF MATHEMATICS

"Proclus hypothesis," formulated in the 5th century AD by the famous Greek philosopher and mathematician Proclus Diadochus (412 - 485), contains the unexpected view on Euclid's "Elements".

According to Proclus, Euclid's goal was not to set forth the geometry itself, but to build a complete theory of regular polyhedra ("Platonic solids"). This theory was set forth by Euclid in the XIII-th, that is, the final book of the "Elements" what in itself is an indirect confirmation of the "Proclus hypothesis".

To solve this problem, Euclid had included the necessary mathematical information into the "Elements." The most curious thing is that he had introduced in the Book II the "golden section," used by him for the creation of geometric theory of the dodecahedron. In Plato's Cosmology, the regular polyhedra had been associated with the "Harmony of the Universe". This means, that Euclid's "Elements" are based on the "harmonic ideas" of Pythagoras and Plato, that is, Euclid's "Elements" are historically first variant of the "Mathematics of Harmony". This unexpected view on the "Elements" leads to the conclusion, which changes our view on the history and structure of mathematics.

As is known, the famous Russian mathematician academician Kolmogorov in the book [56] had identified the two main, that is, "key" problems, which stimulated the development of mathematics at the stage of its origins: the "counting problem" and the measurement problem. However, it follows from "Proclus hypothesis" another "key" problem: the "harmony problem," which underlies Euclid's "Elements".

We can see that three "key" problems—the counting problem, the measurement problem, and the harmony problem—underlie the origins of mathematics (see Figure 8).

The first two "key" problems resulted in the creation of two fundamental notions of mathematics—natural number and irrational number that underlie the Classical Mathematics. The harmony problem, connected with the division in extreme and mean ratio (the "golden ratio") (Proposal II.11 of Euclid's Elements) resulted to the Harmony Mathematics [1]—a new interdisciplinary direction of contemporary science and mathematics.

This approach leads to the conclusion, which is a surprise for many mathematicians. It is found, in parallel with the Classical Mathematics, one more mathematical direction—the Harmony Mathematics—had been developing in ancient science. Similarly to the Classical Mathematics, the Harmony Mathematics has its origin in Euclid's Elements. However, the Classical Mathematics focuses on the axiomatic approach, while the Harmony Mathematics is based on the golden section (Proposal II.11) and Platonic Solids described in Book XIII of Euclid's Elements. Thus, Euclid's Elements is a source of the two independent mathematical directions: the Classical Mathematics and the Harmony Mathematics.

For many centuries, the main focus of mathematicians was on the creation of the Classical Mathematics, which became the Czarina of Natural Sciences. However, the forces of many prominent mathematicians and thinkers—since Pythagoras, Plato and Euclid, Pacioli, Kepler up to Lucas, Binet, Vorobyov, Hoggatt and so forth—were directed towards the development of the basic concepts and applications of the Harmony Mathematics. Unfortunately, these important mathematical directions had been developing separately from one other. The time has come to unite the Classical Mathematics and Harmony Mathematics. This unusual union can lead to new scientific discoveries in mathematics and natural sciences.

A new approach to the mathematics origin (see Figure 8) is very important for school mathematical education. This approach introduces, by very natural manner, the idea of harmony and the golden section into school mathematical education. This provides pupils an access to ancient science and to its main achievement—the har- monic ideas—and to tell them about the most important architectural and sculptural works of the ancient art based upon the golden section (including pyramid of Khufu (Cheops), Nefertiti, Parthenon, Doryphorus, Venus and so on).

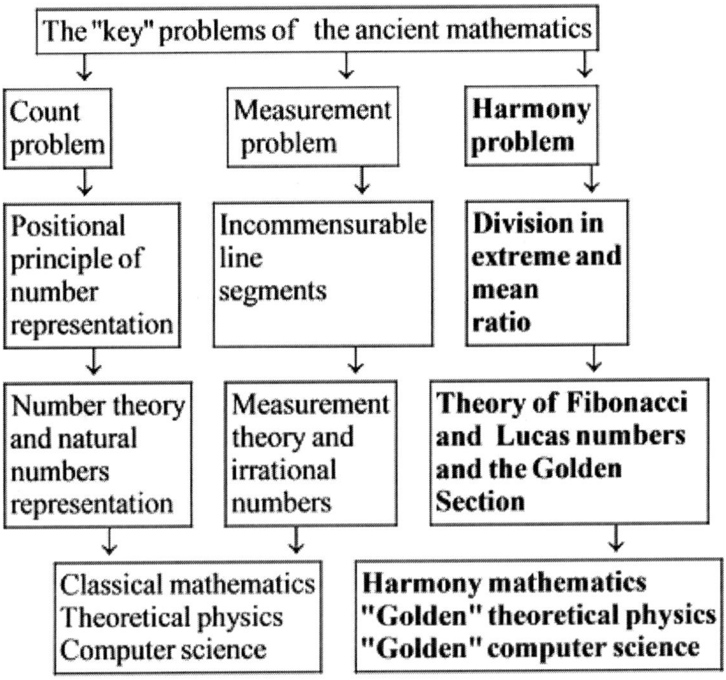

Figure 8. Three "key" problems of the ancient mathematics.

13. HYPERBOLIC FIBONACCI FUNCTIONS AND "BODNAR'S GEOMETRY"

13.1. Classical Hyperbolic Functions

Hyperbolic functions were introduced by the Italian mathematician Vincenzo Riccati in 1757. He first introduced the notation sh and ch. Further investigations of hyperbolic functions were made by Lambert. Hyperbolic functions have the following form:

$$\mathrm{sh}x = \frac{e^x - e^{-x}}{2}; \quad \mathrm{ch}x = \frac{e^x + e^{-x}}{2} \tag{16}$$

The formulas (16) allow us to prove a number of important identities for hyperbolic functions, such as:

$$\mathrm{ch}^2x-\mathrm{sh}^2x=\left(\frac{e^x+e^{-x}}{2}\right)^2-\left(\frac{e^x-e^{-x}}{2}\right)^2=\frac{e^{2x}+2+e^{-2x}}{4}-\frac{e^{2x}-2+e^{-2x}}{4}=1 \quad (17)$$

The interest in the hyperbolic functions in mathematics and theoretical physics has increased dramatically in the 19th century, when the Russian mathematician Nikolai Lobachevsky developed a new type of geometry, based on the hyperbolic functions, and therefore called hyperbolic geometry.

13.2. Hyperbolic Fibonacci Functions

There is an analogy between the hyperbolic functions (16) and Binet's formulas (3), introduced in 19 century. By using this analogy, the Ukrainian mathematicians Alexey Stakhov, Ivan Tkachenko and Boris Rozin have introduced a new class of the hyperbolic functions called hyperbolic Fibonacci functions [17,21].

Hyperbolic Fibonacci sine

$$sFs(x)=\frac{\Phi^x-\Phi^{-x}}{\sqrt{5}} \quad (18)$$

Hyperbolic Fibonacci cosine

$$cFs(x)=\frac{\Phi^x+\Phi^{-x}}{\sqrt{5}} \quad (19)$$

These functions retain, on the one hand, all the known properties of the classical hyperbolic functions (16). In particular, the identity (17) looks as follows:

$$\left[cFs(x)\right]^2-\left[sFs(x)\right]^2=\frac{4}{5}. \quad (20)$$

However, the uniqueness of the hyperbolic Fibonacci functions (18), (19) consists in the fact that they have recursive property, similar to the properties of the Fibonacci numbers. Moreover, the "extended" Fibonacci numbers $F_n\left(n=0,\pm1,\pm2,\pm3,\cdots\right)$ (Table 1), given by Binet's formulas (3), are connected with the hyperbolic Fibonacci functions (18), (19) by the following simple relation:

$$F_n = \begin{cases} sFs(n) & \text{for } n = 2k \\ cFs(n) & \text{for } n = 2k+1 \end{cases} \tag{21}$$

This unique property of the functions (18), (19) is demonstrated on the graphs in Figure 9.

Here the graphs of the hyperbolic sine $y = sFs(x)$ and the hyperbolic cosine $y = cFs(x)$ are represented.

The points on the graph $y = sFs(x)$ correspond to the extended Fibonacci numbers with the even indexes $2n$:

$$F_{2n} = \{\cdots, F_{-8} = -21, F_{-6} = -8, F_{-4} = -3, F_{-2} = -1, F_2 = 1, F_4 = 3, F_6 = 8, F_8 = 21, \cdots\}. \tag{22}$$

The points on the graph $y = cFs(x)$ correspond to the extended Fibonacci numbers with the odd indexes $2n+1$:

$$F_{2n+1} = \{\cdots, F_{-7} = 13, F_{-5} = 5, F_{-3} = 2, F_{-1} = 1, F_1 = 1, F_3 = 3, F_5 = 5, F_7 = 13, \cdots\} \tag{23}$$

13.3. Bodnar's Geometry

Note that the relations (21)-(23) are explaining why the "Fibonacci spirals" appear at the surface of phyllotaxis objects (pine cones, cacti , pineapples, sunflower heads, etc.). This is proved by the Ukrainian researcher Oleg Bodnar in his amazing geometric theory of phyllotaxis, the most common botanical phenomenon. "Bodnar's geometry" [5] shows that the "world of phyllotaxis" is obeying to the laws of the special hyperbolic geometry, which is described by mathematical relationships based on the hyperbolic Fibonacci functions (18) and (19).

"Bodnar's geometry" shows that hyperbolic geometry is much more common in nature than previously assumed. Substantially all wildlife is the epitome of a special type of hyperbolic geometry - "Bodnar's geometry," which uses a special class of hyperbolic functions - hyperbolic Fibonacci functions.

It is this view on the "phyllotaxis geometry" gives us a right to claim that the hyperbolic Fibonacci functions (18) and (19) [17,21] and "Bodnar's geometry" [5] are the fundamental discoveries of modern science.

14. FIBONACCI L-NUMBERS, THE GENERALIZED CASSINI FORMULA AND THE "METALLIC MEANS"

14.1. Fibonacci λ-Numbers

In the end of 20th c. and the beginning of 21 c. the Fibonacci numbers theory [44-46] has been supplemented with the new mathematical results, based on the Fibonacci λ-numbers and the "metallic proportions" [57-63].

Let us give a natural number and consider the following recursive relation:

$$F_\lambda(n+2)=\lambda F_\lambda(n+1)+F_\lambda(n);\;\; F_\lambda(0)=0, F_\lambda(1)=1. \tag{24}$$

The recursive relation (24) "generates" an infinite number of the new numerical sequences, because every number l "generates" its own numerical sequence (24).

For the case $\lambda=1$ the recursive relation (24) is reduced to the following recursive relation:

$$F_1(n+2)=F_1(n+1)+F_1(n);\;\; F_1(0)=0, F_1(1)=1. \tag{25}$$

This recursive relation "generates" the classic Fibonacci numbers:

$$0,1,1,2,3,5,8,13,21,34,\cdots. \tag{26}$$

Based on this fact, we will name the numerical sequences, generated by the recursive relation (24), the Fibonacci l-numbers. Note, that Fibonacci l-numbers are a partial case of the Lucas sequences, introduced in 19th century by the famous French mathematician Eduardo Lucas [64].

For the case $\lambda=2$ the recursive relation (24) is reduced to the recursive relation

$$F_2(n+2)=2F_2(n+1)+F_2(n);\;\; F_2(0)=0, F_2(1)=1, \tag{27}$$

which "generates" Pell numbers:

$$0,1,2,5,12,29,70,\cdots \tag{28}$$

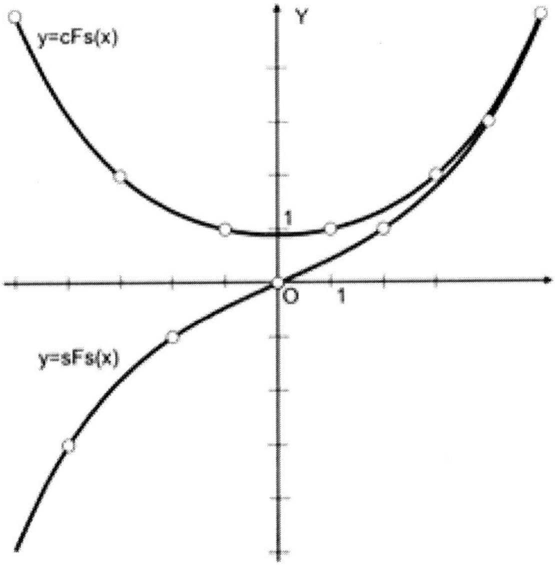

Figure 9. Hyperbolic Fibonacci functions.

14.2. A Generalization of Cassini's Formula for the Fibonacci l-Numbers

The Fibonacci l-numbers have many remarkable properties, similar to the properties of the classic Fibonacci numbers. It is proved that the Fibonacci l-numbers, as well as the classical Fibonacci numbers, can be "expanded" to the negative values of the discrete variable n.

Table 3 shows the four "extended" sequences for the Fibonacci l-numbers, corresponding to the values $\lambda = 1,2,3,4$.

It is proved in [36] that Cassini's formula (2) can be generalized for the case of the Fibonacci l-numbers. The generalized Cassini formula has the following form [36]:

$$F_{\lambda}^{2}(n) - F_{\lambda}(n-1)F_{\lambda}(n+1) = (-1)^{n+1} \tag{29}$$

The most curious thing is that the mathematical result (29) may become as a surprise not only for the "lovers of mathematics," but also for the professional mathematicians specializing in number theory. The Fibonacci l-numbers are a new set of integers (what is the subject of study for number theory) with unique mathematical property (29). This means that within the "mathematics of harmony" [36] a new fundamental mathematical result was obtained. This result has a direct relation to number theory, the oldest mathematical discipline.

14.3. The "Metallic Means"

If we take the ratio of the two adjacent Fibonacci l-numbers $\frac{F_\lambda(n)}{F_\lambda(n-1)}$ and direct the index n to infinity, we get the following result:

$$\lim_{n\to\infty}\frac{F_\lambda(n)}{F_\lambda(n-1)}=\Phi_\lambda, \tag{30}$$

where Φ_λ is a positive root of the following algebraic equation:

$$x^2-\lambda x-1=0. \tag{31}$$

The root Φ_λ has the following form:

$$\Phi_\lambda=\frac{\lambda+\sqrt{4+\lambda^2}}{2}. \tag{32}$$

Note that for the case $\lambda=1$ the formula (32) is reduced to the classical "golden ratio" $\Phi_1=\frac{1+\sqrt{5}}{2}$. This means that the formula (32) gives an infinite number of new mathematical constants called "metallic means" or "metallic proportions" by the Argentinean mathematician Vera W. de Spinadel [57].

If we take in (32) $\lambda=1,2,3,4$, then we get the following mathematical constants, having the following names (according to Vera de Spinadel [57]):

$$\Phi_1=\frac{1+\sqrt{5}}{2}\text{ (golden proportion, }\lambda=1\text{)};\quad \Phi_2=1+\sqrt{2}\text{ (silver proportion, }\lambda=2\text{)};$$

$$\Phi_3=\frac{3+\sqrt{13}}{2}\text{ (bronze proportion, }\lambda=3\text{)};\quad \Phi_4=2+\sqrt{5}\text{ (cooper proportion, }\lambda=4\text{)}.$$

Table 3. The "extended" Fibonacci l-numbers (l = 1, 2, 3, 4).

n	0	1	2	3	4	5	6	7	8
$F_1(n)$	0	1	1	2	3	5	8	13	21
$F_1(-n)$	0	1	–1	2	–3	5	–8	13	–21
$F_2(n)$	0	1	2	5	12	29	70	169	408
$F_2(-n)$	0	1	–2	5	–12	29	–70	169	–408
$F_3(n)$	0	1	3	10	33	109	360	1189	3927
$F_3(-n)$	0	1	–3	10	–33	109	–360	1189	–3927
$F_4(n)$	0	1	4	17	72	305	1292	5473	23184
$F_4(-n)$	0	1	–4	17	–72	305	–1292	5473	–23184

It is clear that the number of the "metallic proportions" (32) is theoretically infinite. The most significant is the fact that for the case l=1 the formula (32) is reduced to the classical "golden ratio." This means that the "metallic proportions" like the "golden ratio" may express some new kinds of "mathematical harmony," which can actually exist in Nature.

15. GAZALE'S FORMULAS AND HYPERBOLIC FIBONACCI L-FUNCTIONS

15.1. Gazale's Formulas

The formulas (24) give the Fibonacci l-numbers recursively. However, the Fibonacci l-numbers can be represented in analytical form through the "metallic proportions" (32). Such formulas have been deduced by the French mathematician Midhat Gazale [58]:

$$F_\lambda(n) = \begin{cases} \dfrac{\Phi_\lambda^n + \Phi_\lambda^{-n}}{\sqrt{4+\lambda^2}} & \text{for } n = 2k+1; \\ \dfrac{\Phi_\lambda^n - \Phi_\lambda^{-n}}{\sqrt{4+\lambda^2}} & \text{for } n = 2k \end{cases} \quad (33)$$

where $k = 0, \pm1, \pm2, \pm3, \cdots$.

Gazale's formulas (23) are a generalization of Binet's formulas (3). A distinctive feature of (33) is that they generate an infinite number of the formulas (33), because each natural number $\lambda = 1, 2, 3, \cdots$ corresponds to the different Gazale's formulas (33). In particular, for the case l = 1 Gazale's formulas (33) are reduced to Binet's formulas (3).

15.2. Hyperbolic Fibonacci l-Functions

Gazale's formulas (33) became a source for the introduction of the hyperbolic Fibonacci l-functions [34,37].

Hyperbolic Fibonacci l-sine

$$sF_\lambda(x) = \frac{\Phi_\lambda^x - \Phi_\lambda^{-x}}{\sqrt{4+\lambda^2}} = \frac{1}{\sqrt{4+\lambda^2}}\left[\left(\frac{\lambda+\sqrt{4+\lambda^2}}{2}\right)^x - \left(\frac{\lambda+\sqrt{4+\lambda^2}}{2}\right)^{-x}\right] \quad (34)$$

Hyperbolic Fibonacci l-cosine

$$cF_\lambda(x)=\frac{\Phi_\lambda^x+\Phi_\lambda^{-x}}{\sqrt{4+\lambda^2}}=\frac{1}{\sqrt{4+\lambda^2}}\left[\left(\frac{\lambda+\sqrt{4+\lambda^2}}{2}\right)^x+\left(\frac{\lambda+\sqrt{4+\lambda^2}}{2}\right)^{-x}\right] \quad (35)$$

It is proved [34,37] that the hyperbolic Fibonacci l-functions (34), (35) coincide with the "extended" Fibonacci l-numbers, given by Gazale's formulas (33), for the discrete values of the variable x = n = 0, ±1, ±2, ±3, ..., that is,

$$F_\lambda(n)=\begin{cases} sF_\lambda(n), & n=2k \\ cF_\lambda(n), & n=2k+1 \end{cases}. \quad (36)$$

It is proved [34,37], that the functions (34) and (35) retain all the known properties of the classic hyperbolic functions (3). For example, the well-known property (17) for the case (34) and (35) takes the following form:

$$[cF_\lambda(x)]^2-[sF_\lambda(x)]^2=\frac{4}{4+\lambda^2}. \quad (37)$$

Note that the hyperbolic functions (34) and (35) are reduced to the hyperbolic Fibonacci functions (18) and (19) for the case $\lambda=1$. For this case, the formula (37) is reduced to the formula (20).

However, a number of the hyperbolic functions, given by (34) and (35), is theoretically infinite because every natural number l = 1, 2, 3, · · · " generates" its own variant of the hyperbolic functions of the kind (34) and (35).

16. HILBERT'S FOURTH PROBLEM AND A NEW CHALLENGE TO THEORETICAL NATURAL SCIENCES

16.1. Hilbert's Problems

In his 23 mathematical problems [65], the famous mathematician David Hilbert (See Figure 10) could not ignore the mathematical problems related to the non-Euclidean geometry. He considered the hyperbolic geometry (Lobachevski's geometry) and elliptic geometry (Riemann geometry) as the geometries closest to the Euclidean geometry. Hilbert's Fourth Problem [65,66], which relates to the non-Euclidean geometry, was formulated by him as follows [67]:

"The more general question now arises: Whether from other suggestive standpoints geometries may not be devised which, with equal right, stand next to Euclidean geometry."

Hilbert's citation contains the formulation of a very important scientific problem, which is of fundamental interest not only for mathematics, but also for all theoretical natural sciences: are there non-Euclidean geometries, which are close to the Euclidean geometry and are interesting from the "other suggestive standpoints?" If we consider it in the context of theoretical natural sciences, then Hilbert's Fourth Problem is about finding NEW HYPERBOLIC WORLDS OF NATURE, which are close to the Euclidean geometry and reflect some new properties of Nature's structures and phenomena.

Unfortunately, the attempts at solving Hilbert's Fourth Problem, made by German mathematician Herbert Hamel (1901) and later by the Soviet mathematician Alexey Pogorelov [68] (1974), have not led to significant progress, as described in Wikipedia's articles "Hilbert problems" [65] and the "Fourth Hilbert problem" [66]. In these articles, the status of the problem is formulated as "too vague to be stated resolved or not" and Pogorelov's solution even is not mentioned. About the same point of view on Hilbert's Fourth Problem is presented in the remarkable book [69]. So, Wikipedia's mathematical community lays the blame for the fact why Hilbert's Fourth Problem is not resolved until now, on David Hilbert, who formulated this problem not clearly enough.

16.2. From the "Game of Postulates" to the "Game of Functions"

According to [41], the reason lies elsewhere. All the known attempts to solve this problem (Hamel, Pogorelov) were in the traditional framework; let's call this the "game of postulates." This "game" started from the works by Gauss, Lobachevski and Bolyai, when Euclid's 5th postulate was replaced by the opposite one. This was the most major step in the development of the non-Euclidean geometry, called "Lobachevski's geometry." The geometry, which changed the traditional geometric representations, is also known as "hyperbolic geometry." This name highlights the fact that this geometry is based on the hyperbolic functions (16).

Figure 10. David Hilbert (1862-1943).

It is important to emphasize that the very name of "hyperbolic geometry" contains another way for the solu- tion of Hilbert's Fourth Problem: searching for the new classes of "hyperbolic functions," which can be the basis for other hyperbolic geometries. Every new class of the hyperbolic functions generates its own new variant of the "hyperbolic geometry." By analogy with the "game of postulates" this way to solve Hilbert's Fourth Prob- lem can be named the "game of functions" [41].

16.3. The Essence of the New Original Solution of Hilbert's Fourth Problem

With the solution of Hilbert's Fourth Problem, described in [37-40], the system of axioms of hyperbolic geometry remains unchanged. But the kind of hyperbolic functions, which describe the basic relationships of hyperbolic geometry, has changed. We use the new classes of the hyperbolic functions—the hyperbolic Fibonacci functions (18) and (19) with the unique mathematical property (21) and the hyperbolic Fibonacci l-functions (34) and (35) with the unique mathematical property (36).

As is known, the classic metric form of Lobachevski's plane with pseudo-spherical coordinates $(u,v), 0<u<+\infty, 0<v<+\infty,$ which has a Gaussian curvature $K=-1$, is given with the following formula [37-40]:

$$(\mathrm{d}s)^2=(\mathrm{d}u)^2+sh^2(u)(\mathrm{d}v)^2, \tag{38}$$

where $\mathrm{d}s$ is an element of length, $sh(u)$ is a classic hyperbolic sine from (16). It is clear that the classical hyperbolic sine plays a key role in the metric form of Lobachevski's plane (38).

Developing the idea of metric form of Lobachevski's plane (38), the following formula for the metric form of Lobachevski's plane, based on the hyperbolic Fibonacci l-functions (34) and (35), has been derived in [37-40]:

$$(\mathrm{d}s)^2=\ln^2(\Phi_\lambda)(\mathrm{d}u)^2+\frac{4+\lambda^2}{4}\left[sF_\lambda(u)\right]^2(\mathrm{d}v)^2, \tag{39}$$

where $\Phi_\lambda=\frac{\lambda+\sqrt{4+\lambda^2}}{2}$ is the "metallic proportion" (32) and $sF_\lambda(u)$ is the hyperbolic Fibonacci l-sine (44). The forms (39) are called the metric l-forms of Lobachevski's plane [25,26].

This formula gives an infinite number of new "Lobachevsky's geometries" ("golden," "silver," "bronze," "copper" and so on to infinity).

Thus, the formula (39) sets an infinite number of the metric forms of Lobachevski's plane. This means that there is infinite number of Lobachevski's geometries, which are based on the "metallic proportions" (32). These new Lobachevski's geometries "with equal right, stand next to Euclidean geometry" (David Hilbert). Thus, the formula (39) can be considered as the partial original solution of Hilbert's Fourth Problem. There is an infinite number of Lobachevski's geometries, described by the formula (39), which are close to Euclidean geometry. Every of these geometries manifest itself in Fibonacci l-numbers (24), which can appear in physical world similarly to "Bodnar's geometry", where the classical Fibonacci numbers appear at the surface of phyllotaxis objects.

16.4. A New Challenge for Theoretical Natural Sciences

Thus, the main result of the research, described in [37-40], is a proof of the existence of an infinite number of hyperbolic functions (34) and (35), based on the "metallic proportions" (32). In addition, each class of hyperbolic functions, corresponding to (34) and (35), "generates" for the given $\lambda = 1, 2, 3, \cdots$ its own "hyperbolic geometry," which leads to the appearance in the "physical world" with specific properties, which depend on the "metallic proportions" (32). The new geometric theory of phyllotaxis, created by Oleg Bodnar [5], is a striking example of this. Bodnar proved that "the world of phyllotaxis" is a specific "hyperbolic world," in which the "hyperbolicity" manifests itself in the "Fibonacci spirals" on the surface of "phyllotaxis objects".

However, the hyperbolic Fibonacci functions (18) and (19), which underlie the "hyperbolic phyllotaxis world," are a special case of the hyperbolic Fibonacci l-functions (34) and (35). In this regard, there is every reason to sup- pose that other types of hyperbolic functions (34) and (34) can be the basis for modeling of new "hyperbolic worlds" that can really exist in Nature. Modern science cannot find these special "hyperbolic worlds", because hyperbolic functions (34) and (35) were unknown for it. Basing on the success of "Bodnar's geometry" [5], one can put forward in front to theoretical physics, chemistry, crystallography, botany, biology, and other branches of theoretical natural sciences the challenge to find new "hyperbolic Nature's worlds", based on other classes of hyperbolic functions given by (34) and (35).

17. FIBONACCI AND "GOLDEN" MATRICES

17.1. Fibonacci Q-Matrices

The Fibonacci Q-matrix [45]

$$Q = \begin{pmatrix} 1 & 1 \\ 1 & 0 \end{pmatrix} \tag{40}$$

is a generating matrix for the Fibonacci numbers and has the following wonderful properties:

$$Q^n = \begin{pmatrix} F_{n+1} & F_n \\ F_n & F_{n+1} \end{pmatrix} \tag{41}$$

$$\det Q^n = F_{n+1}F_{n+1} - F_n^2 = (-1)^n . \tag{42}$$

Note that there is a direct relation between the Cassini formula (2) and the formula (42), which gives the de- terminant of the matrix (41).

17.2. Fibonacci Q_p-Matrices

The article [18] introduces a generating matrix for the Fibonacci p-numbers. It is called the Q_p-matrix $(p = 0,1,2,3,\cdots)$:

$$Q_p = \begin{pmatrix} 1 & 1 & 0 & 0 & \cdots & 0 & 0 \\ 0 & 0 & 1 & 0 & \cdots & 0 & 0 \\ 0 & 0 & 0 & 1 & \cdots & 0 & 0 \\ \vdots & \vdots & \vdots & \vdots & \ddots & \vdots & \vdots \\ 0 & 0 & 0 & 0 & \cdots & 1 & 0 \\ 0 & 0 & 0 & 0 & \cdots & 0 & 1 \\ 1 & 0 & 0 & 0 & \cdots & 0 & 0 \end{pmatrix}. \tag{43}$$

The following properties of the Q_p-matrix (43) are proved in [18]:

$$Q_p^n = \begin{pmatrix} F_p(n+1) & F_p(n) & \cdots & F_p(n-p+2) & F_p(n-p+1) \\ F_p(n-p+1) & F_p(n-p) & \cdots & F_p(n-2p+2) & F_p(n-2p+1) \\ \vdots & \vdots & \ddots & \vdots & \vdots \\ F_p(n-1) & F_p(n-2) & \cdots & F_p(n-p) & F_p(n-p+1) \\ F_p(n) & F_p(n-1) & \cdots & F_p(n-p+1) & F_p(n-p) \end{pmatrix} \tag{44}$$

$$\det Q_p^n = (-1)^{pn}, \tag{45}$$

where $p = 0,1,2,3,\cdots$; and $n = 0,\pm1,\pm2,\pm3,\cdots$

17.3. The "Golden" Matrices

In [33] a special class of the square matrices called the "golden" matrices was introduced. Their peculiarity is the fact that the hyperbolic Fibonacci functions (18) and (19) are elements of these matrices:

$$Q_0(x) = \begin{pmatrix} cFs(2x+1) & sFs(2x) \\ sFs(2x) & cFs(2x-1) \end{pmatrix}; \quad Q_1(x) = \begin{pmatrix} sFs(2x+2) & cFs(2x+1) \\ cFs(2x+1) & sFs(2x) \end{pmatrix} \tag{46}$$

If we calculate the determinants of the matrices (46), we obtain the following unusual identities:

$$\det Q_0(x) = 1; \quad \det Q_1(x) = -1 \tag{47}$$

18. A NEW THEORY OF REDUNDANT CODING BASED ON THE FIBONACCI MATRICES

Let's consider the following coding/decoding method described in [30] and represented in Table 4. Represent the initial message in the form of the square (p + 1) ´ (p + 1)-matrix M with determinant $\det M \neq 0$ and use for its coding the direct Fibonacci Q_p-matrix Q_p^n of the same size and for the decoding the inverse matrix Q_p^{-n} according to Table 4.

It follows from Table 4 that for the given p (p = 0, 1, 2, 3, ···) the Fibonacci coding consists in the multiplication of the initial (p + 1) ´ (p + 1)-matrix M by the coding matrix Q_p^n given by (44). The code matrix E is a result of such matrix multiplication. Then the Fibonacci decoding consists in the multiplication of the code matrix E by the "inverse" matrix Q_p^{-n}. Note that the coding/decoding method is trivial for the case p = 0. However, the coding/decoding methods, corresponding to the cases p>0, represent by themselves some code transformations of the initial message M to the code matrix E and this transformation is useful for some applications.

Error detection and correction is based on the following relation, which connects the determinant of the code matrix and the determinant of the coding matrix $\det Q_p^n = (-1)^{pn}$:

$$\det E = \det M \times \det Q_p^n = \det M \times (-1)^{pn} \tag{48}$$

Note that the coding/decoding method, given by Table 4, ensure infinite variants of the Fibonacci coding/decoding, because every Fibonacci coding matrix Q_p^n and its inverse matrix Q_p^{-n} $(p = 1, 2, 3, \cdots; n = \pm 1, \pm 2, \pm 3, \cdots)$ "generate" their own Fibonacci coding/decoding method.

A new error-correcting theory based on the Fibonacci Q_p-matrices [30] has the following advantages in comparison to the algebraic error-correcting codes:

1) The Fibonacci coding/decoding method is reduced to matrix multiplication, that is, to the well-known algebraic operation, which is realized very well in modern computers.
2) The main practical peculiarity of the coding method is the fact that large information units, in particular, matrix elements, are objects of error detection and correction.
3) The simplest Fibonacci coding/decoding method (p = 1) can guarantee the restoration of all "erroneous" code (2 ´ 2)-matrices having "single", "double" and "triple" errors.
4) The potential correction ability of the method for the simplest case p = 1 i exceeds essentially the potential correction ability of all well-known algebraic error-correcting codes. This means that the new coding theory based on matrix approach [30] is of great practical importance for modern computer science.

19. THE "GOLDEN" TERNARY MIRROR-SYMMETRICAL ARITHMETIC

For the first time, the "golden" ternary mirror-symmetrical arithmetic was described in the article "Brousentsov's ternary principle, Bergman's number system and ternary mirror-symmetrical arithmetic", published in "The Computer Journal" in 2002 [19]. The article caused a positive reaction from the Western computer com- munity.

The prominent American mathematician Donald Ervin Knuth (see Figure 11) congratulated the author with the publication of the article [19]. In his letter, he informed the author about his intention to include a de- scription of the "golden" mirror-symmetrical arithmetic into the new edition of the book "Art of Computer Pro- gramming".

Table 4. Fibonacci coding/decoding method.

Coding	Decoding
$M \times Q_p^n = E$	$E \times Q_p^{-n} = M$

Figure 11. American mathematician Donald Knuth.

Why the article [19] caused the interest of Donald Knuth? It is proved in [19] that any integer N (positive or negative) may be represented as follows:

$$N = \sum_i b_i \Phi^{2i}, \tag{49}$$

where $b_i \in \{\bar{1},0,1\}$ is a ternary numeral, Φ^{2i} is the weight of the ith digit, $\Phi = \frac{1+\sqrt{5}}{2}$ is the golden proportion.

Note that the base of the number system (49) is the square of the "golden ratio," that is,

$$\Phi^2 = \frac{3+\sqrt{5}}{2} \approx 2.618. \tag{50}$$

This means that the number system (49) belongs to the category of the number systems with irrational bases.

The abridged ternary digital record of the number system (49) has the following form:

$$N = b_m b_{m-1} \cdots b_i \cdots b_2 b_1 b_0 \cdot b_{-1} b_{-2} \cdots b_{-i} \cdots b_{-(m-1)} b_{-m} . \quad (51)$$

It is proved in [19], that the ternary digital record (51) for any integer N consists of two parts, the left hand part $b_m b_{m-1} \cdots b_i \cdots b_2 b_1$ and right hand part $b_{-1} b_{-2} \cdots b_{-i} \cdots b_{-(m-1)} b_{-m}$, symmetrically located relatively to the 0th digit b_0.

Table 5 presents the ternary digital records of the initial natural numbers in the number system (49). As fol- lows from Table 5, all ternary records of positive integers have the property of the so-called "mirror symmetry." The essence of this property is that the left part $b_m b_{m-1} \cdots b_i \cdots b_2 b_1$ of the ternary code (51) is a mirror reflection of the right part $b_{-1} b_{-2} \cdots b_{-i} \cdots b_{-(m-1)} b_{-m}$; in this connection, $b_i = b_{-i} \ (i = 1, 2, 3, \cdots, m)$. It is proved in [19] that all integers (positive and negative) have the property of "mirror symmetry" at their representation in the ternary number system (49). That is why, the ternary system (49) is named in [19] the ternary mirror-symmetrical number system.

We note one more unusual property of the ternary mirror-symmetrical number system (49). In traditional po- sitional number systems (decimal, binary), the increase of a number is realised by the extension of digital re- cording in the direction of the higher digits. As follows from Table 5, for the case of the mirror-symmetrical ternary recording (51), the extension of ternary digital recording is realized by the extension of the ternary digi- tal recording mirror-symmetrically to the left and to the right simultaneously.

It is proved in [19] that the property of "mirror symmetry" remains valid at the realization of arithmetical op- erations, that is, the results of these operations are always represented in the mirror-symmetric form. This means that this property can be used in computers to control the arithmetical operations.

20. INTERNATIONAL CONGRESS ON MATHEMATICS OF HARMONY

This Congress was held in October 2010 at Mechnikov's Odessa National University (Ukraine). https://sites.google.com/site/ harmonymath kongress/.

Table 5. Ternary mirror-symmetric digital recordings of natural numbers.

i	3	2	1	0	−1	−2	−3
N/Φ^i	Φ^3	Φ^2	Φ^1	Φ^0	Φ^{-1}	Φ^{-2}	Φ^{-3}
0	0	0	0	0	0	0	0
1	0	0	0	1.	0	0	0
2	0	0	1	−1.	1	0	0
3	0	0	1	0.	1	0	0
4	0	0	1	1.	1	0	0
5	0	1	−1	1.	−1	1	0
6	0	1	0	−1.	0	1	0
7	0	1	0	0.	0	1	0
8	0	1	0	1.	0	1	0
9	0	1	1	−1.	1	1	0
10	0	1	1	0.	1	1	0

This Congress gathered together leading experts in the field of the "golden ratio," Fibonacci numbers and their applications not only from Ukraine, Russia, Belarus, but also from other countries and continents (USA, Germany , Chile, South Africa) (see Figure 12).

Point 7 of the Congress Resolution claims the following:

"The publication of the book AP Stakhov "The Mathematics of Harmony. From Euclid to Contemporary Mathematics and Computer Science "(World Scientific, 2009) is one of the most important events in this area. The book is the result of author's 40-year research in this area. Stakhov's book gained a widespread international recognition. One of the leading U.S. experts in the field of harmony and the "golden section" Prof. Scott Olsen wrote in his review: "Stakhov's ideas Stakhova so significant that his new book may well change not only the consideration of the history of mathematics, but also the future development of mathematics in its applications to theoretical natural sciences."

Figure 12. Participants of the international congress on the mathematics of harmony.

21. CONCLUSIONS

The following conclusions follow from this study:

1) The first conclusion touches on the question of the origin of mathematics and its development. This conclusion can be unexpected for many mathematicians. We assert that since the Greek period, the two mathematical doctrines—the Classical Mathematics and the Harmony Mathematics—begun to develop in parallel and independent one another. They both originated from one and the same source—Euclid's Elements, the greatest mathematical work of the Greek mathematics. Geometric axioms, the beginnings of algebra, theory of numbers, theory of irrationals and other achievements of the Greek mathematics were borrowed from Euclid's Elements by the Classical Mathematics. On the other hand, a problem of division in extreme and mean ratio (Theorem II.11) called later the golden ratio, and a geometric theory of regular polyhedrons (Book XIII), expressing the Harmony of the Cosmos in Plato's Cosmology, were borrowed from Euclid's Elements by the Mathematics of Harmony. We affirm that Euclid's Elements were the first attempt to reflect in mathematics the main scientific idea of the Greek science, the idea of Harmony. According to Proclus, the creation of geometric theory of Platonic Solids (Book XIII of Euclid's Elements) was the main purpose of Euclid's Elements.

2) The second conclusion touches on the development of number theory. We affirm that new constructive definitions of real numbers, based on Bergman's system (11) and the codes of the golden p-proportions (12), overturn our ideas about relationship between rational and irrational numbers [3]. A special class of irrational numbers—the golden ratio and golden p-proportions, following from Pascal's triangle, becomes a base of new number theory—the "golden" number theory [24], because all rest of real numbers can be reduced to them by using the definitions (11) and (12). New property of natural numbers Z-property (15), following from this ap- proach, is a confirmation of a fruitfulness of this approach [24].

3) The third conclusion touches on the development of hyperbolic geometry. We affirm that a new class of hyperbolic functions, the hyperbolic Fibonacci functions (18) and (19) [17,21] and the hyperbolic Fibonacci λ- functions (34) and (35) [34,37], can become inexhaustible source for the development of hyperbolic geometry. We affirm that the formulas (34) and (35) give an infinite number of hyperbolic functions similar to the classical hyperbolic functions, which underlie Lobachevski's geometry. This affirmation can be referred to

one of the main mathematical results of the Mathematics of Harmony. The original solution of Hilbert's Fourth Problem [37-41] confirms a fruitfulness of this approach to hyperbolic geometry.

4) The fourth conclusion touches on the applications of the Mathematics of Harmony in computer science. We affirm that the Mathematics of Harmony is a source for the development of new information technology— the "Golden" Information Technology based on the Fibonacci p-codes (9), Bergman's system (11), codes of the golden p-proportions (12), and "golden" ternary mirror-symmetrical arithmetic [19].

5) The fifth conclusion touches on the applications of the Mathematics of Harmony in modern mathematical education. We affirm that the Mathematics of Harmony should become a base for the reform of modern mathematical education on the base of the ancient idea of Harmony and golden section. Such an approach can increase an interest of pupils in studying mathematics because this approach brings mathematics and natural sci- ences together. A study of mathematics turns into fascinating search of new mathematical regularities of Nature.

6) The sixth conclusion touches on the general role of the Mathematics of Harmony in the progress of contemporary mathematics. We affirm that the Mathematics of Harmony can overcome a contemporary crisis in the development of the 20th century mathematics that resulted in the weakening of the relationship between mathematics and theoretical natural sciences [70]. The Mathematics of Harmony is a true "Mathematics of Nature" incarnated in many wonderful structures of Nature (pine cones, pineapples, cacti, heads of sunflowers and so on), and it can give birth to new scientific discoveries. A new geometric theory of phyllotaxis ("Bodnar's geometry") [5] is a brilliant confirmation of this fact.

ACKNOWLEDGEMENTS

A. P. Stakhov, (U.S.) for the English edition of this article and valuable remarks.

REFERENCES

1. P. Stakhov, "The Mathematics of Harmony: From Euclid to Contemporary Mathematics and Computer Science," World Scientific (International Publisher), Singapore City, 2009. [Citation Time(s):7]
2. S. Olsen, "The Golden Section: Nature's Greatest Secret," Walker (Publishing Company), New York, 2006.
3. P. Stakhov, "The Mathematics of Harmony: Clarifying the Origins and Development of Mathematics," Congressus Numer- antium, Vol. 193, Utilitas Mathematica, Winnipeg, 2008, pp. 5-48.
4. E. M. Soroko, "Structural Harmony of Systems," Nauka i Tekhnika, Minsk, 1984.
5. O. Y. Bodnar, "The Golden Section and Non-Euclidean Geometry in Nature and Art," Svit, Lvov, 1994.
6. S. V. Petoukhov, "Metaphysical Aspects of the Matrix Analysis of Genetic Code and the Golden Section," Metaphysics: Cen- tury XXI, BINOM, Moscow, 2006, pp. 216-250.
7. "Academician Mitropolsky's Commentary on the Scientific Research of the Ukrainian Scientist Doctor of Engineering Sciences Professor Alexey Stakhov," In: A. Stakhov, Ed., The Mathematics of Harmony: From Euclid to Contemporary Mathematics and Computer Science, World Scientific, Singapore, 2009.
8. P. Stakhov, "The Golden Section in the Measurement Theory," Computers & Mathematics with Applications, Vol. 17, No. 4-6, 1989, pp. 613-638.
9. P. Stakhov, "Application of Natural Redundancy of Fibonacci's Number Systems for Computer Check," Automation and Computer Engineering, Vol. 6, 1975, pp. 80-87.
10. P. Stakhov, "The Golden Ratio in Digital Technology," Automation and Computer Technology, Vol. 1, 1980, pp. 27-33.
11. P. Stakhov, "The Golden Section and Modern Harmony Mathematics," Applications of Fibonacci Numbers, Vol. 7, Kluwer Academic Publishers, Dordrecht, 1998, pp. 393-399.
12. P. Stakhov, "Introduction into Algorithmic Measurement Theory," Soviet Radio, Moscow, 1977.
13. P. Stakhov, "Algorithmic Measurement Theory," Nauka, Moscow, 1979.
14. P. Stakhov, "Codes of the Golden Proportion," Radio and Communication, Moscow, 1984.
15. P. Stakhov, "Noise-Tolerant Codes. Fibonacci Computer," Znanie, Moscow, 1989.
16. Stakhov, V. Massingue and A. Sluchenkova, "Introduction into Fibonacci Coding and Cryptography," Osnova, Kharkov, 1999.
17. P. Stakhov and I. S. Tkachenko, "Hyperbolic Fibonacci Trigonometry," Reports of the Ukrainian Academy of Sciences, Vol. 208, No. 7, 1993, pp. 9-14.
18. P. Stakhov, "A Generalization of the Fibonacci Q-Matrix," Reports of the National Academy of Sciences of Ukraine, Vol. 9, 1999, pp. 46-49.
19. P. Stakhov, "Brousentsov's Ternary Principle, Bergman's Number System and Ternary Mirror-Symmetrical Arithmetic," The Computer Journal, Vol. 45, No. 2, 2002, pp. 221-236.
20. P. Stakhov, "Hyperbolic Fibonacci and Lucas Functions: A New Mathematics for the Living Nature," ITI, Vinnitsa, 2003.
21. Stakhov and B. Rozin, "On a New Class of Hyperbolic Functions," Chaos, Solitons & Fractals, Vol. 23, No. 2, 2004, pp. 379-389.
22. Stakhov and B. Rozin, "The Golden Shofar," Chaos, Solitons & Fractals, Vol. 26, No. 3, 2005, pp. 677-684.
23. Stakhov, "The Generalized Principle of the Golden Section and Its Applications in Mathematics, Science, and Engineering," Chaos, Solitons & Fractals, Vol. 26, No. 2, 2005, pp. 263-289.
24. P. Stakhov, "The Generalized Golden Proportions and a New Approach to Geometric Definition of a Number," Ukrainian Mathematical Journal, Vol. 56, No. 8, 2004, pp. 1143-1150.
25. Stakhov, "Fundamentals of a New Kind of Mathematics Based on the Golden Section," Chaos, Solitons & Fractals, Vol. 27, No. 5, 2006, pp. 1124-1146.

26. Stakhov and B. Rozin, "The 'Golden' Algebraic Equations," Chaos, Solitons & Fractals, Vol. 27, No. 5, 2006, pp. 1415-1421.
27. Stakhov, "The Generalized Principle of the Golden Section and Its Applications in Mathematics, Science, and Engineering," Chaos, Solitons & Fractals, Vol. 26, No. 2, 2005, pp. 263-289.
28. Stakhov and B. Rozin, "Theory of Binet Formulas for Fibonacci and Lucas p-Numbers," Chaos, Solitons & Fractals, Vol. 27, No. 5, 2006, pp. 1162-1177.
29. Stakhov and B. Rozin, "The Continuous Functions for the Fibonacci and Lucas p-Numbers," Chaos, Solitons & Fractals, Vol. 28, No. 4, 2006, pp. 1014-1025.
30. Stakhov, "Fibonacci Matrices, a Generalization of the 'Cassini Formula', and a New Coding Theory," Chaos, Solitons & Fractals, Vol. 30, No. 1, 2006, pp. 56-66.
31. P. Stakhov, "The Generalized Golden Proportions, a New Theory of Real Numbers, and Ternary Mirror-Symmetrical Arithmetic," Chaos, Solitons & Fractals, Vol. 33, No. 2, 2007, pp. 315-334.
32. Stakhov and B. Rozin, "The 'Golden' Hyperbolic Models of Universe," Chaos, Solitons & Fractals, Vol. 34, No. 2, 2007, pp. 159-171.
33. Stakhov, "The 'Golden' Matrices and a New Kind of Cryptography," Chaos, Solitons & Fractals, Vol. 32, No. 3, 2007, pp. 1138-1146.
34. P. Stakhov, "Gazale Formulas, a New Class of the Hyperbolic Fibonacci and Lucas Functions, and the Improved Method of the 'Golden' Cryptography," Academy of Trinitarizam, Moscow, 2006.
35. P. Stakhov and B. N. Rozin, "The Golden Section, Fibonacci Series and New Hyperbolic Models of Nature," Visual Mathematics, Vol. 8, No. 3, 2006. https://eudml. org/doc/ 256745
36. Stakhov, "A Generalization of the Cassini Formula," Visual Mathematics, Vol. 14, No. 2, 2012.
37. Stakhov, "On the General Theory of Hyperbolic Functions Based on the Hyperbolic Fibonacci and Lucas Functions and on Hilbert's Fourth Problem," Visual Mathematics, Vol. 15, No. 1, 2013. http://www.mi.sanu.ac.rs/vismath/pap.htm
38. Stakhov and S. Aranson, "Hyperbolic Fibonacci and Lucas Functions, 'Golden' Fibonacci Goniometry, Bodnar's Geometry, and Hilbert's Fourth Problem, Part I. Hyperbolic Fibonacci and Lucas Functions and 'Golden' Fibonacci Goniometry," Applied Mathematics, Vol. 2, No. 1, 2011, pp. 74-84. http://www.scirp.org/journal/am/
39. Stakhov and S. Aranson, "Hyperbolic Fibonacci and Lucas Functions, 'Golden' Fibonacci Goniometry, Bodnar's Geometry, and Hilbert's Fourth Problem, Part II. A New Geometric Theory of Phyllotaxis (Bodnar's Geometry)," Applied Mathematics, Vol. 2, No. 2, 2011, pp. 181-188.
40. Stakhov and S. Aranson, "Hyperbolic Fibonacci and Lucas Functions, 'Golden' Fibonacci Goniometry, Bodnar's Geometry, and Hilbert's Fourth Problem, Part III. An Original Solution of Hilbert's Fourth Problem," Applied Mathematics, Vol. 2, No. 3, 2011, pp. 283-293.
41. P. Stakhov, "Hilbert's Fourth Problem: Searching for Harmonic Hyperbolic Worlds of Nature," Applied Mathematics and Physics, Vol. 1, No. 3, 2013, pp. 60-66.
42. S. Blackburn, "Harmony of Spheres: The Oxford Dictionary of Philosophy," Oxford University Press, Oxford, 1994, 1996, 2005.
43. V. Dimitrov, "A New Kind of Social Science. Study of Self-Organization of Human Dynamics," Lulu Press, Morrisville, 2005.
44. N. N. Vorobyov, "Fibonacci Numbers," Nauka, Moscow, 1961.
45. V. E. Hoggat, "Fibonacci and Lucas Numbers," Houghton-Mifflin, Palo Alto, 1969.
46. S. Vajda, "Fibonacci & Lucas Numbers, and the Golden Section: Theory and Applications," Ellis Horwood limited, Chichester, 1989.
47. G. D. Grimm, "Proportionality in Architecture," ONTI, Leningrad Moscow, 1935.
48. V. P. Shenyagin, "Appeal to the International Scientific Community: On the Appropriateness of Nomination of Professor Stakhov AP to Award the Abel Prize in Mathematics in 2014," Academy of Trinitarizm, Moscow, 2013.

49. S. Olsen, "Professor Alexey Stakhov Is an Absolute Genius of Modern Science (in Honor of Alexey Stakhov's 70th Birthday)," Academy of Trinitarizm, Moscow, 2009.
50. E. M. Soroko, "A Review on the Book 'The Mathematics of Harmony: From Euclid to Contemporary Mathematics and Com- puter Science'," In: A. Stakhov, Ed., Academy of Trinitarizm, Moscow, 2010.
51. S. K. Abachiev, "The Mathematics of Harmony through the Eyes of Specialist in History and Methodology of Science," Academy of Trinitarizm, Moscow, 2007.
i. Zenkin, "The Mistake by Georg Cantor," Problems of Philosophy, Vol. 2, 2000, pp. 165-168.
52. D. Polya, "Mathematical Discovery," Nauka, Moscow, 1970.
53. G. Bergman, "A Number System with an Irrational Base," Mathematics Magazine, Vol. 31, 1957, pp. 98-119.
54. P. Stakhov, "Codes of the Golden Proportion, or the Number Systems for Future Computers?" Technology for Young People, Vol. 7, 1985, pp. 40-44.
55. N. Kolmogorov, "Mathematics in Its Historical Development," Nauka, Moscow, 1991.
56. V. de Spinadel, "The Family of Metallic Means," Visual Mathematics, Vol. 1, No. 3, 1999. http://members.tripod.com/vismath/
57. M. J. Gazale, "Gnomon. From Pharaohs to Fractals," Princeton University Press, Princeton, 1999.
58. J. Kappraff, "Connections. The Geometric Bridge between Art and Science," 2nd Edition, World Scientific, Singapore, 2001.
A. Tatarenko, "The Golden T_m-Harmonies' and D_m?Fractals," Academy of Trinitarism, Moscow, 2005. http://www.trinitas.ru/rus/doc/0232/009a/02320010.htm
59. H. Arakelyan, "The Numbers and Magnitudes in Modern Physics," Armenian Academy of Sciences, Yerevan, 1989.
60. V. P. Shenyagin, "Pythagoras, or How Everyone Creates His Own Myth," The Fourteen Years after the First Publication of the Quadratic Mantissa's Proportions. Academy of Trinitarism, Moscow, 2011. http://www.trinitas.ru/rus/doc/0232/013a/02322050.htm
61. S. Falcon, "Angel Plaza, On the Fibonacci k-Numbers," Chaos, Solitons & Fractals, Vol. 32, No. 5, 2007, pp. 1615-1624.
62. E. Lucas, "The Theory of Simply Periodic Numerical Functions," American Journal of Mathematics, Vol. 1, No. 2, 1878, pp. 184-240, 289-321. http://www.fq.math.ca/ Books/Complete/simply-periodic.pdf
63. "Hilbert's Problems," from Wikipedia, the free Encyclopedia. http://en.wikipedia.org/wiki/Hilbert's_problems
64. "Hilbert's Fourth Problem," From Wikipedia, the free Encyclopedia. http://en.wikipedia.org/wiki/Hilbert's_fourth_problem
65. "Lecture 'Mathematical Problems' by Professor David Hilbert," 1900. http:// aleph0.clarku.edu/~djoyce/hilbert/problems.html#prob4
66. V. Pogorelov, "Hilbert's Fourth Problem," Nauka, Moscow, 1974.
67. H. Yandell, "The Honors Class-Hilbert's problems and Their Solvers," A. K. Peters/CRC Press, Boston, 2003.
68. M. Kline, "The Loss of Certainty," Oxford University Press, New York, 1980.

CHAPTER 13

Scholz's First Conjecture: A Brief Demonstration

José M. Sautto[1], Agustín Santiago[2], Carlos N. Bouza[3], Verónica Campos[1]

[1]*Campus Costa Chica, Universidad Autónoma de Guerrero, Guerrero, México*

[2]*Facultad de Matemáticas, Universidad Autónoma de Guerrero, Acapulco, México*

[3]*Facultad de Matemática y Computación, Universidad de la Habana, Ciudad de La Habana, Cuba*

ABSTRACT

This paper presents a brief demonstration of Schulz's first conjecture, which sets the upper and lower limits on the length of the shortest chain of addition. Two methods of the upper limit are demonstrated; the second one is based on the algorithm of one of the most popular methods for obtaining addition chains of a number, known as the binary method.

Keywords: Addition Chain, Exponentiation, Short Chain, Scholz's Conjecture, Binary Method

1. INTRODUCTION

With the development of the Internet and adding chain applications in the development of cryptography—which permits safe handling of data over the Internet—the theme began with the publication in 1937 of Arnold Scholz's paper [1] , which defines a minimal addition chain, along with his three famous conjectures. In 1939 [2] , Alfred Brauer gave a strong impetus to this issue, gaining importance in this area. During the last decades of the last century deterministic methods flourished. The most popular were the binary method and the window method [3] -[5] . Heuristic methods began to emerge in the 70s and toward the end of the century, began to dominate in the literature [6] [7] . This is the second paper we write on the theme [8] and in both present simple demonstrations, the third and first conjecture, we use deterministic algorithms. Our intention is to build a framework for the development of intelligent methods for generating addition chains.

2. BASIC DEFINITIONS

The first conjecture presented by Scholz was: $m+1 \le l(n) \le 2m$; for the n which satisfy:

$$2^m + 1 \le n \le 2^{m+1}; m \ge 1. \tag{1}$$

In addition to setting maximum and lower bounds for the minimum length of addition chains, this conjecture induces a partition of our study space, in this case the natural numbers. The bounded sets defined by Schulz $2^m + 1 \le n \le 2^{m+1}$ for $m \ge 1$, exclude the numbers 1 and 2. For $m = 1$ the possible values of n correspond to the set $\{3,4\}$; from there, if we increase m by one, the possible values of n are doubled. The point of this partition is that we can delimit our study space, in this case the natural numbers, to set general properties on addition chains.

We will begin our discussion with the following definitions:

Definition 2.1. Let $S_e = \{a_i\}$ be a finite sequence of natural numbers. We will call it an addition chain of a natural number if it satisfies:

1) I. $1 = a_0 < a_1 < \cdots < a_r = e$.
2) II. $a_i = a_j + a_k$, for some (j,k) and for each step i, $0 \le k \le j < i$, with $0 < i \le r$.

Definition 2.2. Let $S_e = \{a_i\} = \{1 = a_0, a_1 = 2, \cdots, a_r = e\}$ be an addition chain of a number e, the highest index of the sequence r is called length of the chain S_e, and it is denoted by $l(S_e)$.

Definition 2.3. The minimum length of all addition chains of a natural number e is denoted by $l(e)$, that is:

$$l(e) = \min\{l(S_e) \mid S_e \text{ is an addition chain of } e\}.$$

Definition 2.4. Let us consider the family G_i of natural subsets defined by:

$$G_i = \begin{cases} i = 0, G_0 = 1; \\ i \geq 1, G_i = \{n \in N \mid 2^{i-1} + 1 \leq n \leq 2^i\}. \end{cases}$$

The set G_i will be called ith generation of natural numbers.

Definition 2.5. For every ith generation with $i \geq 2$, we will define:

$G_i^p = \{x \mid x = 2y; y \in G_{i-1}\}$ even numbers of G_i.

$G_i^m = \{x \mid x = 2y - 1; y \in G_{i-1}\}$ odd numbers of G_i.

Definition 2.6. For every $n \in N$, we will define the nth dominant chain as:

$S_{2^n}^n = \{a_i\}$ defined by $a_i = 2^i, 0 \leq i \leq n.$

3. IMPORTANT PROPERTIES

Proposition 3.1. The dominant chains are of maximum growth. That is to say, for every x if $S_x = \{b_i\} \neq S_{2^r}^r = \{a_i\}$ with $l(S_x) = r$; then $x < 2^r$.

Proof. As $\{b_i\} \neq \{a_i\}$ then there exists $1 < j \leq r$ (since the first two values of any addition chain are the same) in such a way that $b_j \neq a_j$ as $a_j = 2 * a_{j-1} > a_{j-1} + a_k = b_j$ with $k < j-1$, as $b_{j-1} = a_{j-1}$; $b_{j-1} + b_k$ $b_j =$; $k < j-1$ because if the first term of the sum had an index less than $j-1$, at most it could be b_{j-1}, which would imply that the sequence would not be an addition chain because it does not strictly increase, and as $k < j-1$ and the sequence strictly increases we have that: $b_j < a_j$. If it was the last term of the sequence then the inequality is true, and if it was not the last term from there $b_{j+i} \ll a_{j+i}$, for $i = 1, \cdots, r-j$ at each increment of the difference between the terms of the sequences increases. Hence $x = b_r < a_r = 2^r$.

Proposition 3.2. Dominant chains are addition chains defined on the numbers of the form $e = 2^n$, of length $l(S_{2^n}^n) = n$.

Proof. By definition $S_e^n = \{a_i\}$ defined by $a_i = 2^i, 0 \le i \le n$ from where:

$$S_{2^n}^n = \{a_0 = 2^0 = 1, a_1 = 2^1 = 2, \cdots, a_n = 2^n\}.$$

From the definition of dominant chain: $a_0 = 2^0 = 1$, from where its first element is 1and the last one is $a_n = 2^n$, in addition for every i the following is true: $a_{i-1} = 2^{i-1} < 2^i = a_i$ from where $0 < i < n$, therefore it increases and ends in $a_n = 2^n$, which proves the first property of addition chain.

Let us have a look at the second property, that is:

$$a_i = 2^i = 2^{i-1} + 2^{i-1} = a_{j=i-1} + a_{k=i-1},$$

clearly $0 \le k \le j < i$, with $0 < i \le r$, so it satisfies the second property and therefore they are addition chains.

Finally, since $a_n = 2^n$ is an addition chain of 2^n, the Proposition (3.1) assures that any other chain of that length is of a number $x < 2^n$, which implies that it is the only chain of length n de 2^n. This proves that $l(S_{2^n}^n) = n = l(2^n)$.

To underline this last fact we will state the following in this corollary:

Corollary 3.3. The numbers of the form $e = 2^n$ have $S_{2^n}^n$ as minimum length chain, whose length is $l(S_{2^n}^n) = n$. This chain is unique.

Proof. The Proposition (3.2) assures that $S_{2^n}^n$ is addition chain of 2^n, and the Proposition (3.1) states that any other chain of that length different from $S_{2^n}^n$ makes $x < 2^n$ true, which guarantees the uniqueness of $S_{2^n}^n$.

Another important result:

Corollary 3.4. If $l(x)=r$ then $x \le 2^r$.

Proof. If $l(x)=r$ implies that there exists S_x in such a way that $l(S_x)=r$ if $S_x \ne S_{2^r}^r$ by the Proposition (3.1), we have that $x<2^r$ and if $S_x = S_{2^r}^r$ by the Proposition (3.2) we know that $x=2^r$ from where $x \le 2^r$.

Proposition 3.5. If $x \in G_i$, then $l(x) \ge i$.

Proof. If we assume that (3.5) is not true, then there exists $x \in G_i$ in such a way that $l(x) \le i-1 < i$; from where if $l(x)=i-1$ there exists S_x in such a way that $l(S_x)=i-1$. The Corollary (3.4) assures that $x \le 2^{i-1}$, and this implies that $x \notin G_i$, which contradicts our hypothesis, from which we conclude that if $x \in G_i$ then $l(x) \ge i$.

Proposition 3.6. If $z \in G_i^p$, then $l(z) \le l(x)+1$ for any $x \in G_{i-1}$.

Proof. As $z \in G_i^p$, then there exists $x \in G_{i-1}$ in such a way that $z=2x$. Let $S_x = \{a_0 = 1, \cdots, a_r = x\}$ be a minimum length chain of x, from this one we will build an addition chain of z.

$S_z = \{a_0 = 1, \cdots, a_r = x, a_{r+1} = z = 2x\}$, it is clearly an addition chain of z, of length $l(x)+1$, which proves that

$$l(z) \le l(x)+1.$$

Proposition 3.7. If $x \in G_i^p$ and $x < 2^i$, then $z = x+1 \in G_i^n$ and $l(z) \le l(x)+1$.

Proof. As x is an even number and lower than the upper limit of G_i, that is $x<2^i$; then $x \le 2^i - 2$ from where $z = x+1 \le 2^i - 2 + 1 = 2^i - 1 \rightarrow z \in G_i^n$, which guarantees that $z \in G_i^n$.

Let $S_x = \{a_i\}$ be a minimum length chain $l(x)$, then its first term is 1 and its last term is x, from where if we add as the last term $z = x+1$ to that sequence, now this sequence is an addition chain of z, that is: $S_z = \{a_i\} \cup \{x+1=z\}$, it is an addition chain of Z of length $l(x)+1$, from where $l(z) \le l(x)+1$.

Proposition 3.8. If $z \in G_i^n$, then $l(z) \le l(x)+2$ for any $x \in G_{i-1}$.

Proof. By definition $G_i^n = \{x \mid x = 2y-1; \in G_{i-1}\}$, odd numbers G_i, from where: $G_i^n = \{2^{i-1}+1, 2^{i-1}+3, \cdots, 2^{i-1}\}$, from these numbers only the first $z = 2^{n-1}+1$ does not have an even numbered predecessor in G_i, an addition chain of this number is given by $S_z = S_{2^{n-1}}^{n-1} \cup \{2^{i-1}+1\}$, whose length is $l(S_z) = n-1+1 = n$, this length is minimum since if there exists another chain of z with a length lower than n according to the Corollary (3.4), we have that $z \le 2^{n-1}$ and clearly $z > 2^{n-1}$, from where we conclude that it is minimum, that is $l(z = 2^{i-1}+1) = n$. For the rest of the elements of G_i^n, according to the Proposition (3.7), there exists a $y \in G_i^p$ in such a way that $l(z) \le l(y)+1$. Now, the Proposition (3.6) assures $\exists x \in G_{i-1}$ in such a way that $l(y) \le l(x)+1$, from where $l(z) \le l(y)+1 \le l(x)+1+1 = l(x)+2$ for any $x \in G_{i-1}$.

Proposition 3.9. For every $i \ge 2$; $\wp = \{G_i^p, G_i^n\}$ is a partition of G_i.

Proof. We have to demonstrate that for every $i \ge 2$ we will always have:

a) $G_i = G_i^p \cup G_i^n$.

b) $G_i^p \cap G_i^n = \emptyset$.

For $i = 2$ we have $G_{2-1} = G_1 = \{2\}$ according to the definition of G_i from where $G_2^p = \{4\}$ and $G_i^n = \{3\}$ and $G_2 = \{n \in N \mid 2^{2-1}+1 = 2^1+1 = 3 \le n \le 2^2 = 4\} = \{3,4\}$, from where we clearly see that a) and b) are true.

Proposition 3.10. The number 3 only has an addition chain of length equal to 2.

Proof. The only addition chain of the number 3 is: $S_3 = \{a_0 = 1, a_1 = 2, a_2 = 3\}$, we will demonstrate that it is an addition chain and that it is unique.

It is an addition chain since it begins with 1, it has increasing terms and it ends with the number 3, from where it satisfies the property I of the definition of the chain addition.

Clearly $a_1 = a_0 + a_0 = 1 + 1 = 2$ and $a_2 = a_1 + a_0 = 2 + 1 = 3$, which proves part II of the definition $a_k = a_j + a_i; 0 \le i \le j < k$.

We will demonstrate now that it is unique. Let us suppose that it is not, then there exists $S_3^+ \neq S_3$, which is chain of the number 3. Let $S_3^+ = \{b_0 = 1, b_1 = 2, \cdots, b_p = 3\}$ be an addition chain different from S_3, the definition of chain forces $b_0 = 1$ and the property II when $k = 1$, and $0 \le i \le j < k = 1$, the only possible value of i and j is zero, from where $b_1 = 2$. For b_2 the possible values of (i, j) are three, which are presented in Table 1.

The only possible value for b_2 is 3, according to the definition of addition chain it would be the last term of the sequence. However, it is equal to S_3, which contradicts the fact that S_3^+ is different from S_3, from where we conclude that it is unique. As $b_2 = 3; l(3) = 2$.

4. DEMONSTRATION OF SCHOLZ'S FIRST CONJECTURE

In terms of the given definitions, Scholz's first conjecture is equivalent to:

$m + 1 \le l(n) \le 2m$; for the n which satisfy: $2^m + 1 \le n \le 2^{m+1}; m \ge 1$.

If we make a change of variables $i = m + 1$ we will have:

$i \le l(n) \le 2i - 2$ for the n which satisfy: $2^{i-1} + 1 \le n \le 2^i; i \ge 2$.

As $G_i = \{n \in N \mid 2^{i-1} + 1 \le n \le 2^i\}$ the conjecture is now:

$i \le l(n) \le 2i - 2$; for the $n \in G_i, i \ge 2$.

The new formulation would be:

Theorem 4.1. If $x \in G_i$, then $i \le l(x) \le 2i - 2$; for every $i \ge 2$.

Proof. The Proposition (3.5) guarantees the first part of the inequality, that is: if $x \in G_i$, then $i \le l(x)$.

We will now demonstrate the second part of the inequality.

The Proposition (3.6) guarantees us that if $z \in G_i^p$, then $l(z) \leq l(x)+1$, for any $x \in G_{i-1}$. The Proposition (3.8) guarantees us that if $z \in G_i^p$, then $l(z) \leq l(x)+2$, for any $x \in G_{i-1}$. The Proposition (3.9) proves that $G_i = G_i^p \cup G_i^n$, from where if k_{i-1} is another upper bound of $\{l(x) \mid x \in G_{i-1}\}$, we have that for every i, $k_i \leq k_{i-1}+2$ is true, it is the upper bound of $\{l(x) \mid x \in G_i\}$, that is $l(x) \leq k_{i-1}+2$; for every $x \in G_i$.

Let k_2 be the upper bound of G_2; then

$$k_i = k_2 + 2(i-2);\ i \geq 2. \tag{2}$$

As $G_2 = \{3,4\}$ and $4 = 2^2$, its minimum addition chain is $l(4=2^2)=2$, according to the Corollary (3.3) and the 3 according to the Proposition (3.10) $l(3)=2$, from where the upper bound of G_2 is 2. This is $k_2 = 2$, and substituting this value of k_2 in (2) we will have:

$$k_i = k_2 + 2(i-2) = 2 + 2(i-2) = 2(i-2+1) = 2(i-1) = 2i-2,$$

which proves that if $x \in G_i$, then $l(x) \leq 2i-2$; for $i \geq 2$. This demonstrates the second part of the inequality.

An alternative way of deriving the upper bound established in Schulz's first conjecture is obtainable by using the binary analysis methods for constructing the addition chains [3] . It is expressed in the following algorithm.

5. BINARY METHOD

Input: an integer: $e = (e_0, \cdots, e_{m-2}, e_m)$

Output: Addition chain $U = \{1, 2, \cdots, e\}$

Table 1. Sequence of addition chain for possible values of (i,j).

i	j	$b_i + b_j = b_2$	Comment
0	0	$b_0 + b_0 = 1 + 1 = 2$	The sequence would not be strictly increasing
0	1	$b_0 + b_1 = 1 + 2 = 3$	Possible value
1	1	$b_1 + b_1 = 2 + 2 = 4$	The last value must be 3, there can't be higher numbers

Start:

$$1: x = 1; U = \{1\}$$

for i=1 to m

$$x = 2 * x$$

$$U = U \cup \{x\}$$

If $e_i == 1$ then: $x = x + 1 \wedge U = U \cup \{x\}$

End

Therefore the length of the obtained chain depends on the length of the binary expression of the number and the quantity of ones that it has, as for each one, without taking into account the first one, is added another number to the output chain; for example see Table 2.

The output chain has the same number of elements that the number of bytes of the expression, in base 2, of the input number e, plus the quantity of ones of the binary expression, minus one, which is at the beginning of the binary expression. In this case: 77 = 1001101, the length of the binary expression is of 7 bytes and the number of ones minus the first one is 3. Hence the length of the output chain is 7 + 3 = 10; U = {1, 2, 4, 8, 9, 18, 19, 38, 76, 77}. Therefore the length of the addition chain is 9. This fact is expressed in the next proposition, as follows:

Proposition 5.1. The length of the addition chain generated by the binary method of a number e is equal to the last sub-index of the binary expression $e = e_0, e_1, \cdots, e_m$, plus the quantity of ones it has, minus one. That is $m + p - 1$, where p is the number of ones that the binary expression e has.

Proof: Take $e \in \mathbb{N}$; and its binary expression $e = e_0, e_1, \cdots, e_m$. The algorithm starts with a one and eliminates e_0, it is equal to one, and adds it to the addition chain U. For each element from e_1 to e_m that is for m elements we have added one more to U. Then $U = \{u_0, u_1, \cdots, u_m, u_{m+1}, u_{m+2}, \cdots, u_{m+p-1}\}$ has m plus $p - 1$, where p is quantifies the number of e_i's equal to one. Therefore, according to Definition 2.2 the length of the addition chain is equal to the last sub-index of the chain, in this case $m + p - 1$.

Table 2. Algorithm fot the binary method.

Description	Example
Input: integer: e	77
Write $e=(e_0,e_1,\cdots,e_m)$	$77=1001101$
Output: Addition Chain U	$U=\{1,2,\cdots,e\}$
$x=1;U=\{1\};i=1$	$x=1$; $U=\{1\};i=1$
While $i<m$ $x=2*x$; $U=U\bigcup\{x\}$; $i=i+1$	While $i<7$ $e_1=0$; $x=2*1=2$; $U=\{1,2\}$; $i=1+1=2$
As $i<m$ $x=2*x$; $U=U\bigcup\{x\}$; $i=i+1$	As $i=2<7$ $e_2=0$; $x=2*2=4$; $U=\{1,2,4\}$; $i=3$
Repeat: Step $4:x=2*x$ Add x at the end of the list U As $e_i==1$ than $x=x+1$ $U=U\bigcup\{x\}$; $i=i+1$	As $i=3<7$ $x=4*2=8;U=\{1,2,4,8\}$ As e_3 is $=1$; $x=8+1=9$ $U=\{1,2,4,8,9\}$; $i=3+1=4$
As $i<m;x=2*x$ Add x at the end of the list U If $e_i==1$ then $x=x+1$ Add x the list of U; $i=i+1$	As $i=4<7$ $e_4=1$; $x=9*2=18$; $U=\{1,2,4,8,9,18\}$ As e_4 is $=1$; $x=18+1=19$ and $U=\{1,2,4,8,9,18,19\}$; $i=4+1=5$

As $i < m$: $x = 2 * x$ Add x at the end of the list U; $i = i + 1$	As $i = 5 < 7$; $e_5 = 0$; $x = 19 * 2 = 38$ $U = \{1, 2, 4, 8, 9, 18, 19, 38\}; i = 5 + 1 = 6$
e_i's equal to one.	As $i = 6 < 7$ $e_6 = 1$; $x = 38 * 2 = 76$ $U = \{1, 2, 4, 8, 9, 18, 19, 38, 76\}$ As e_6 is $= 1$; $x = 76 + 1 = 77$ $U = \{1, 2, 4, 8, 9, 18, 19, 38, 76, 77\}; i = 6 + 1 = 7$
As $i = m \nless m$ stop the output is: U	As $i = 7 \nless 7$ stop and the output is: $U = \{1, 2, 4, 8, 9, 18, 19, 38, 76, 77\}$

As m is the last sub-index of the binary expression, and p is the number of ones in the expression. Note that we have that $U=\{u_0,u_1,\cdots,u_m,\cdots,u_{m+p-1}=e\}$. Then from Definition 2.2, we derive that the length of the chain is the last of the sub-indexes, which is $m+p-1$.

Q.E.D.

Note that the numbers belonging to the generation G_i, for $i>2$ have has binary expression with length equal to n, minus the superior limit, which has $n+1$ elements with 1 and n with zeros, which corresponds to the superior limit of the generation, as it is observed in the following Table 3.

Note that the upper bound of the generation is given by 2^n; its minimum addition chain is n, because of Proposition 3.2. Then we do not take i into account. The maximum expected length of a chain, belonging to the generation n, is given by the number whose binary expression is equal to n, corresponding to $e=2^n-1$, one less than the superior limit, with binary expression: $2^n-1=(e_0=1,e_1=1,\cdots,e_{n-1}=1)$. The obtained addition chain will have n – 1 components, as the number of ones of the expression is n, where the addition chain is given by $U=\{u_0=1,u_1=2,u_3=3,\cdots,u_{2n-2}=e\}$. Hence, the length of this chain is $2n-2$, which is the longest generated by the method in that generation. Then is proved the following result:

Corollary 5.1. The length of the longest addition chain generated by the binary method in G_i corresponds to $e=2^i-1$ that has as length $l(2^i-1)=2i-2$.

Theorem 5.1. The binary method for the generation of addition chains proofs the right hand side of the First Schulz's Conjecture that is:

If $n\in G_i$, then $l(n)\leq 2i-2$; for every $i\geq 2$.

Proof. If $n\in G_i$ then n has the binary expression $n=(e_0=1,e_1,e_2,\cdots,e_{i-1})$, where at least a component is different form e_0 is equal to 1, because if all are zeros, it will be the superior limit of G_{i-1}. The generation i has $i-1$ elements whose binary expression has two ones, for these elements, according to Proposition 5.1, its generated addition chain is of length equal to i, as we add only another value and the first value of the generated chain has as sub-index $i=i-1+1$. The rest will have longer chains. The longest corresponding to a $n=2^i-1$ has length $2i-2$ due to the results of corollary.

Table 3. Example binary sequence.

Generation	Limits of the partition		Limits of the partition expressed in binary		Size of the partition
G_0	1	1	1	1	1
G_1	2	2	10	10	1
G_2	3	4	11	100	2
G_3	5	8	101	1000	4
G_4	9	16	1001	10000	8
G_5	17	32	10001	100000	16
G_6	33	64	100001	1000000	32
G_7	65	128	1000001	10000000	64
G_8	129	256	10000001	100000000	128
G_9	257	512	100000001	1000000000	256
G_{10}	513	1024	1000000001	10000000000	512
			....		
G_n	$2^{i-1}+1$	2^i	$1,0,\cdots,0,1$	$1,0,\cdots,0$	2^{n-1}

This fact proofs that all the chains generated with this method are not larger than $2i-2$. Hence, the minimum chain must be smaller or equal to these values. This last result completes the proof.

REFERENCES

1. Scholz, A. (1937) Aufgabe 253. Jahresbericht der Deutschen Mathematiker-Vereingung, 47, 41-42.
2. Brauer, A.T. (1939) On Addition Chains. Bulletin of the American Mathematical Society, 45, 736-739.
3. Knuth, D.E. (1969) The Art of Computer Programming. Vol. 1: Fundamental Algorithms. Second Printing, Addison- Wesley Publishing Co., Reading.
4. Kunihiro, N. and Yamamoto, H. (2000) New Methods for Generating Short Addition Chains. IEICE Transactions on Fundamentals, E83-A, 60-67.
5. Koc, C.K. (1995) Analysis of Sliding Window Techniques for Exponentiation. Computers & Mathematics with Applications, 30, 17-24.
6. Cruz-Cortes, N., et al. (2008) An Artificial Immune System Heuristic for Generating Short Addition Chains. IEEE Transactions on Evolutionary Computation, 12, 1-24.
7. Bos, J. and Coster, M. (1990) Addition Chain Heuristics. Advances in Cryptology, 435, 400-407.
8. Sautto-Vallejo, J.M., Agustín, S.-M., Carlos, B.-H. and Verónica, C.-G. (2013) Scholz's Third Conjecture: A Demonstration for Star Addition Chains. Applied Mathematics, 4, 1-12.

INDEX

G

H

L

M

N

O

P